Emerging Mobile and Web 2.0 Technologies for Connected E-Government

Zaigham Mahmood
University of Derby, UK & North West University Potchefstroom, South Africa

A volume in the Advances in Electronic
Government, Digital Divide, and Regional
Development (AEGDDRD) Book Series

Information Science
REFERENCE
An Imprint of IGI Global

Managing Director:	Lindsay Johnston
Production Editor:	Jennifer Yoder
Development Editor:	Vince D'Imperio
Acquisitions Editor:	Kayla Wolfe
Typesetter:	Thomas Creedon
Cover Design:	Jason Mull

Published in the United States of America by
Information Science Reference (an imprint of IGI Global)
701 E. Chocolate Avenue
Hershey PA 17033
Tel: 717-533-8845
Fax: 717-533-8661
E-mail: cust@igi-global.com
Web site: http://www.igi-global.com

Library of Congress Cataloging-in-Publication Data

CIP Data - Pending
ISBN 978-1-4666-6082-3 (hardcover)
ISBN 978-1-4666-6083-0 (ebook)
ISBN 978-1-4666-xxxx-x (print &perpetual access)

This book is published in the IGI Global book series Advances in Electronic Government, Digital Divide, and Regional Development (AEGDDRD) (ISSN: 2326-9103; eISSN: 2326-9111)

British Cataloguing in Publication Data
A Cataloguing in Publication record for this book is available from the British Library.

All work contributed to this book is new, previously-unpublished material. The views expressed in this book are those of the authors, but not necessarily of the publisher.

For electronic access to this publication, please contact: eresources@igi-global.com.

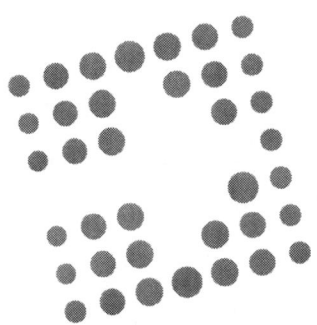

Advances in Electronic Government, Digital Divide, and Regional Development (AEGDDRD) Book Series

Zaigham Mahmood
University of Derby, UK & North West University, South Africa

ISSN: 2326-9103
EISSN: 2326-9111

MISSION

The successful use of digital technologies (including social media and mobile technologies) to provide public services and foster economic development has become an objective for governments around the world. The development towards electronic government (or e-government) not only affects the efficiency and effectiveness of public services, but also has the potential to transform the nature of government interactions with its citizens. Current research and practice on the adoption of electronic/digital government and the implementation in organizations around the world aims to emphasize the extensiveness of this growing field.

The **Advances in Electronic Government, Digital Divide & Regional Development (AEGDDRD)** book series aims to publish authored, edited and case books encompassing the current and innovative research and practice discussing all aspects of electronic government development, implementation and adoption as well the effective use of the emerging technologies (including social media and mobile technologies) for a more effective electronic governance (or e-governance).

COVERAGE

- Digital Democracy
- E-Citizenship
- Electronic & Digital Government
- ICT Adoption in Developing Countries
- ICT within Government & Public Sectors
- Knowledge Divide
- Public Information Management
- Regional Planning
- Urban & Rural Development
- Web 2.0 in Government

IGI Global is currently accepting manuscripts for publication within this series. To submit a proposal for a volume in this series, please contact our Acquisition Editors at Acquisitions@igi-global.com or visit: http://www.igi-global.com/publish/.

Titles in this Series

For a list of additional titles in this series, please visit: www.igi-global.com

Revolutionizing the Interaction between State and Citizens through Digital Communications
Sam B. Edwards, III (Green Mountain College, USA) and Diogo Santos (Federal University of Maranhao, Brazil)
Information Science Reference • copyright 2015 • 327pp • H/C (ISBN: 9781466662926) • US $195.00 (our price)

Emerging Issues and Prospects in African E-Government
Inderjeet Singh Sodhi (St. Wilfred's Post Graduate College, India)
Information Science Reference • copyright 2015 • 338pp • H/C (ISBN: 9781466662964) • US $175.00 (our price)

Advanced ICT Integration for Governance and Policy Modeling
Peter Sonntagbauer (Cellent AG, Austria) Kawa Nazemi (Fraunhofer IGD, Germany) Susanne Sonntagbauer (Cellent AG, Austria) Giorgio Prister (Major Cities of Europe, Italy) and Dirk Burkhardt (Fraunhofer IGD, Germany)
Business Science Reference • copyright 2014 • 414pp • H/C (ISBN: 9781466662360) • US $200.00 (our price)

E-Governance and Social Inclusion Concepts and Cases
Scott Baum (Griffith University, Australia) and Arun Mahizhnan (National University of Singapore, Singapore)
Information Science Reference • copyright 2014 • 300pp • H/C (ISBN: 9781466661066) • US $205.00 (our price)

Design, Development, and Use of Secure Electronic Voting Systems
Dimitrios Zissis (University of Aegean, Greece) and Dimitrios Lekkas (University of Aegean, Greece)
Information Science Reference • copyright 2014 • 270pp • H/C (ISBN: 9781466658202) • US $195.00 (our price)

Digital Access and E-Government Perspectives from Developing and Emerging Countries
Peter Mazebe II Mothataesi Sebina (University of Botswana, Botswana) Kgomotso H. Moahi (University of Botswana, Botswana) and Kelvin Joseph Bwalya (University of Botswana, Botswana & University of Johannesburg, South Africa)
Information Science Reference • copyright 2014 • 356pp • H/C (ISBN: 9781466658684) • US $195.00 (our price)

Technology Development and Platform Enhancements for Successful Global E-Government Design
Kelvin Joseph Bwalya (University of Botswana, Botswana & University of Johannesburg, South Africa)
Information Science Reference • copyright 2014 • 511pp • H/C (ISBN: 9781466649002) • US $235.00 (our price)

IT in the Public Sphere Applications in Administration, Government, Politics, and Planning
Zaigham Mahmood (University of Derby, UK & North West University, South Africa)
Information Science Reference • copyright 2014 • 359pp • H/C (ISBN: 9781466647190) • US $200.00 (our price)

www.igi-global.com

701 E. Chocolate Ave., Hershey, PA 17033
Order online at www.igi-global.com or call 717-533-8845 x100
To place a standing order for titles released in this series, contact: cust@igi-global.com
Mon-Fri 8:00 am - 5:00 pm (est) or fax 24 hours a day 717-533-8661

To Rehana Zaigham Mahmood

Table of Contents

Chapter 12

Detailed Table of Contents

Section 1
Service Orientation and Virtualisation Technologies for Connected Government

> *Muthu Ramachandran, Leeds Metropolitan University, UK*
> *Zaigham Mahmood, University of Derby, UK & North West University Potchefstroom, South Africa*
> *Pethuru Raj, IBM India, India*

Connected Government suggests provisioning of a government's services to its citizens using the Web and communications technologies employing the latest software development paradigms and related methodologies. This also requires appropriate integration of, and interaction between, software applications and e-services developed by various government departments as well as the other influencing sectors of the society such as commerce. This is especially so as the connected government (c-government) applications require open, flexible, interoperable, collaborative, and integrated architecture to provide services for the emerging technologies such as mobile, cloud, and big data. This, in turn, suggests a robust and standard mechanism to develop such applications and services. In this context, Service-Oriented Architecture (SOA) is an attractive approach to adopt. SOA has already been proven successful in providing such a framework for delivering software applications as services with flexibility and multi-platform and multi-channel integration that are necessarily required for c-government application offerings. This chapter provides a discussion of the SOA paradigm and the associated citizen and administrative requirements. The chapter also presents a service-oriented architectural framework based on a set of evaluated application characteristics that support newer technologies. A number of service-component models have also been proposed that provide required customisation, reuse, flexibility, and extensibility. In the context of the proposed overall service-oriented architecture, a large-scale sub-system that the authors term "e-Taxservice" has been used as a case study. The study has a service design that has been validated against a set of key service quality attributes.

Chapter 2

Miloš Milutinović, Belgrade University, Serbia
Marijana Despotović-Zrakić, Belgrade University, Serbia
Konstantin Simić, Belgrade University, Serbia
Mihajlo Anđelić, Belgrade University, Serbia

Modern information and communication systems can process massive amounts of data automatically and efficiently, but human beings have a limited cognitive capacity. Organizations that handle data on a larger scale need to adjust and streamline their operations in order to cope with this complexity. The e-Government information systems present problems that are on an entirely different scale, with communication streams between citizens and government easily dwarfing those in the private sectors. The information flows in e-Government are further constrained by established processes and practices that are hard to change and strict privacy concerns. A solution to problems of complexity and inefficiency of data manipulation in e-Government is needed. This chapter analyzes models and techniques of data categorization and visualization that can be employed in the context of e-Government. Methods of categorization, metadata, and ontologies in particular are explored for use in such an environment. A simple government ontology framework is developed as a starting point for introduction of ontologies into the e-Government context and the information is structured in such a way to allow easy correlation and navigation between concepts. A simple but intuitive visual representation of information and their relations is developed to facilitate better understanding of complex topics.

Chapter 3

Mohammed Al-Husban, Southampton Solent University, UK
Carl Adams, University of Portsmouth, UK

Efficient public service delivery is a primary task of public administration within any governance model. The main theme of modern governance implies an integrated, effective, and citizen-centric practices of government and administration as a prerequisite for a long-term positive development of the economy. Electronic public service delivery via e-government portal has become a convenient means for the customers—citizens and businesses—to fulfil their requirements. However, the quality of service delivery is heavily based on the level of integration of the services between different partners in the back office. Service integration requires good governance among partners in agencies in various departments and sometimes at different government levels. This chapter provides an interoperability integration framework that connects closely coordinated services based on Service-Oriented Architecture, Enterprise Service Bus, and Web services. The proposed framework is presented as an attempt to align the organizational structures and processes of different government departments while reducing implementation and ownership costs. The framework is applied to a realistic case example of integrating three different public services, namely applying for a Tourism Agency License, applying for a Vocational License, and applying for No Criminal Record Certificate, in a highly interoperable manner and a high level of adaptability to existing government policies and priorities.

Section 2
Web 2.0 Technologies and E-Participation for Next Generation E-Government

Chapter 4

This chapter presents a discussion on e-readiness, Web 2.0, social media, mobile/wireless technologies, and other Information and Communication Technologies (ICTs) that can help to facilitate the attainment and sustenance of an e-ready environment necessary to enhance e-governance in Nigeria. The chapter aims to clearly articulate the necessary steps to be taken to provide all stakeholders with a blueprint of areas and factors on which to focus. An assessment of how e-ready the Nigerian government and its citizens currently are and the requirements necessary for further steps to be taken (such as policies, programmes, and processes to be put in place, infrastructures to be acquired, and training provisions to equip Nigerian citizens and government officials with the capacity to benefit from and sustain the use of acquired e-technologies) are also presented. Specific ways by which Nigeria can harness the various emerging technologies (social media, Web 2.0, and mobile/wireless technologies) are highlighted. If employed appropriately, these technologies can help to provide improved processes, increased efficiency, improved transparency, and citizen's effective participation and involvement in governance to further improve the lives of Nigerian citizens.

Chapter 5

E-Government is an evolving field with continually changing practice and priorities. It is also a global phenomenon, from the richest and most technologically developed nations to the poorer and less technologically developed countries, involving a range of latest Information and Communication Technologies (ICT) and diverse methodologies. In such a dynamic field spanning all sectors of the governments and societies, it is difficult for e-government researchers and practitioners to identify the trends in the e-government activity and learn from previous cases and experiences. In this context, the aim of this chapter is to present an in-depth evaluation of e-government practice and research since 2007, to provide insight on research practicalities and emerging issues in e-government activity, and to identify the trends and technologies. The chapter also focuses on the current mobile and Web 2.0 technologies and examines the practicalities of using mobile technologies in various countries such as USA, Canada, UK, Austria, Japan, and others, as well as the practicalities of Web 2.0 technologies in some domains such as government, regulation, cross-agency cooperation, law enforcement, etc. This chapter presents a framework based on the mobile and Web 2.0 technologies in the context of e-government activity. In addition, the authors propose a framework for a government-people relationship. We hope to make a contribution for researchers, practitioners, policy makers, and people interested in e-government by providing a base of the e-government domain knowledge, practice, and framework. Additionally, the chapter illustrates how the implementation of mobile and Web 2.0 technologies support connected e-government.

Chapter 6

Maria Moloney, Escher Group Ltd., Ireland
Gary Coyle, Escher Group Ltd., UK

The evolving model of the Future Internet has, at its heart, the users of the Internet. Web 2.0 and Government 2.0 initiatives help citizens communicate even better with their governments. Such initiatives have the potential to empower citizens by giving them a stronger voice in both the traditional sense and in the digital society. Pressure is mounting on governments to listen to the voice of the public expressed through these technologies and incorporate their needs into public policy. On the other hand, governments still have a duty to protect their citizens' personal information against unlawful and malicious intent. This responsibility is essential to any government in an age where there is an increasing burden on citizens to interact with governments via electronic means. This chapter examines this dual agenda of modern governments to engage with its citizens, on the one hand, to encourage transparency and open discussion, and to provide digitally offered public services that require the protection of citizens' private information, on the other. In this chapter, it is argued that a citizen-centric approach to online privacy protection that works in tandem with the open government agenda will provide a unified mode of interaction between citizens, businesses, and governments in digital society.

Section 3
Mobile Technologies for Smarter and Sustainable Mobile Government

Chapter 7

Pethuru Raj, IBM India, India

There are hordes of data-driven, context-aware, and people-centric applications and services for smarter environments such as smarter homes, governments, buildings, cities, and organizations. With the exponential growth of smart phones, there are service repositories and application stores in remote mobile clouds. Similarly, with the ceaseless advancements in the device ecosystem and in the IT field, government-specific applications will flourish and be deployed and maintained in special cloud stores, platforms, and infrastructures to be found, bound, and used by any input/output devices for a variety of everyday personal and professional purposes. Smart, sustainable, intuitive, and citizen-aware services can be dynamically created from the ground up as well as orchestrated or choreographed out of multiple atomic and discrete software services. Such composite services are directly fulfilling government activities. Thus, clouds emerge as the most common and minimum requirement for not only producing and stocking services but also for hosting application platforms. Further, clouds facilitate provisioning and renting out their configurable and customizable assets on demand. Through self-service portals, the cloud usage is to pick up fast in the days to unfold. In this chapter, the authors write about how cloud adoption is to ring in delectable transformations for worldwide governments as well as their citizens, that is, how governments can accomplish more with less, how people can experience high quality, technology-sponsored digital living, how the cloud idea becomes a centre of attraction for more ingenuity towards newer and nimbler service conceptualization, concretization, and delivery.

Chapter 8

Olalekan Samuel Ogunleye, Meraka Institute, South Africa
Jean-Paul Van Belle, University of Cape Town, South Africa

Mobile technology has played a crucial role in facilitating democratic change in many of the developing countries. Many countries have attempted to implement Mobile Government (m-government), which is a form of electronic government, using mobile and other latest technologies such as social media as the most fundamental infrastructure for implementing such changes. However, m-government projects' scalability and sustainability are amongst the key issues relating to the use of Information and Communication Technologies (ICT). This chapter attempts to discuss the scalability and sustainability of m-government projects in the context of developing countries. The aim is to provide a broader understanding of the inherent issues surrounding scalability and sustainability of m-government projects: in general terms and also in relation to mobile phone-based projects for governments' service delivery. In order to understand these issues, definitions of these two concepts are provided and various e-government maturity models are discussed. This is then followed by an overview of the challenges of scaling up and sustaining the m-government projects in developing countries, and lastly, an elaboration of how sustainability and scalability can be achieved is also presented.

Chapter 9

Rodrigo Sandoval-Almazan, Universidad Autónoma del Estado de México, Mexico
Yaneileth Rojas Romero, Universidad Autónoma del Estado de México, Mexico

The mobile government has become a reality in a large majority of countries around the world. However, the use of mobile apps (small software programs for use on mobile devices) to link government Websites and information is a recent trend that is becoming of interest to citizens and public officials. The uses, advantages, and disadvantages have recently become a study field for several scholars around the globe. The mobile government is not new for e-government scholars; however, the explosion of apps and the increase of smart phones have created a new trend in the mobile government field. In order to understand these phenomena in the Mexican society, the authors have gathered data from different sources: government departments, business enterprises, and citizen organizations. Based on this information, they analyze the impact of apps across the country and suggest a classification method that can be used for a better understanding of this new field. In this chapter, the authors discuss five small case studies, which they consider good examples to follow by different government organizations. To accomplish this objective, they divide this chapter into seven main sections. After the introduction, the authors provide a literature review, describe the method of study and classify the apps, discuss the findings with the model application, present the case studies for government apps, discuss ideas for future research on government apps, and then in the final section, they present final remarks and conclusions of the investigation.

Section 4
Social Computing and Data Modelling for Connected Services for Inclusive Government

Chapter 10

Connected Government requires different government organizations to connect seamlessly across functions, agencies, and jurisdictions in order to deliver effective and efficient services to citizens and businesses. In the countries of the European Union, this also involves the possibility of delivering cross-border services, which is an important step toward a truly united Europe. To achieve this goal, European citizens and businesses should be able to interact with different public administrations in different Member States in a seamless way to perceive them as a single entity. Interoperability, which is a key factor for Connected Government, is not enough in order to achieve this result, since it usually does not consider the social dimension of organizations. This dimension is at the basis of co-operability, which is a form of non-technical interoperability that allows different organizations to function together essentially as a single organization. In this chapter, it is argued that, due to their unique capacity of coupling several technologies and processes with interpersonal styles, awareness, communication tools, and conversational models, the integration of social computing services and tools within inter-organizational workflows can make them more efficient and effective. It can also support the "learning" process that leads different organizations to achieve co-operability.

Chapter 11

It was not long ago when Information and Communication Technologies (ICT) were not ubiquitous and Web 2.0 was the stuff of science fiction. However, these technologies are now here to stay, and local governments should learn how to make the most of them. In this chapter, the situation of emerging ICT in Spain in general and for Spanish e-government in particular is described. Next, the results of an empirical study based on a longitudinal quantitative survey are shown. The survey was carried out in 2005, before the advent of Web 2.0, and again in 2012. In the survey, the Chief Information Officers (CIO) of Spanish municipalities express their opinions on critical success factors that may enhance or hinder the effectiveness, connectivity, and transparency of their strategies for a connected government (c-government). The comparative findings reveal that political issues set off, then and now, local e-government success and failure, whereas ICT-based issues, once very important for these CIOs, have been downgraded in their minds. Therefore, the emergence of social media, mobile technologies, Web 2.0, and connected government has not had a truly significant role in the quest for e-government success on their own, but in combination with other factors. The chapter also discusses the related factors.

Chapter 12

Abhishek Roy, The University of Burdwan, India
Sunil Karforma, The University of Burdwan, India

In the current climate of global economic decline, the developing countries are facing severe challenges in maintaining an efficient administration within an affordable budget. If this economic slowdown continues, there will be serious difficulties which will hamper the socio-economic development of the entire region. To respond to the situation, the governments must reduce budget expenses and still maintain efficiency and openness. To do so, the administration must deploy ICT-based mechanisms to fulfil the desired objectives. In this chapter, the authors present the development of a multifaceted electronic card-based secured e-governance mechanism to attempt to redress the inherent issues and explore new dimensions of interdisciplinary research. The proposed system will also act as the all-purpose electronic identity of the Citizen and hopefully replace the existing identity instruments such as Voter Card, Permanent Account Number Card, Driving License, Ration Card, Below Poverty-Line Card, Employment Card, Health Card, Insurance Card, etc. Moreover, this electronic instrument will also enable Citizen to perform financial transactions. Clearly, the authentication procedure of the proposed mechanism must also exist otherwise the intruders will be able to breach the system and execute their ill intentions. To ensure appropriateness of security features of the mechanism, the authors have also implemented a user authentication technique using object-oriented modelling of RSA digital signature algorithm for a Government-Citizen (G2C) type of e-governance. For better management of such a huge amount of sensitive information, the authors also discuss data modelling techniques used during user authentication of the proposed model.

Preface

OVERVIEW

With the emergence of Web 2.0, mobile technologies, and social media (Facebook, Twitter, YouTube, etc.), societies are becoming more online and better connected. Similarly, electronic government (e-government) that previously relied on older Information and Communications Technologies (ICT) is now making use of the newer technologies and evolving into what is beginning to be referred to as the *connected government* (c-government) or *Government 2.0*. Various departments of the government (federal and local) that simply develop and advertise e-services but are otherwise not digitally integrated and not fully connected with other vital sectors of the society (such as education, health, finance, etc.) are no longer acceptable anymore. Citizens, who have access to inexpensive mobile devices, often free of cost cloud storage, and free use of social media, expect governments and their ministries and departments to utilize emerging technologies. They would like their governments to be more visibly online at all times, better connected with the governed, and more responsive to the citizens' needs. They now expect e-governance to be more open, efficient, effective, and fully transparent. Thus, e-participation of the citizens in the running of the affairs of the government and the production of the relevant e-services that reflect the citizens' opinions and voices are now high on the agenda and a necessary minimum requirement.

With this understanding, governments that are at a higher level of c-government maturity are beginning to employ latest methodologies such as service-oriented or component-driven architectures for more efficient and seamless development and integration of Web-based services. These governments are keen to be seen as using the emerging technologies such as Web 2.0, social media, mobile technologies, and cloud computing for better provision and access of such e-services.

In this context, this book, *Emerging Mobile and Web 2.0 Technologies for Connected E-Government*, considers the various dimensions of the connected government and connected e-governance and presents the current situation in the form of status reports, development methodologies, practical examples, best practices, case studies, and the latest research. The present volume is a collection of 12 chapters authored by well-known academics and industry practitioners from around the world. The book will serve as a reference text in the subject areas of connected e-government and e-governance as well as mobile government (m-government).

BOOK OBJECTIVE

This book, *Emerging Mobile and Web 2.0 Technologies for Connected E-Government,* aims to serve as a reference text and presents latest research focusing on the use of the latest ICTs, in particular Web 2.0, social media, and mobile technologies, as well as the case studies. The objective is to understand the use and effectiveness of such technologies and best practices that provide successful strategies towards developing a connected government that is more responsive to the needs of the general public and is engaged in full e-participation of the citizens. To this end, the current volume presents reports and discussions on the following:

- Current research and new ideas for integrated e-government
- Frameworks and strategies for connected e-government
- Novel practices for the use of Web 2.0 and mobile technologies
- E-governance, e-inclusion, e-democracy, and m-government
- Best practices, practical suggestions, and recommendations
- Case studies highlighting practical experiences.

TARGET AUDIENCE

This volume, *Emerging Mobile and Web 2.0 Technologies for Connected E-Government,* is a reference text aimed at several groups of readers, including the following:

- University students, lecturers, and researchers interested in the field of Government 2.0 and the latest Web 2.0, communication, and mobile technologies
- Information systems specialists, technology experts, and practitioners in the field of ICT, Web 2.0, social media, and mobile technologies
- Decision makers, managers, and directors in government departments, public administration, and the business sector responsible for offering e-services
- Project managers and information systems architects tasked with the development of e-services and management of connected e-government projects.

BOOK ORGANISATION

Emerging Mobile and Web 2.0 Technologies for Connected E-Government is organised in 4 sections with a total of 12 chapters, authored by 23 academics and practitioners from around the world, as follows:

- **Section 1**: "Service Orientation and Virtualisation Technologies for Connected Government." There are three contributions in this section. The first chapter discusses the Service-Oriented Architecture (SOA) approach for developing Web-based applications, the second chapter focuses on data categorization and virtualization techniques to enhance data management in e-government projects, and the next chapter presents a connected service delivery framework for interoperable government.

- **Section 2**: "Web 2.0 Technologies and E-Participation for Next Generation E-Government." This section also has three chapters. The first of these discusses e-readiness of the Nigerian nation considering Web 2.0, ICT infrastructure, and training provision. The second contribution focuses on mobile and Web 2.0 technologies for connected e-government, and the third chapter looks at the next generation of e-government in terms of the e-participation and data protection agenda.
- **Section 3**: "Mobile Technologies for Smarter and Sustainable Mobile Government." This section of the book also comprises three chapters. The first contribution presents ideas for a smarter governance employing latest technologies such as mobile and cloud technologies, whereas the second chapter focuses on the scalability and sustainability of mobile government projects. The final contribution in the section presents a case study of Mexican mobile government by considering e-services as mobile applications.
- **Section 4**: "Social Computing and Data Modelling for Connected Services for Inclusive Government." This final section also consists of three chapters. The first chapter suggests the importance of and mechanisms for cross-boundary delivery of e-services. The second contribution presents an analysis of critical success factors in terms of traditional and newly emerging technologies. The final contribution of this section of the book presents data modelling of a multifaceted electronic-card-based secure e-governance system.

CHAPTER DESCRIPTIONS

Detailed abstracts of the book chapters appear in the *Detailed Table of Contents* section. Here, very brief summaries of chapters' content are presented.

Chapter 1 is titled "Service-Oriented Architecture for Developing Web-Based Applications for Connected Government." Authored by Ramachandran, Mahmood, and Raj, the chapter looks into the use of Service-Oriented Architecture (SOA) paradigm for the development of open and flexible applications for an integrated and collaborative e-government provision. The concept of SOA is explained together with the associated technologies such as XML, WSDL, UDDI, and SOAP. The chapter also discusses a number of service component models that can be usefully employed to develop e-government applications. To illustrate the use of service orientation, a large-scale system, e-Taxservice, is presented as a case study.

Chapter 2 is developed by Milutinovic, Despotovic-Zrakic, Simic, and Andelic. Titled "Enhancing Data Management in E-Government Using Data Categorization and Visualization Techniques," it analyses models and techniques for data categorization and visualization that can be utilised in the context of e-government developments. Various methods of categorization, metadata, and ontologies are explored, and a simple government ontology framework is developed as a starting point for introduction of ontologies into the e-government context. The aim is to correctly structure the information to allow easy correlation and navigation between concepts. The chapter also presents an intuitive visual representation of information to facilitate better understanding of the topic.

Chapter 3, co-authored by Al-Husban and Adams, is titled "Connected Service Delivery Framework: Towards Interoperable Government." This contribution presents an interoperability integration framework that connects closely coordinated services based on SOA, Enterprise Service Bus, and Web services. The aim is to align the organizational structures and processes of different government departments while reducing implementation and ownership costs. To illustrate the effectiveness of the framework, it is applied to a realistic case example of integrating three different public services in a highly interoperable manner with a high level of adaptability, the services being the applications for a Tourism Agency License, a Vocational License, and a No Criminal Record Certificate.

Chapter 4 is contributed by Ikponmwosa and titled "Web 2.0, ICT Infrastructure, and Training Provision for E-Government Readiness in Nigeria." It presents a discussion on Web 2.0, social media, mobile/wireless, and other ICTs that can usefully help towards e-government readiness in the context of Nigeria. The chapter aims to articulate the necessary steps required (e.g. policies, programmes, processes, and technology infrastructures) for e-readiness assessment including the training provision for the citizens. The chapter highlights specific opportunities for the nation to harness the emerging technologies (such as social media, Web 2.0, and mobile/wireless technologies) to provide improved processes, increased efficiency, better transparency, and citizens' effective participation.

Chapter 5 is titled "A Base of Knowledge, Mobile, and Web 2.0 Technologies for Connected E-Government." In this contribution, the authors, Yusuf and Adams, present an in-depth evaluation of e-government practice and research since 2007 to provide insight on emerging issues, trends, and technologies (including mobile and Web 2.0 technologies). Discussing the practicalities of Web 2.0 technologies in domains such as government, regulation, cross-agency cooperation, and law enforcement, the chapter also presents a framework based on mobile and Web 2.0 technologies emphasizing government-people relationship to illustrate how the implementation of such latest technologies can successfully support connected e-government.

Chapter 6 is developed by Moloney and Coyle. Titled "Next Generation E-Government: Reconciling the E-Participation and Data Protection Agendas," this contribution focuses on the active participation of citizens in the affairs of the government and the responsibility of the government to ensure citizens data confidentiality and protection. The chapter examines this dual agenda of modern governments to engage with its citizens to encourage transparency and open discussion. The chapter argues that a citizen-centric approach to online privacy protection that works in tandem with the open government agenda will eventually provide a unified mode of interaction between citizens, business organizations, and government departments in a digital society.

Chapter 7 is authored by Puthuru Raj. Titled "Mobile and Cloud Technologies for Smarter Governance," this chapter looks into the context-aware people-centric applications and services for smarter environments such as smarter homes and efficient governments. The author suggest that such services can be dynamically created, orchestrated, and choreographed out of multiple atomic and discrete Web-based software services and hosted in cloud environments that facilitate provisioning and renting out already configurable and customizable resources on demand. The chapter discusses the related technologies including service-oriented architecture with respect to services development and subsequent adoption and migration to cloud environments.

Chapter 8 is titled "Scalability and Sustainability of M-Government Projects Implementation in Developing Countries." Co-authored by Ogunleve and Van Belle, this contribution discusses the role of mobile and Web 2.0 technologies such as social media in the implementation of mobile government (m-government) in the context of developing nations. The chapter aims to provide an understanding of the inherent issues surrounding scalability and sustainability of m-government projects and applications. Various e-government maturity models are also discussed and guidelines presented with respect to the challenges of scaling up and sustaining m-government projects for a more effective connected e-government.

Chapter 9 is titled "The Case of Mexican Mobile Government: Measurement and Examples." Co-authored by Sandoval-Almazan and Romero, it looks at the development of mobile government in Mexico in the wake of the increase in the use of smart phones and mobile apps available to the citizens. Based on the data gathered from government departments, business enterprises, and citizen organizations, the chapter analyses and measures the impact of mobile apps on the uptake of connected government and suggests a classification approach. Forty-seven applications are classified and the findings are discussed in terms of five case studies to understand the effect of mobile technology for further developments in this context.

Chapter 10 is developed by Walter Castlenovo. The contribution titled "Social Computing and Co-operation Services for Connected Government and Cross-Boundary Services Delivery" discusses the delivery of cross-border services, across geographical boundaries between co-operating countries, as the next step towards truly global government. Considering the case of the EU member states, the chapter focuses on the need for seamless interoperability and social dimension of organizations as the basis of co-operability in the light of the unique nature of coupling of several technologies and processes with interpersonal styles – also considering the communication tools and conversational models employed within the inter-organizational workflows.

Chapter 11 is jointly authored by Claver-Cortes, Juana-Espinosa, and Valdes-Conca. Titled "Emerging and Traditional ICT as Critical Success Factors for Local Governments: A Longitudinal Analysis," it describes the situation of emerging ICT in Spain and for Spanish e-government and presents the results of an empirical study based on a longitudinal quantitative survey carried out in 2005 to discuss the critical success factors that may enhance or hinder the effectiveness, connectivity, and transparency of a connected government. The findings establish an emphasis on political issues as triggers, both positive and negative, of the local e-government success. The chapter also discusses the effects of the emergence of social media, mobile technologies, and Web 2.0.

Chapter 12 is titled "Data Modelling of a Multifaceted Electronic Card-Based Secure E-Governance System." Developed by Roy and Karforma, this chapter presents the development of a multifaceted electronic card-based secured e-governance mechanism. It is suggested that the proposed smart card system can act as an all-purpose electronic identity of the citizens, replacing existing identity instruments such as Voter Card, Driving License, Employment Card, Health Card, Insurance Card, etc. as well as an instrument to conduct financial transactions. To address the security issues, data modelling techniques and authentication procedures are also discussed and implemented using object-oriented modelling of the RSA digital signature algorithm.

Zaigham Mahmood
University of Derby, UK & North West University Potchefstroom, South Africa
31 January 2014

Acknowledgment

The editor acknowledges the support and efforts of a number of colleagues. First and foremost, I would like to thank the contributors to this book, the 23 authors and co-authors from academia as well as industry from around the world who collectively developed and submitted a total of 12 chapters. Without their efforts in developing quality chapters conforming to the required guidelines and meeting the strict deadlines, this text would not have been possible. Their names and brief biographical notes are listed in a separate section in this book.

Secondly, my grateful thanks are due to the members of the advisory and editorial board of this book, who willingly volunteered their time in reviewing the book chapters and providing further advisory and editorial support. Their names and affiliations also appear in a separate section in this book.

Finally, I would like to thank members of my immediate family, especially Rehana Mahmood, for her encouragement and moral support and for happily allowing me time to devote to the development of this 4th in a series of reference books on electronic government.

Thank you all.

Zaigham Mahmood
University of Derby, UK & North West University Potchefstroom, South Africa
31 January 2014

Section 1
Service Orientation and Virtualisation Technologies for Connected Government

Chapter 1
Service-Oriented Architecture for Developing Web-Based Applications for Connected Government

Muthu Ramachandran
Leeds Metropolitan University, UK

Zaigham Mahmood
University of Derby, UK & North West University Potchefstroom, South Africa

Pethuru Raj
IBM India, India

ABSTRACT

Connected Government suggests provisioning of a government's services to its citizens using the Web and communications technologies employing the latest software development paradigms and related methodologies. This also requires appropriate integration of, and interaction between, software applications and e-services developed by various government departments as well as the other influencing sectors of the society such as commerce. This is especially so as the connected government (c-government) applications require open, flexible, interoperable, collaborative, and integrated architecture to provide services for the emerging technologies such as mobile, cloud, and big data. This, in turn, suggests a robust and standard mechanism to develop such applications and services. In this context, Service-Oriented Architecture (SOA) is an attractive approach to adopt. SOA has already been proven successful in providing such a framework for delivering software applications as services with flexibility and multi-platform and multi-channel integration that are necessarily required for c-government application offerings. This chapter provides a discussion of the SOA paradigm and the associated citizen and administrative requirements. The chapter also presents a service-oriented architectural framework based on a set of evaluated application characteristics that support newer technologies. A number of service-component models have also been proposed that provide required customisation, reuse, flexibility, and extensibility. In the context of the proposed overall service-oriented architecture, a large-scale sub-system that the authors term "e-Taxservice" has been used as a case study. The study has a service design that has been validated against a set of key service quality attributes.

DOI: 10.4018/978-1-4666-6082-3.ch001

INTRODUCTION

Electronic Government or e-Government aims at a citizen centred vision of a government that provides effective governance, increased transparency, improved management, effective processes and efficient services through the use of the Internet and ICT (Information and Communication Technologies). It is about harnessing the information revolution to improve the lives of citizens and businesses and to improve the efficiency of government policies and implementations (Borras, 2004). In this context, e-Government is an enabler for better governance, where technology is used as a strategic tool to modernise structures, processes, regulatory frameworks, human resources and the culture of public administrations to increase public value (Centeno et al., 2004). Ultimately, the goal is to enhance interaction between three important components of a society: citizens (i.e. general public), government (including other agencies and employees) and the business sector.

Connected Government or c-Government takes a step further and attempts to ensure a much better interaction between the government and the governed through the use of newer computing technologies such as Web 2.0, social media and various mobile technologies and devices. This is sometimes also referred to as Government 2.0.

Interaction between the government departments and the citizens is carried out through the provision of electronic services (e-services) over the Internet that are then consumed by the general public, other government agencies and various other sectors of the society. These services implement the required functionality for the users to make use of e.g. a citizen can use these services to file income tax return electronically or purchase an item online or cast his/her vote using mobile telephone, etc.

These services are software applications developed by governmental agencies as well as the other sectors of the society e.g. financial sector, health organisations etc. Often, there is a need to integrate such applications e.g. certain border agency services to be linked to the police enforcement applications. In this case, it is important that the integration is seamless and interface such that the output from one application is easily and correctly read by another application without further human intervention. It then becomes imperative that the design of such software applications conforms to certain set standards that all agencies, especially the federal and local governments, adhere to. Service Oriented Architecture (SOA) is one design approach that can be effectively used. Component based design is another and there are many variations of these. The core benefit of such architectural styles is that the basic programming unit is a *service* or a *component* that is fully cohesive but functionally independent and self-contained. These units can be built with interactions with latest technologies and devices that can be easily constructed and modified as required; besides, these units can be orchestrated and linked relatively easily.

In this chapter, we first present the concept of SOA and then discuss an SOA based framework for Web applications development. The chapter also proposes a number of service component models that provide required customisation, reuse, flexibility and extensibility. To illustrate the effectiveness of the suggested technologies, a large scale sub-system that we term as e-Tax service has been used as a case study for services design which has been validated against a set of key service quality attributes. In the chapter, e-government and c-government terms are sometimes used interchangeably.

SERVICE ORIENTED ARCHITECTURE (SOA)

Service Oriented Architecture (SOA) is an architectural style for developing and integrating large applications. It is an organisational and technical framework to enable an organisation to deliver self-describing and platform independent business functionality (Cartwright et al., 2006) providing a way of sharing functions, typically, business functions, in a widespread and flexible way, that is necessarily required when integrating applications developed by various governmental agencies. The idea is that functionality is developed as a number of independent components, called *services* and then services connected in the correct way to form bigger complete applications. SOA is, therefore, a broad, standalone and standards based framework in which services are built, deployed, managed and orchestrated in pursuit of an agile and resilient IT infrastructure (Mahmood, 2007). Furthermore, this architecture aims to provide enterprise business solutions that can extend or change *on demand*. SOA is being certainly established as a disruptive and transformative business technology for the booming ICT and software development domain.

The distinct and decisive factors and facets of SOA are that it is extremely simple, supple, extensible and above all, process-aligned. Due to its extreme flexibility and adaptability, several business behemoths and IT powerhouses have created, demonstrated and sustained their own service-oriented architectures and frameworks. For example, CISCO has successfully formulated the service-oriented networking architecture (SONA) to closely and compactly acquaint and associate their products, skills and services with the blooming service-orientation concepts.

In SOA, the business and technical processes are implemented as *services*. Each service represents a certain functionality that maps explicitly to a step in a business process (Groves, 2005). In this context, a service is an independent and totally cohesive software component that can be reused by another software component or accessed via a standard-based interface over a network e.g. the Internet. An important aspect of service-orientation is the separation of service *interface* (the WHAT) from its implementation or *content* (the HOW). The interface provides service identification, whereas, the content provides business logic. Zimmermann et al. (2004) suggests three levels of abstractions within SOA:

- **Operations:** Units of functions operating on received data, having specific interfaces and returning structured responses.
- **Services:** Logical groupings of operations.
- **Business Processes:** Actions or activities to perform specific business goals by invoking multiple services.

In terms of service-orientation, we can envisage three types of services (Hohpe, 2002):

- **Infrastructure Services:** To include security, management and monitoring.
- **Business-Neutral Services:** To include service brokers and notification, scheduling and workflow services.
- **Business Services:** To include services based on the business domain e.g. credit card validation, address verification and inventory checks.

According to Erl (2007), services are governed by the following basic and core principles:

- Services are autonomous and self-contained.
- Services share a formal interface, called *contract*, which is platform independent.
- Services are loosely coupled.
- Services abstract underlying logic – underlying logic is invisible to outside world.
- Services are composable, allowing logic to be represented at different levels of granularity.
- Services are reusable and stateless.

Figure 1. Publish, find, bind and execute paradigm of SOA (Mahmood, 2007a)

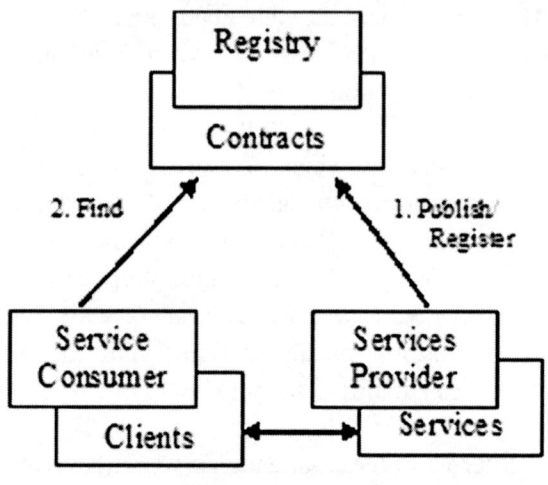

SOA uses a *publish-find-bind-execute* paradigm. The main components, as shown in Figure 1, include:

- **Service Providers:** Components (made available to *consumers*) that execute business functions using given inputs and producing outputs.
- **Service Consumers:** components that use services provided by service *providers* and published in a service registry
- **Service Registry:** a repository containing service descriptions for service consumers to know how services may be accessed.

Service Providers build services and offer them via an intranet or Internet. They *register* services with service brokers and *publish* them in distributed registries. Each service has an interface, known as *contract* and functionality, which is kept separate from the interface. The *Service Consumers* search for services (based on some criteria) - when found, a dynamic *binding* is performed. In this case, the service provides the consumer with the *contract* details and an *endpoint* address. The consumer then *invokes* the service.

Services are often implemented as Web Services (WS) which are applications accessed through using a network e.g. Internet. These are delivered using technologies such as eXtensible Mark-up Language (XML), Web Services Description Language (WSDL), Simple Object Access Protocol (SOAP) and Universal Description Discovery and Integration (UDDI). Thus, Web services provide the necessary standards to implement service oriented architectures: SOAP, WSDL and UDDI.

SOAP defines the structure of the messages and the protocol used for the standardized exchange of information in heterogeneous networks. It can be employed as the messages exchange protocol in a service-oriented architecture. Moreover, the service consumers do not know the necessary message format for the invocation of a service. Therefore, the service provider must provide the interface of the service. WSDL is a language employed by the service providers to describe their services in an interoperable manner. By employing SOAP and WSDL, the *invoke* operation, presented in the design model of a service-oriented architecture, can be applied. To implement the *publish* and *invoke* operations, another standard UDDI is defined. The fundamental underlying technology is the XML.

Many proprietary SOA tools and frameworks have also been produced for the development of web services and implementation of SOA. Majority of these are difficult to use and do not necessarily deliver the business benefits claimed. They lack vital capabilities like configuration control or testing prior to deployment. Hohpe [5, 8] believes the next generation tools will provide facilities for testing and debugging as well as provide support for monitoring and management. For a review of a number of such products from vendors such as BEA, Eclipse, IBM and CapeClear, refer to Mahmood (2007b).

SOA FOR E-GOVERNANCE

SOA has been successful in providing a framework for delivering software applications as connected services with flexibility and multi-platform and multi-channel integration that is required for connected government applications.

The primary requirement of such applications and e-services is the seamless interoperability in order to ensure spontaneous interactions among them. Generally, c-government services are highly disparate and distributed and hence the goal of interoperability gains a greater momentum in the context of effective and dynamic governance. A typical federal government has to have multiple and geographically distributed ministries, departments, agencies, and wings of federal governments. There would be multiple IT services providers assisting and automating different obligations of governments towards their citizens in a systematic and streamlined fashion. With the powerful emergence of the service paradigm, government tasks are being presented as a dynamic pool of reusable, composable, and accessible services. The role and relevance of service oriented architecture in taking government services to their authenticated and authorised users is constantly growing. IT providers are increasingly leaning upon the raging service concepts, tools, platforms, and infrastructures in order to elegantly conceptualize, concretize and deliver new-generation government services to people.

Indeed, as per the service idea, everything is a service so that for developing multifaceted government applications, all the participating and contributing services are carefully chosen, picked up from different service repositories being maintained by service developers and providers, and programmatically composed of the chosen services to form a well-intended and defined applications with all the functional as well as non-functional requirements e.g. quality of service (QoS) attributes embedded therein. In this model, the design and development of an application is concentrated on the smart combination of highly connected and distributed services on a network to form larger applications. There are numerous integration standards, platforms, patterns, practices to enable services assemblage to quickly drive well-formulated applications. As services bring in a clear-cut separation between interfaces and implementations, the goals of service modifiability, substitutability, replaceability, transparency, technology-agnostism and sustainability are eventually fulfilled without any major hitch or hurdle. That is, the service provider must offer a description explaining the necessary steps to invoke the offered services.

Referring to Figure 1, the objective of the publish-find-bind-execute architecture is to offer effective and flexible services in a network. To implement service-oriented applications in a normalized way, we use composite services: multiple services connected together. In fact, this technology proposes standards for the message exchanges and the service descriptions.

As mentioned before, Web Services (WS) are delivered using technologies such as XML, WSDL, SOAP and UDDI. This approach, which uses the services technologies, represents a good solution vis-à-vis a lot of the problems of c-government applications. It makes it possible to accomplish the long-standing interoperability requirement amongst a myriad of government organizations and administrations. Services offer a flexible environment allowing the much-needed cooperation between various governmental administrations. The result of the cooperation is in the form of e-services intended to accomplish citizens' needs. The following sub sections mention the citizens' requirements as well as the administrative requirements with respect to the provision of e-services and effective e-governance.

Citizens Requirements

The citizens' requirements define what a citizen wishes to obtain through a c-government application portfolio. Some of the well recognized requirements, taken from Sellami and Jamaiel (2007) are listed below.

Multiple Access Channels: The citizens want to have multiple ways of, and options for, obtaining e-services. This requirement can be met by the separation of the presentation layer from the application layer in the service architecture - whereas the presentation layer is responsible for providing the access interface to the services, the application layer presents business logic. Such a separation enables the presentation layer to have several forms (Web servers, mobiles and embedded servers), without having to carry out any modification at the applicative i.e. logic level. In this context, mobile-based access (mobile applications and browsers), web browsers, IVR systems, dashboards, public kiosks, smartphones and tablets, laptops, information and web appliances, wearables, etc. are the popular input and output devices for accessing and using content, data, services and applications.

Information Portals: Portals are turning out to be the most sought-after medium and mechanism for informing delivery and all kinds of correct and relevant details to citizens in a timely manner. The validated and verified information is made available to people via such kinds of portal solutions. There are multiple technologies and tools facilitating the design, development and sustainability of portals as the one-stop place for all kinds of services being made available to various categories of people and sections of the society.

Service Registry and Repository: For mobile services, there are several cloud-based services and content and data repositories across the globe. Similarly, for governments also, cloud-based repositories are becoming essential for empowering people to conceive newer services based on the available and authenticated e-services. Crafting government-centric and people-centric services

out of one or many discrete, atomic and basic services is the main motto behind the service repositories. However, in this case, the citizens need to have access to latest technologies such as mobile and hand held processing devices to access the cloud services.

Service Orchestration and Delivery Platform: Service composition through service orchestration or choreography is an important factor for the unprecedented success of the service paradigm across business verticals including governments. We need an adaptive platform for simplifying and streamlining service assemblage in order to accelerate the process of deriving service composites linking various services over networks using disparate devices and technological systems.

Service Security: All kinds of user identity and access control methods and security mechanisms need to be in place in order to restrict user authentication and authorization so that services cannot be misused or compromised and accessed by the unauthorised users. Similarly, the data privacy, confidentiality and security while data is in transit, persistence or usage, need to be taken care of for service integrity and assurance and citizens' trust. Data privacy and trust are of vital importance.

Authenticity of C-Government Applications: The citizens would like to be sure that they are connected to the government-provided applications belonging to the administration in question. For this, while connecting to an e-service or application, the citizens will need to receive a digital certificate proving the identity of the owner of the application. To implement this, we can use the SSL protocol which allows the authentication of a server – and this is just one of the many mechanisms.

Privacy Constraints: The general public want and need to be confident that the information they will provide over the network (e.g. identity card number, salary, social security number, etc) will be used for purposes it is intended for and by only the relevant departments who will have access to such information. Also, that the data and information will remain secure.

Data Confidentiality: The information sent by the member of general public has to remain confidential. The data sent to the presentation layer will have to be encrypted with the public key contained in the digital certificate of the concerned administration or by some other robust mechanism. For this, one solution is to employ the SSL protocol which makes it possible to implement encrypted sessions for the data exchange.

An Acquittal: After achieving the benefits from a service or the successful use of it, the consumer must be able to prove it, if necessary. At the end of the execution of a service, the application will have to respond to the citizen by a digitally signed document attesting the fulfilment of the service in question.

Administrative Requirements

In this section, we describe the constraints relating to the administrative requirements (Sellami & Jmaiel, 2007).

Interoperability: In general, each governmental administration is responsible for the development of its own information systems. Often, there are multiple heterogeneous devices and technologies in the form of hardware, transport protocols, programming languages, platforms, etc. resulting in enhanced complexity for administrators. This added complexity, induced through multiplicity and heterogeneity, can be cooled down by the smart leverage of the service principles and practices. Indeed, ensuring interoperability allows the cooperation and interaction between heterogeneous systems seamless and spontaneous since the underlying concept is technology-neutral.

Data Restriction: The access to confidential customer and corporation data is hugely restricted everywhere. Even in the governments too, the access is restricted considering the scope of possible misuse. The services with switches provide an appropriate solution for this requirement, while maintaining the autonomy of the c-government applications.

Citizen Authentication: The government administration must be able to authenticate and verify a citizen requesting to use a service. This is realized by using the *username/password* mechanism and other better authentification techniques such IAM (Identity and Access Management) during the communication between the administration and the citizen.

Administration Authentication: A government administration must be able to authenticate another administration which requires to use a service and to authenticate itself near the other administrations. Since, we are using the SOA approach; the different administrations are communicating using the SOAP transportation protocol. So, their authentication can be assured through a digital signature contained in the exchanged SOAP messages. The signature will be carried out in accordance with the Web-Services-Security standards.

Requestors' Integrity: The administration will need to be assured of the integrity of the received requests, i.e. it must check that they have not been deteriorated. A digital signature applied to a SOAP message makes it possible to the governmental administration to be sure that the data contained in the latter was not modified or compromised. This signature will also be inserted in the SOAP message in accordance with the WS-Security specifications.

Filtered Services Access: Some administrations may have the right to invoke a service while others may not have such rights. As a solution, we can allocate to each service different sets of *username/password*. To invoke a service, the consumer must provide the correct pair of authentication codes as an input.

Data Confidentiality: The data contained in the application's data layer and exchanged between the different administrations must remain private and confidential. Concerning the data layer, it has to be protected via a firewall to filter the connections requests by their IP address. Moreover, the data contained in the exchanged SOAP messages will need to be encrypted in order to ensure the confidentiality. Just as for the digital signature, encryption should be carried out following the WS-security specifications.

Services with Switches: Services enable devices and applications to talk with each other over any network. All kinds of dependencies and deficiencies are being eliminated with the leverage of service concepts while designing, developing, deploying and delivering service-based applications. However, there is, generally, a considerable lack of confidence in automated systems. As services bring in more aggressive automation, there are various options being explored in order to embed as much resoluteness in accessing data and services. Services with switches are being recommended as the best way forward in curbing all the illegal activities and curtailing the unauthenticated access. These services can be switched on or off at will by the chief administrator at the organisation owning the service. When switching off a service, it will be automatically replaced by a notification service whose role is to inform the administration that someone invoked the switched off service. At this point, there can be two options.

- **The Service is Activated (Switched On):** In this case, the access to the data base by the service is automatic. The event will be recorded in a log file. The administrator will be able to consult the execution result of this service through a follow-up monitoring device.
- **The Service is Deactivated (Switched Off):** When the service needs to access the data layer a notification will be sent to the administrator of the organisation. In the same way, this event will be recorded in a journal. Then, the employee will need to respond to the received request manually. He plays the role of an intermediary between the other administrations and the data layer.

The services with switches constitute a solution vis-à-vis the requirement and this is not to give an automatic access to the databases. With this solution, the services offered by an administration can be activated or deactivated according to their needs. Thus, an administration which wishes to offer an autonomous c-government application will be able to do it and others will be able to create applications where a human intervention is necessary.

Summary

In summary, the SOA architectural style has the following features:

- It is based on the design of the services which mirror real-world business activities comprising the enterprise (or inter-enterprise) business processes. Here, a government is treated as a complex enterprise.
- Service representation utilises business descriptions to provide the context (i.e. business process, goal, rule, policy, service interface, and service component) and implements the services using service orchestration, following some kind of Business Process Engineering (BPE).
- It places unique requirements on the infrastructure. It is recommended that implementations use open standards to realize interoperability and location transparency.
- Implementations are environment-specific and they are constrained or enabled by context and must be described within that context.
- It requires strong governance of service representation and implementation.

SOA is based on three main concepts: services, interoperability through an enterprise service bus (ESB), and loose coupling. An enterprise service bus is the infrastructure, which enables high interoperability between distributed systems for services. It makes it easier to distribute business processes over multiple systems using different platforms and technologies. Loose coupling is the concept of reducing system dependencies – also it gives rise to strong cohesion. SOA enables developers to reuse existing functionality to create new applications based on existing software services by packaging them differently and adding new services as required. The cost of developing new applications via SOA is therefore relatively low, as many well tested components and software modules would already exist. This is technically known as 'orchestration' or 'aggregation', a method in which new business processes and applications are combined and built from existing services, as illustrated in Figure 2.

Service Identification and Classification

As indicated above, there has to be a service registry and repository that hold interfaces for various services and the implementation logic. Instead of developing every service from the ground up, picking up the services that are already tested, tried, and trusted goes a long way in reducing the

developmental complexity and time. Services can be subscribed from other services too. The idea is to identify those services that are the possible candidates and compose them programmatically to create and sustain e-government applications and services. Following are the various categories of the services to be identified:

- Government to Citizen (G2C) Services.
- Government to Business (G2B) Services.
- Government to Employee (G2E) Services.
- Government to Government (G2G) Services.
- Shared Services.
- Informational Services.
- Interactive Services.
- Transactional Services.
- Integrated Services.

Government to Citizen (G2C) Services: These services provide information and facilitate electronic transitions such as paying bills, applying for passports and financial benefits, making appointments and renewing licenses. These services are about giving citizens the convenience of choosing when and where they access public services and how they consume such services. Governments

Figure 2. The steps for service-oriented e-governance application implementation (Wauters, P., Declercq K., et al., 2011)

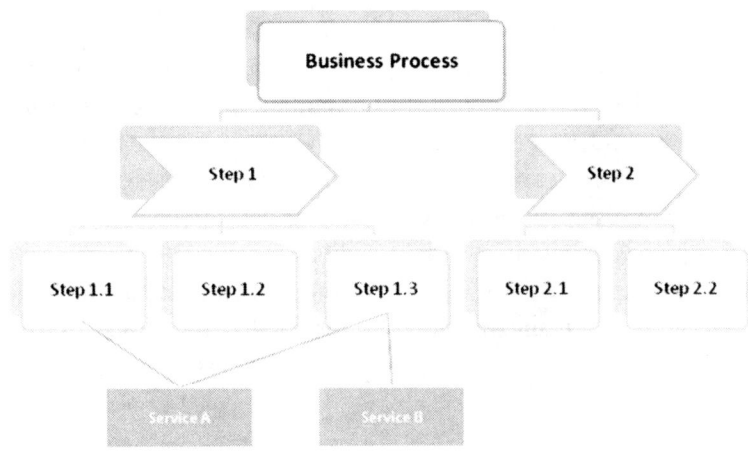

have to proactively formulate newer citizen-centric schemes and services as an obligation in order to provide extra care, convenience and choice.

Government to Business (G2B) Services: These services include providing information, such as industry standards, supplier directories and ratings, and conducting transactions, such as electronic quotations and company registrations. G2B is about making interactions, transactions, and communication faster, clearer and easier for businesses. Businesses need to invest in establishing newer types of industries and state-of-the-infrastructures in order to create new opportunities for citizens. Thus, the governmental support for employment generation via a direct and indirect support to industrial sectors is a must.

Government to Employee (G2E) Services: These services relate to transactions between employees and their ministries or agencies and include information management (via intranets), knowledge management (via some kind of content management systems) and collaborative and communication management (using e-mail, messaging systems, etc). All kinds of employee benefits need to reach out to the authorised and authenticated employees in time. In this context, IT is a great enabler for governments. Newer Web 2.0 and mobile technologies are proving highly beneficial.

Government to Government (G2G) Services: G2G services or transactions and interactions between ministries or agencies could include the provision of central services and information. For a variety of reasons, government agencies need to interact more closely considering the welfare of their people. Cooperation at the government level brings in more security, safety and prosperity. Service oriented systems for establishing a beneficial linkage among worldwide governments is being projected as the most appropriate investment for government. Just like B2B interactions, G2G connectivity is a grandiose enabler of many things.

Shared Services: To provide reliable and efficient e-services, a government agency will typically need to provide customer support, collect payment online, and authenticate the users. Customer service is a capability that will be needed for all e-services, to guide and support users. Services are becoming shared in order to fulfil their utilization rates substantially.

Informational Services: These include general information, made available to the general public in the form of instructions, information required by the users in the normal course of life, and information that pertains to the individual user and is likely to be private. As indicated above, information portals are the primary sources for conveying verified and validated information to all the stakeholders.

Interactive Services: These services are more sophisticated and formal interactions between citizens and government in which communication is conducted through email, online feedback, and the like, often using mobile and Web 2.0 technologies, as well. Interactive services also include the ability to search for records, download forms and applications, and to submit them. There are many applications that are extremely interactive, collaborative and transaction-centric. The service paradigm is being prescribed as the perfect approach for creating and sustaining interactive, informative and intuitive services.

Transactional Services: A government can provide a wide range of services through online transactions, such as enabling users to apply for housing, pay traffic fines, or apply for permits. It is at this stage that e-government becomes functionally interactive and therefore especially useful to the public. Transaction is an important criterion for e-commerce and e-business applications. For governments too, transaction has become an important feature and hence in the days to come, there will be much more of government-sponsored citizen-centric transaction services.

Integrated Services: With correct integration, services become seamless and client focused. Government services are clustered along common needs and linked for ease of use. Several atomic, discrete, and basic services are combined together to derive complex and integrated services that directly supports government processes thereby easily accomplishing all kinds of process innovation, excellence, orchestration, mining, simulation, management, control, integration, and good governance.

Identification of E-Forms: Electronic forms (e-forms) help the government departments to maintain a structured and streamlined process. Also, with e-forms, government departments process forms more efficiently, timely and securely, helping achieve cost savings and meeting the demands of an online department. Structure (metadata) of e-forms normally includes: name, data type, length, context, etc. Constructing electronic forms from metadata promotes the reuse of metadata across forms, which in turn reduces form authoring costs, and facilitates harmonization across forms. The c-governance initiatives within the government employ XML as a core technology to enable data interoperability, exchange, and reuse. The major types of e-forms are:

- Information Gathering Forms.
- Transaction Processing Forms.
- Regulatory Compliance Forms.

A FRAMEWORK FOR C-GOVERNMENT IMPLEMENTATION

The main purpose of a c-government is to maximise interaction between the governments and people with the effective use of Information and Communication Technologies (ICT). However, the current state of the art of ICT has improved tremendously with the attraction of social media and popularity of cloud computing technologies. Therefore, the current implementation of c-government technologies needs to be redeveloped and/or re-engineered to a service-oriented paradigm. ITU (2009) has proposed four dimensions of e-Government implementation framework as shown in Figure 3, to help implement ICT based services. This is also helpful for making ICT implementations systematically from a set of four views: Infrastructure, Policy, Governance, and Outreach.

Infrastructure: Dimension is the key to effective implementation of e-government services with the use of all available ICT facilities such as traditional and emerging technologies as well as supporting multitude of mobile devices. In the current trend, this should also include green energy for ICT and related devices to save electricity costs as the government is the promoter of green energy across the sectors.

Policy: With respect to e-government refers to promoting ICT businesses and to support business effectiveness with the use of ICT strategies

Figure 3. The four dimensions of e-government implementation strategy

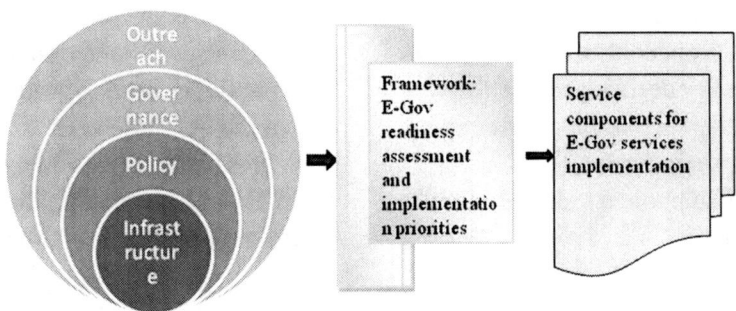

and protocols. Example of a Policy is to promote e-voting system by a forthcoming election or to provide an e-tax system for filing income tax returns.

Governance: Refers to performance of public administration and government departments efficiently. Improved governance can only be achieved with e-government initiatives and its implementation. The e-governance performance includes quality of governance with respect to time, cost, compliance, procedures, standards, process, effectiveness and consistency of the procedures adopted across different workflows and re-engineering.

Outreach: Is the fourth dimension of e-governance strategy which refers to integrated services provided by a government to businesses and citizens. It also includes services in supply and demand situation. SOA based implementation plays a major role in providing integrated, outreach on-demand, customisable, one-stop services to end-users.

The measurement of e-Government framework is based on the use of emerging technologies, a multitude of newer hand held devices (e.g. mobile telephones, netbooks, etc.), and multi-channels of communications. In addition, it also places emphasis on the width of services, depth of services, services coverage, and quality of services. Carromeu et al. (2010) have developed a number of WebApps based on Software Components and Software Product Line Engineering (SPLE) for Brazilian Government with especially established LEDES (Software Engineering Lab). They have also reported to have realised the importance of flexibility gained with component based approach to service design. Component based service development supports key design quality attributes that was set in our requirements for c-government applications (as shown in Figure 3).

A set of service components should include at least the following:

- Outreach:
  - E-Government Services to Citizens
  - E-Government Services to Businesses
  - E-Government Services to Interagency
- Governance:
  - E-Government Services to IT-Backoffice
  - E-Government services to Re-Engineering Public Processes
  - E-Government Services to Inter-Government Departments
  - E-Government Services to Local Governments
- Policy:
  - E-Government Services on Laws on E-Government Security (E-Governance principles)
  - E-Government Services for National Security Policies
  - E-Government Services for National Computer Forensics
  - E-Government Services for National Crimes Unit & Its Data Access for Secured Analysis
  - E-Government Services for Institutional Models of E-Government
- Infrastructure:
  - E-Government Services for Mobile Infrastructures
  - E-Government Services for Social Media & Emerging Technologies
  - E-Government Services for Broadband Infrastructure

C-Government services need to be designed as autonomous, self-composable and recoverable and thus providing service availability to customers, dynamically. These need to be re-configurable; thus providing service integration and support for multi-channels, scalable (multi-devices and platforms), adoptable (for use on heterogeneous devices) and reusable. One of the main aims of this chapter is to propose a SOA based c-government solution that is now presented in the following section.

SOA APPROACH FOR C-GOVERNMENT APPLICATIONS

As mentioned before, c-government applications require open, flexible, interoperable, collaborative and integrated architecture to provide services to various stakeholders such as citizens (e-health, e-tax, e-national security, e-pension, e-payment, e-education and training, e-work and employment, e-funding, etc), business organisations and vendors (e-procurement), and government agencies (both inter and intra governments). Therefore, designing architecture for c-government applications pose a design challenge.

Software architectural design needs to be verified, assessed and evaluated for its quality before its development and deployment. There are well known generic design criteria such as flexibility, maintainability, testability, reusability, etc. The key to achieving good architectural quality for a system being developed is to extract key characteristics of this system from various sources such as non-functional requirements, customer requirements, existing systems, experts, research literatures (Goteza, 2009; Sellami & Jamiel, 2007; Wauters, 2001; Open Group, 2009) and so on. This chapter identifies such characteristics from various perspectives supporting emerging technologies which are shown in Figure 4. These c-government application characteristics are the backbone for developing service-oriented architecture and components. For example, c-government applications need to work and interface with multi-platform and multi-vendor applications and therefore these need to be designed for interoperability. Multi-channel delivery of services has been recommended to deliver c-government services through various and emerging communication channels such as mobile phones, digital TV, call centres, kiosks, emails, PCs, teleconferencing, Web, and webinars - some of these are now being increasingly used by some governments, at present.

Figure 4. Characteristics of e-government applications

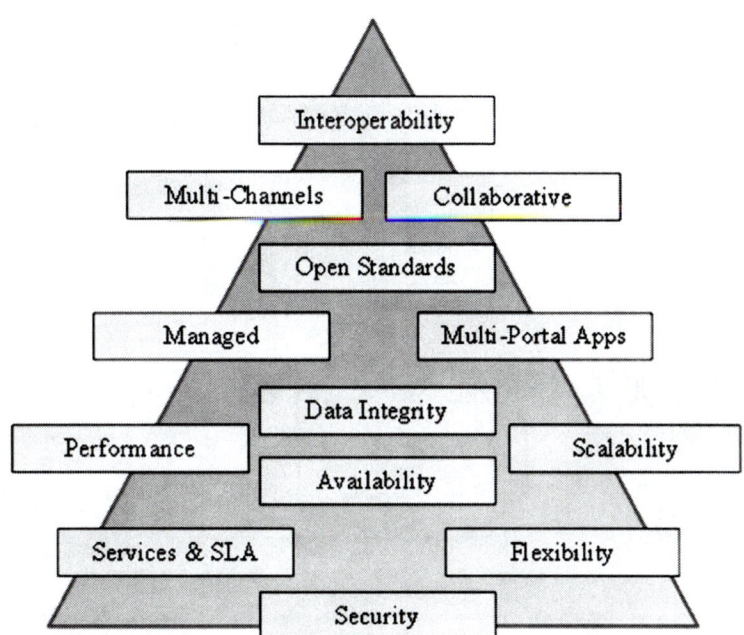

Each characteristic shown in Figure 4 represents not only the application attribute but it also provides a set of key design criteria and quality attributes for developing architecture that support emerging technologies for e-government. For example, the design criteria interoperability is paramount for c-government applications as they are multi-platform and multi-channel based. Interoperability is supported in the service-oriented architecture design for e-government applications by means of a service bus and loose coupling of the other components which is shown in Figure 5.

These services can be made available as stand alone, integrated, componentised, web based service component, composite service (a set of interconnected services), virtualised services (e.g. cloud based), and dynamically re-configurable services. The architecture presented in this chapter is then based after a critical review and analysis of a number of existing architectures for c-government applications. As mentioned before, the SOA based architecture consists of four distinct levels of abstraction layers which are connected and communicated by messages through a core communication channel known as a *service bus* or a central bus. These layers are: 1) a *business* layer with a dedicated set of services, 2) an *orchestration* layer with a set of services where new services can be composed, 3) an *e-government* (or c-government) layer that supports integration of services, government departments and local governments, and 4) an *e-business* layer that supports new businesses and integration of data. The SOA based architecture for c-government services then ensures that it achieves the expected service-oriented design factors such as customisation, cost-effectiveness, availability, etc.

Figure 5. Service-oriented architecture for e-government

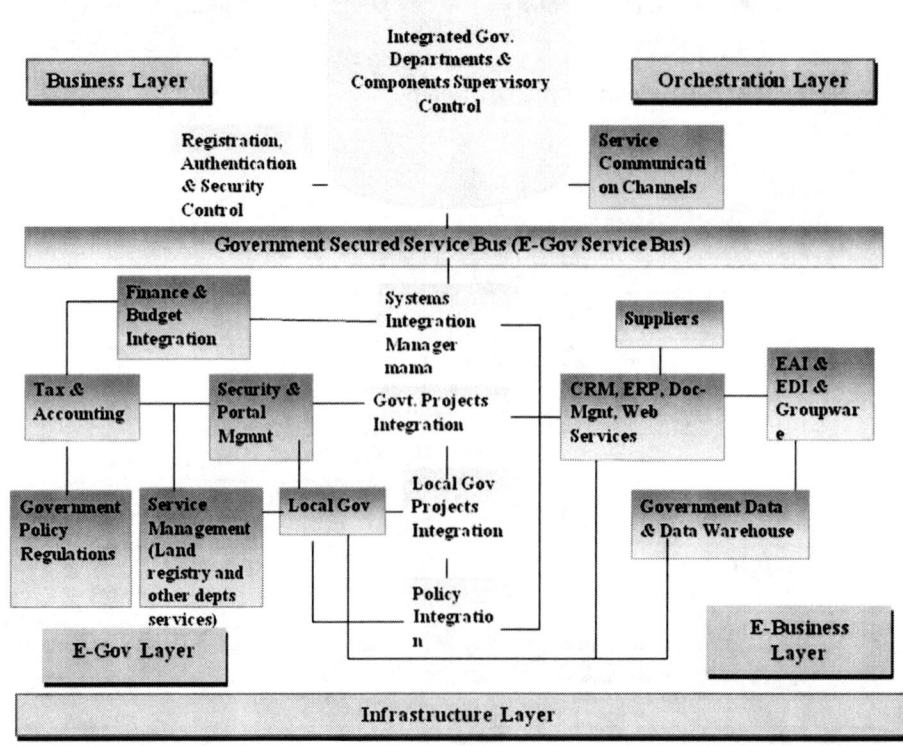

Referring to Figure 5, at the business and orchestration layers provide high level service composition based on new business perspective and policies (both political and economical factors). Mostly, the customisation and the new business needs arise from these two key variables. The sub-systems such as registration control, security control, integrated services for e-government applications control, and communications channels help to achieve customisation at a higher level of abstraction without affecting underlying business logic services. These are communicated and connected to layers below using a concept of service bus known as e-government (or c-government) secured service bus. The layer below the business layer provides services from various e-government departments, and external suppliers (e-business layer).

A SERVICE COMPONENT MODEL FOR C-GOVERNMENT APPLICATIONS

Here, we present another approach to developing c-government applications. As stated before, e-government applications require open, flexible, interoperable, collaborative and integrated architecture to provide services. These services can be made available as stand alone, integrated, componentised, web based service component, composite service (a set of interconnected services), virtualised services (cloud based), and dynamically re-configurable services. This vision is similar to the Open Group's Service Integration Maturity Model (OSIMM Open Group, 2009). The OSIMM provides:

- A process roadmap for attaining key practices with metrics.
- Seven levels of maturity to improve.
- A quantitative model for assessing current practices and to improve with recommended practices.

In order to achieve the recommended process and key practices, we need to have interoperability, reconciliations and resiliency between systems, where interface linkages are appropriate to other services such as those of emerging technologies interfaces (e.g. *IServiceInterfaceFacebook*, *IServiceInterfaceLinkedIn*, *IServiceContractGovRegulations*). Refer to Figure 6. Similarly, such reconciliations services must be automated for the sake of cost-efficiency. Design flexibility must be maintained where there is a need for dynamic re-configuration of services (e.g. when a *TaxService* encounters an unknown identity of a person

Figure 6. Service component model for e-government applications

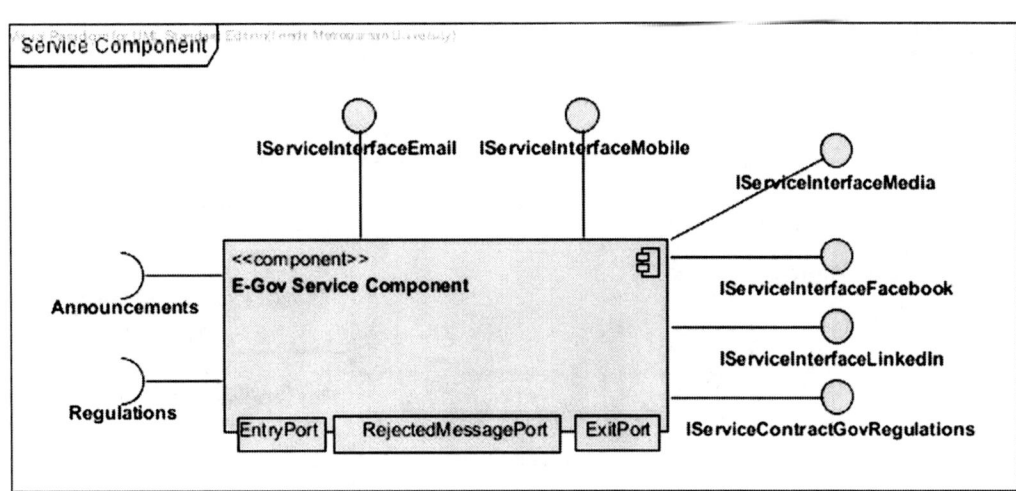

or a work place). In this scenario, at run-time, the *TaxService* could look for identification of a person and re-configure itself to check against all social media portals and local governments portals. Similarly, an *ImigrationService* could encounter similar issues where national integrity can be protected. Therefore, designing systems with component based services can play a major role in utilising emerging technologies more effectively.

To develop an integrated c-government service-oriented architecture system, it is critical that service analysts and software engineers identify and address the key challenges when implementing c-government services. Some of the relevant questions are listed below:

- How to adopt a new process and to identify the scope of the business services to be developed and supported?
- How data integrity will be achieved and secured?
- How data will be protected when needed to support management decisions?
- How systems will fit together to support service choreography and orchestration layers?
- How systems will fit together to support the enabling emerging technologies (system architecture needs to be service-orient-ed thereby any new emerging technologies in the future should easily be integrated more cost-effectively than with traditional system architectures)?
- How data and services will be safeguarded and secured to ensure the integrity and re-configurability of service operations and personal data (information assurance)?
- How new services can be composed and re-configured at run-time?

With this background, we now propose an integrated service component model for an e-TaxService as shown in Figure 7. This is highly customisable component model providing a type of services known as *TaxRulesBands* that is a required service requiring variable parameters based on two dynamic variable factors viz: economic and political. This component provides services on *ISelfTaxReturn*, *BusinessTaxReturn*, and interfaces with the revenue department, *InterfaceRevenueDept*.

In addition, this component model also provides highly secured interfaces known as ports. There are three types of port as it is a service component viz: Entry Port, Rejected Message Port (RMP), and Exit Port.

Figure 7. Service component model for e-TaxService

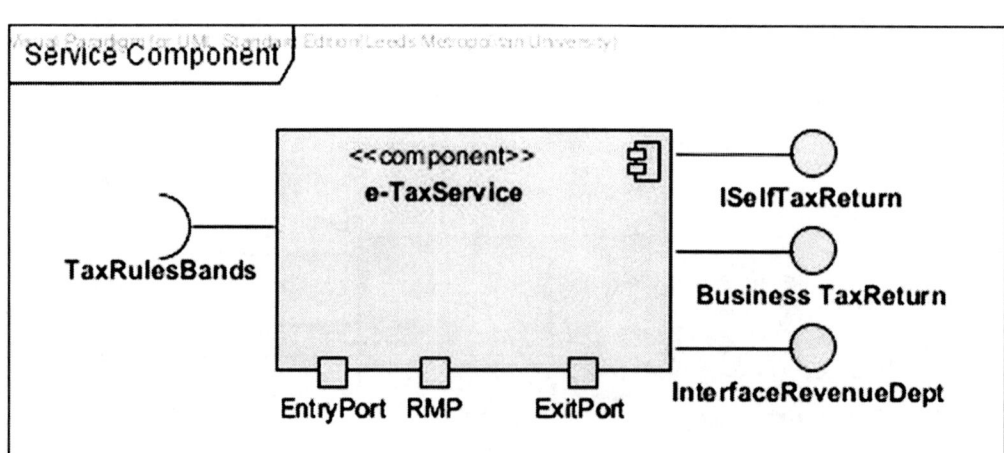

CONCLUSION

C-Government applications need to be developed to suit modern technologies (e.g. Web 2.0, social media, cloud computing, enterprise computing, mobile technologies, etc) for providing services dynamically, cost-effectively and in an easy to access manner. This chapter has suggested the use of a service oriented architectural (SOA) framework for the development of such applications. It is based on a set of evaluated application characteristics that support emerging technologies including web services and mobile technologies. The SOA concept has been presented and discussed how it can be beneficially used to achieve seamless integration of large applications developed by various governmental and other agencies including the business sector. The chapter has also proposed the use of service component models that provide the required customisation, reuse, flexibility, interoperability and extensibility. Based on the service oriented architecture (SOA) framework, a large scale sub-system known as e-*Taxservice* has been used as a case study with service design which has been validated against a set of key service quality attributes.

REFERENCES

Borras, J. (2004). International Technical Standards for E-Government. *Electronic Journal of E-Government, 2*(2).

Carromeu, C., et al. (2010). *Component-based Architecture for e-Gov Web Systems Development.* Paper presented at the 17th IEEE International Conference and Workshops on Engineering of Computer Based Systems (ECBS). Oxford, UK.

Cartwright, I., & Doernenburg, E. (2006, May). Time to jump on the SOA bandwagon. *IT Now*.

Centeno, C., Van Bavel, R., & Burgelman, J. C. (2004). *E-Government in the EU in the Next Decade: The vision and Key Challenges*. Seville, Spain: Inst for Perspective Technological Studies.

Erl, T. (2005). *Core principles for service-oriented architectures*. Retrieved Oct 2013, from www.looselycoupled.com/opinion/2005/erl-core-infr0815.html

Goteza, J. et al. (2009). Cross-National Interoperability and Enterprise Architecture. *Informatica, 20*(3), 369–396.

Groves, D. (2005, September 11). Successfully planning for SOA. *BEA Systems Worldwide*.

Hohpe, G. (2002, May). Stairway to Heaven. *Software Development*.

ITU. (2009). *E-government implementation toolkit: A Framework for e-Government Readiness and Action Priorities*. Retrieved from http://www.itu.int/ITU-D/cyb/app/docs/eGovernment%20toolkitFINAL.pdf

Mahmood, Z. (2007a). Service Oriented Architecture: Potential Benefits and Challenges. In *Proceedings of WSEAS Int Conference*. WSEAS.

Mahmood, Z. (2007b). Service oriented architecture: tools and technologies. In *Proceedings of WSEAS Int Conference*. WSEAS.

Mahmood, Z. (2007c). Service Oriented Architecture: A New Paradigm for Enterprise Application Integration. In *Proceedings of WSEAS Int Conference*. WSEAS.

Open Group. (2009). *OSIMM*. Retrieved Oct 2013, from https://www2.opengroup.org/ogsys/jsp/publications/PublicationDetails.jsp?publicationid=12450

Sellami, M., & Jamiel, M. (2007). *A Secured Service-Oriented Architecture for E-government in Tunisia*. Retrieved from www.redcad.org/PDFs/C33.pdf

Wauters, P., Declercq, K., van der Peijl, S., & Davies, P. (2011). *Study on cloud and service-oriented architectures for e-government.* FP7 report ref. Ares(2012)149022 - 09/02/2012.

Zimmermann, O., Krogdahl, P., & Gee, C. (2004, June 2). *Elements of service-oriented analysis and design.* IBM Corporation.

ADDITIONAL READING

Agostino, D. (2012). The Effectiveness of Social Software for Public Engagement.

Alio, J., Ibrahim, M., Pickton, D., & Bassford, M. (2009). A systems-based media effectiveness framework for e-marketing communications. In, Systems Conference, 2009 3rd Annual IEEE. IEEE, pp. 150-55.

Barnes, D., The service oriented architecture: more than just another TLA?, *British Computer Society.*

(2009). *BEAN, B.* Questions For Evaluating Social Media Effectiveness.

Bertot, J. C., Jaeger, P. T., & Grimes, J. M. (2010). Using ICTs to create a culture of transparency: E-government and social media as openness and anti-corruption tools for societies. *Government Information Quarterly*, 27(3), 264–271. doi:10.1016/j.giq.2010.03.001

Blecher, L. (2011). *Measuring social media: Free tools to quantify your social influence.*

Chang, A.-M., & Kannan, P. (2008). *Leveraging Web 2.0 in government.* IBM Center for the Business of Government.

Chua, A. Y., Goh, D. H., & Ang, R. P. (2012). Web 2.0 applications in government web sites: Prevalence, use and correlations with perceived web site quality. *Online Information Review*, 36(2), 175–195. doi:10.1108/14684521211229020

Clark, L., SOA gathers pace in the enterprise, *Computer Weekly, UK*, 26 Sept 2006.

Colan, M., Service-oriented architecture expands the vision of web services – part1, *IBM Corporation*, 21 April 2004.

Collins, S. (2009). Government 2.0, e-Government and Culture. In J. Gotze, & C. B. Pedersen (Eds.), *State of the Eunion: Government 2.0 and Onwards.* AuthorHouse.

Danis, C., Bailey, M., Christensen, J., Ellis, J., Erickson, T., Farrell, R., & Watson, W. A. K. I. T. (2009). Mobile applications for the next billions: a social computing application and a perspective on sustainability. Strengthening the Role of ICT in Development:309.

Delone, W. H. (2003). The DeLone and McLean model of information systems success: a ten-year update. *Journal of Management Information Systems*, 19(4), 9–30.

DeLone, W. H., & McLean, E. R. (1992). Information systems success: the quest for the dependent variable. *Information Systems Research*, 3(1), 60–95. doi:10.1287/isre.3.1.60

DiMaio, A. (2009). Government 2.0: A gartner definition. Industry Research, ID (G00172423).

Eggers, W. D. (2007). *Government 2.0: Using technology to improve education, cut red tape, reduce gridlock, and enhance democracy.* Rowman & Littlefield.

European Commission. (2003). *The Role of E-Government for Europe's Future.* Brussels: European Economic and Social Committee Report, European Parliament.

Fowler, M. (2002). *Patterns of enterprise application architecture.* Addison Wesley.

Freeman, R. J., & Loo, P. (2009). Web 2.0 and E-government at the Municipal Level. In, Privacy, Security, Trust and the Management of e-Business, 2009. CONGRESS'09. World Congress on. IEEE, pp. 70-78.

Gruen, N. (2009). *Engage: Getting on with Government 2.0 Report of the Government 2.0 Taskforce*. Australian Government Information Management Office.

Hohpe, G. Developing Software in a service-oriented world, *Whitepaper, ThoughtWorks Inc.*, Jan 2005. Johnson, B., The benefits of service oriented architecture, *Objectsharp Consulting*. Retrieved Oct 2013, from http://objectsharp.com/cs/blogs/bruce/pages/235.aspx

Kes-Erkul, A., & Erkul, R. E. (2009). Web 2.0 in the Process of e-participation: The Case of Organizing for America and the Obama Administration.

Knorr, E. & Rist, O., 10 Steps to SOA, *Info World - San Mateo*, Vol. 27, Issue 45, pp. 23-35, Nov 2005.

Kodali, R. R. What is service oriented architecture? *JavaWorld.com*, 13 June 2005. Retrieved Oct 2013, from http://www.javaworld.com/javaworld/jw-06-2005/jw-0613-soa.html

Kuzma, J. (2010). Asian government usage of Web 2.0 social media. European Journal of ePractice (9), 1-13.

Magro, M. J., Ryan, S. D., Sharp, J. H., & Ryan, K. A. (2009). Using social networking for educational and cultural adaptation: an exploratory study.

Mahmoud, Q. H., Service-oriented architecture and web services: the road to enterprise application integration, *Sun Microsystems Inc.,* April 2005.

Misra, H. Understanding SOA Perspective of e-Governance in Indian Context: Case Based Study in User Interfaces (Ed Matrai R), Interopen, May 2010, available at: www.interopen.com

O'reilly, T. (2007). What is Web 2.0: Design patterns and business models for the next generation of software. Communications & strategies (1), 17.

Osimo, D. (2008). Web 2.0 in government: Why and how. Institute for Prospectice Technological Studies (IPTS), JRC, European Commission, EUR 23358.

Overall, D., Have we been there before?, *Opinions, Computer Weekly, UK*, 11 April 2006

Oxford Dictionary Online, -. (2013).

Raeth, P., Smolnik, S., Urbach, N., & Zimmer, C. (2009). *Towards Assessing the Success of Social Software in Corporate Environments* (p. 662). AMCIS.

Retrieved Feb. 2007, from www.bcs.org/server.php?show= ConWebDoc.3041

Saran, C., SOA will fail without governance: warns Gartner, *Computer Weekly, UK*, 12 Sept 2006.

Sivarajah, U., & Irani, Z. (2012). Exploring The Application Of Web 2.0 In E-Government: A United Kingdom Context. In, Transforming Government Workshop.

Solis, B. (2010). Engage: The complete guide for brands and businesses to build, cultivate, and measure success in the new web. Wiley. com. Sonic Software Solutions, Service oriented architecture. Retrieved Oct 2013, from www.sonicsoftware.com/solutions/service_oriented_architecture/index.ssp

Steinhuser, M., Smolnik, S., & Hoppe, U. (2011). *Towards a Measurement Model of Corporate Social Software Success-Evidences from an Exploratory Multiple Case Study* (pp. 1–10). IEEE.

Stewart, D. W., & Pavlou, P. A. (2002). From consumer response to active consumer: measuring the effectiveness of interactive media. *Journal of the Academy of Marketing Science, 30*(4), 376–396. doi:10.1177/009207002236912

Veljković, N., Bogdanović-Dinić, S., & Stoimenov, L. (2005). Building E-Government 2.0-A Step Forward in Bringing Government Closer to Citizens. e-government 1 (1989).

Zhang, W., Johnson, T. J., Seltzer, T., & Bichard, S. L. (2010). The Revolution Will be Networked The Influence of Social Networking Sites on Political Attitudes and Behavior. *Social Science Computer Review*, 28(1), 75–92. doi:10.1177/0894439309335162

KEY TERMS AND DEFINITIONS

Component Based Design: Designing software as components provides a self-dependent unit of abstraction with two types of interfaces for connecting and composing with other components. CBSE supports reuse and productivity gains.

Data Integrity: This refers to the validity of data. Since, data can be compromised in a number of ways, various authentication and security mechanisms have been devised to minimize the threats to data integrity.

E-Government: This refers to the provisioning of a government's functionality and its services using the web and communications technologies including the latest software development and deployment paradigms and related methodologies such as Web 2.0 and mobile technologies.

Flexibility: This is often used term that describes a property of software systems. It is the ability of software to change to be modified easily in response to different user and system requirements or conditions. And, if the software fails, it fails in a predictable fashion.

Granularity: It is a characteristic that refers to the extent to which a system contains separate components (like granules). The more components in a system or the greater the granularity; the more flexible it is and so higher the granularity. Similarly, the components that are more specific are more *fine* grained.

Interoperability: This refers to the ability of software and hardware on different machines from different vendors to share data. Interoperability can become an issue if the vendors or products use particular or specific standards. Using open standards is the answer.

Multi-Channel: This refers to medium of communication between parties and is a delivery of e-government services through various and emerging communication channels such as mobile phones, digital TV, call centres, kiosks, emails, PCs, teleconferencing, Web, and webinars.

Resource Management: This refers to the efficient and effective deployment of an organization's resources when they are needed. Such resources may include financial resources, inventory, human skills, production resources as well as formation technology resources.

REST: *Representational State Transfer* is an architectural style for large-scale software design. REST emphasizes scalability of component interactions, generality of interfaces, independent deployment of components, and intermediary components to reduce interaction latency, enforce security, and encapsulate legacy systems.

Scalability: This is a one of the often sought property of software and hardware systems that can improve performance and speed by adding or removing more processors, memory, and other resources. Scalability can be *horizontal* or *vertical*.

Service Oriented Architecture (SOA): This is an architectural style for developing and integrating large applications. The basic building block is a *service* that is independent and self contained. The application system is a collection of such services linked in the correct way.

SLA: The acronym stands for *service-level agreement*. It is a contractual agreement on the level of service to be provided by a service provider to a customer. The level of service refers to the availability, performance, robustness and other quality of service attributes.

SOA Standards/Open Standards: These provide a unified framework for designing service oriented software systems. SOA standards are becoming much more evident. There are four very important SOA architecture standards from The Open Group that provide the basis for business to create their SOA solutions. These are: SOA Reference Architecture; Open Service Integration Maturity Model (OSIMM); SOA Governance; and SOA Ontology.

SOAP: The *Simple Object Access Protocol* is a lightweight XML-based messaging protocol used to encode the information in Web service request and response messages before sending them over a network. SOAP messages are independent of any operating system or protocol and may be transported using a variety of Internet protocols, including SMTP, MIME, and HTTP.

Web Service Security (WS-Security): This is a W3C specification that sets the standards for protecting web services from potential security attacks. WS-Security is an extension to SOAP to apply security to web services.

Web Services: This is a distributed computing model that allows application to application communications. This also allows developers to add web services to their applications much more easily. Web services do not provide GUIs, instead these shares business logic, data, and process through a programmatic interface across a network thus establishing distributed computation.

WSDL: The *Web Services Description Language* is an XML-formatted language used to describe Web services' capabilities as collections of communication endpoints capable of exchanging messages. WSDL is an integral part of UDDI, an XML-based worldwide business registry. WSDL is the language that UDDI uses. WSDL was developed jointly by Microsoft and IBM.

XML: The acronym stands for *Extensible Mark-up Language*. Defined by W3C, it is a language for specifying document styles. An XML file looks similar to an HTML file but it allows the use of user defined tags and is much more flexible with additional facilities.

Chapter 2
Enhancing Data Management in E-Government Using Data Categorization and Visualization Techniques

Miloš Milutinović
Belgrade University, Serbia

Marijana Despotović-Zrakić
Belgrade University, Serbia

Konstantin Simić
Belgrade University, Serbia

Mihajlo Anđelić
Belgrade University, Serbia

ABSTRACT

Modern information and communication systems can process massive amounts of data automatically and efficiently, but human beings have a limited cognitive capacity. Organizations that handle data on a larger scale need to adjust and streamline their operations in order to cope with this complexity. The e-Government information systems present problems that are on an entirely different scale, with communication streams between citizens and government easily dwarfing those in the private sectors. The information flows in e-Government are further constrained by established processes and practices that are hard to change and strict privacy concerns. A solution to problems of complexity and inefficiency of data manipulation in e-Government is needed. This chapter analyzes models and techniques of data categorization and visualization that can be employed in the context of e-Government. Methods of categorization, metadata, and ontologies in particular are explored for use in such an environment. A simple government ontology framework is developed as a starting point for introduction of ontologies into the e-Government context and the information is structured in such a way to allow easy correlation and navigation between concepts. A simple but intuitive visual representation of information and their relations is developed to facilitate better understanding of complex topics.

DOI: 10.4018/978-1-4666-6082-3.ch002

INTRODUCTION

One of the key issues in the internal work of any organization that deals with internal and external information flows are the difficulties with manipulating, processing, analyzing, and understanding the information in the system. This is an issue for both the humans and the machines that take part in the process. For the machines, it manifests itself as a problem of interoperability and understanding data in different formats, while for the humans the issue becomes the availability of right information at the right time and choosing the appropriate representation of said information. This is especially significant in an e-Government context. Every government represents a complex, all-encompassing system, composed out of a number of hierarchically ordered entities that communicate between themselves and with citizens, private organizations, companies, foreign institutions, etc. Modern technologies have also contributed to an increase in the amount of citizen-government communication, leading to a significant upsurge in the complexity and volume of information in e-Government systems.

E-Governments attempt to improve their efficiency by introducing new technologies and digitalizing and virtualizing existing processes. This can engender problems typical for environments that experience rapid introduction of new technologies without a general plan – problems relating to the standardization of protocols, processes, data, centralization of services, and integration and interoperability of diverse systems.

Interoperability is a critical requirement in this context, and it refers to a property that enables different systems and organizations to work together in order to achieve common goals. Interoperability is achieved by applying common standards and work practices in existing processes or restructuring them to comply with external requirements in order to enable exchange of information and knowledge (Interoperability Solutions for European Public Administrations,

2010). Interoperability requires a certain level of compatibility between systems that take part in the exchange of information in order to assure its proper interpretation. Technological diversity often makes this difficult and required compatibility can be achieved by applying abstractions to hide complexities and implementation details.

A number of constraints that influence government integration and interoperability are defined in (Scholl & Klischewski, 2007) and include constitutional/legal, jurisdictional, collaborative, organizational, informational, managerial, cost, technological, and performance constraints. A similar, but more specific list of barriers to achieving interoperability on a wider scale using modern information systems and technologies is presented in (Landsbergen Jr. & Wolken Jr., 2001) and includes:

- **Political:** Citizen privacy concerns, agency statutory authority ambiguities, and openness to public scrutiny.
- **Organizational:** Lack of mutual trust between agencies, lack of experience in developing interoperable systems, and lack of awareness of possibilities for data sharing.
- **Economic:** Lack of resources and "low-bid" procurement methods that favour approaches which are cost-effective in the short run.
- **Technical:** Hardware/software incompatibility, reliance on private contractors, and data-sharing standards.

Although interoperability depends on a number of factors, the core of the problem is the standardization of underlying data formats and developing methods for data management, interlinking, access authorization and presentation appropriate to entities or individuals utilizing the data. The solution applied to this problem will define other, higher-level technical, economic, organizational and political problems and barriers to introducing interoperability.

The issue of structuring information in e-Government is also a critical one when considering the processes of analyzing the current situation and decision-making. Government employees and decision-makers need to have a holistic view of the entire situation in order to make valid decisions. Decision making in modern society is heavily dependent on cooperation with other government units, as well as with other external entities that control necessary resources (both material and non-material). Decision-makers are also unlikely to posses all of the required knowledge, so the process needs to be decentralised. Attempts are made to include the citizens into an interactive decision-making process in order to generate better policies, and bridge a democratic gap between government and its citizens (Klijn & Koppenjan, 2000). Information resources used in the decision-making process need to provide clear insight into the problem and describe all relations with other external concepts that either contribute to the problem or are influenced by the proposed measures.

By describing the informational resources in the system using a rich, descriptive, standard set of attributes, additional semantics (meaning) can be attached to them. These attributes represent metadata, data about data, and can facilitate better, more efficient information use at any level. Machines can use the strictly defined relationships between information objects to apply higher-level logic in their processing. They can also provide better browsing and searching capabilities and generate results that are more relevant to user queries. Using standard attributes and objects inherently facilitates interoperability, and governments can utilize them to describe different sources of data maintained by different organizational units, mix them, and have them presented appropriately. For an example, a regional government could access data about tourist facilities, regional specifics, policies and strategies, and government-funded research results about water quality, then combine it all, and compare it with other regions in order to create a new policy for water park development. In order to achieve this, a sufficiently complex meta-model needs to be defined, as well as the appropriate controlled vocabularies for values describing the individual information resources. There are different methods of achieving this, including the use of ontologies, an extremely powerful tool for description of any concepts and their relations.

When all resources are appropriately semantically annotated, existing processes can be simplified and made easier to understand, and novel methods of presentation can be applied. Additionally, applying visual methods of presentation like graphs and trees can abstract extraneous details to allow focusing on the desired topic. In the domain of web technologies, it is especially easy to find free visualization libraries of high quality that can be plugged in into existing systems and fed relevant information sources. When dealing with unannotated or weakly annotated information sources, applying visualization techniques can be very hard and the results might amount to displaying a number of resource "islands," disconnected from each other and the wider context. On the other hand, when there is an abundance of semantic attributes and relations between resources, various forms of presentation can be selected. Certain resources can be abstracted, elements and relations filtered or highlighted, and different perspectives generated in order to observe a certain aspect of the resource set being analyzed.

This chapter attempts to describe a general model for handling documents and other information resources in e-Government. Visualization techniques that can be applied to annotated documents are also explored, with the goal of creating a powerful environment for exploration of semantic relations that would serve as a basis for decision-making. Special attention is given to interactions between government and citizens for their frequency, large scale, and impact on citizen satisfaction, but other domains in which governments operate are also considered.

The structure of this chapter is as follows. In the second part, the concept of classification is explained and relevant methods are listed with an emphasis on using ontologies. In the third part of the chapter, different classification methods are compared from the perspective of their applicability in e-Government systems. Model for classification of information resources in government is defined, based on an ontology of government concepts, which was judged as being the most suitable for the context of e-Government. The fourth part gives a short overview of the advantages of quality visualization techniques for large amounts of data. In the fifth part of the chapter, methods of visualization compatible with classification models previously given are presented. The visualizations are designed to augment the existing system and their role is explained. Finally, in the sixth chapter, a discussion of presented model is given, along with some possibilities for future extension and directions for future research.

DOCUMENT CLASSIFICATION

The process of finding relevant information buried in a sea of documents and resources can be a daunting task for a government employee. The pattern of accessing a complex information base consists of two phases: the locator phase and the navigation phase (Sacco, 2000). In the first, locator phase, the user attempts to find the desired information by formulating a query and inspecting the result list. In the following, navigation phase, the user uses the set of retrieved items as a starting point for further exploration. Problems can arise from the fact that the users are not always certain what precisely they are looking for, nor what the information content at their disposal is. Users' queries might be imprecise resulting in too large or too small result sets. This makes it difficult for government employees to understand the wider picture, indicating that there is a need for a system that would organize the information in some useful, efficient way.

Accordingly, when dealing with massive amounts of data encapsulated in the form of documents or other resources, a method for browsing and discovery of relevant information needs to be provided. An elementary but not very efficient solution can be the application of brute force on the available data. User-inputted terms or patterns can be used to perform a full-content search on all of the documents. This approach is valid in some contexts and can always be provided at least at some scale, complementing other methods that will first narrow the result set to a more manageable size. This narrowing still remains a problem: a full-content search on a large set of documents can take time and consume a large amount of computing resources, potentially slowing down the system for other users as well. The results themselves often do not satisfy the needs of the user for several reasons and additional, manual browsing is required. Instances of queried terms will be discovered even if they are only mentioned marginally in irrelevant documents or in different contexts. Unless there is some special, linguistic logic integrated into the system, synonyms, related terms, and wrongly spelled words will be ignored. Another point that needs to be considered is the format of available resources. Even if the content is mainly comprised of text, it can still be packed into documents of different formats that need to be compatible with the system performing the search. Resources that are represented in the form of images, audio, or video data can only be searched using specific, often experimental, semantically weak, resource-hungry, and unreliable methods like comparing the colours or patterns in an image.

Classification and Metadata

Clearly, some form of classification of resources that would facilitate the entire process of browsing and discovery is needed. Broadly speaking, classification is the process of categorizing information resources in accordance to some pre-existing hierarchy or scheme. If such a scheme is well defined and applied in a wider context, it will ensure that different parties can manage the complying resources consistently. The definition of the applied schema is called a meta-model: a model that defines data structures and relations describing data and resources in some domain. The values stored in these structures represent metadata since they are, in accordance to the metadata definition, "data about data." A more relevant term for these values is "descriptive metadata" or "metacontent," which refers to metadata that exists to describe individual instances of data content. There are two perspectives on the focus of metadata (Burnett, Ng, & Park, 1999). The first states that metadata records characterization and relationships of the source data; it identifies location of information objects and facilitates collocation of subject content. This perspective is indeed aligned with outlined needs of government employees, but the other, more technical perspective needs to be taken into account. The other perspective states that metadata is any information that supports effective use of data, including data management, data access, and data analysis (data mining, visualization). Since these perspectives are not incompatible, metadata should both describe source data in order to support its discovery and also facilitate its effective use in information systems.

By defining or selecting an appropriate, sufficiently descriptive meta-model, more complex and detailed search queries can be defined, leading to better results. When defining data attributes to be associated with objects in the model, an existing vocabulary for metadata can be used. General-purpose vocabularies provide the highest level of interoperability between different systems but might need to be expanded in order to satisfy the specific needs of a domain in which they are used. One of well-known vocabularies is Dublin Core that defines a set of 15 properties applicable to information resources, including such as "title," "subject," and "description," with an intention of facilitating their discovery (Weibel, 2005a). Values that individual documents will have for

these attributes are well suited to be used for classification purposes. In the case of the Dublin Core vocabulary, the "subject" attribute is the most suitable choice, since it is defined as the topic of the resource, containing keywords or phrases that describe the subject or content of the resource, including controlled vocabularies or classification schemes (Dublin Core Metadata Initiative, 2012).

Some of the attributes defined by the meta-model can only take values from a strictly defined set; a good example are attributes that contain time values: for instance, months can only take 12 different, specific values. Other attributes can have values freely inputted by the user. The attributes that describe the topic of the resource, e.g. the subject attribute in the Dublin Core vocabulary, are of special interest for purposes of classification of information resources. If the users of a system freely fill in such attributes, many problems typical for cases when there is a lack of standardization can emerge. Some users might prefer certain synonyms for description of topics, others might use terms that overlap with other terms, adding to the confusion. In order to perform an efficient, all-encompassing classification of all information resources in a domain, some kind of a controlled vocabulary for document subjects is required. This is the main concern of subject-based classification.

Subject-Based Classification

There are several methods of subject-based classification, the main being controlled vocabularies, taxonomies, thesauri, faceted classifications and ontologies. Many of these methods are often freely combined with other mechanisms, adjusted, and expanded, and often the names of the methods are used loosely to denote different approaches. Still, the basic definitions of listed methods are as follows (Garshol, 2004):

Controlled Vocabulary: The basic method where only discrete terms are defined, but not relations between them. The users do not input their own terms but select one or more from the vocabulary and therefore cannot make mistakes, define meaningless or too broad terms, or use synonyms.

Taxonomies: Denotes a hierarchical system of classification. By ordering and grouping terms, basic relations between them can be expressed. The new concept of hierarchy, however, makes no changes to the metadata attributes used to describe information resources and only defines relations that help the user when selecting the appropriate terms. Advanced relations like similarities and differences between terms, synonyms and antonyms cannot be expressed using this method.

Thesauri: Allows more statements to be made about subjects. Types of relations defined by the ISO specification are BT (Broader Term), SN (scope note, a description of a term), USE (a reference to a preferred synonym), TT (Top Term, the root of the hierarchy), and RT (Related Term).

Faceted Classification: Divides all terms into "facets" that can be considered as different, orthogonal axes, or sides of a multidimensional cube. Same terms cannot be present on different sides of the cube at the same time. Within sides, the terms can be ordered using other methods like thesauri or taxonomies. Resources are described by applying one term from every side.

Ontologies: Represents an open vocabulary, a model to describe the world using types, attributes, and relations. Of special interest is the technology known as topic maps that represents an ontology framework for information retrieval. Topic maps define three constructs for description of subjects: names, occurrences and associations.

Taxonomies and thesauri are often incapable or at least inadequate of representing many complex conceptual domains. However, when several distinct hierarchies (taxonomies, thesauri, or other methods) are combined to represent different dimensions or attributes, faceted classification is the result, and their expressiveness grows accordingly. A study (Pratt, Hearst, & Fagan, 1999) has shown that users found more answers and their satisfaction was higher when using a faceted

interface to perform browsing, compared to other methods. Another research (Uddin & Janecek, 2007) focused on academic staff and students, has shown that they were more successful in finding relevant results and that the possibilities of switching from one facet to another, combining the facets, previewing the results, and navigating using breadcrumbs were regarded as the most useful features.

Ontologies

Ontologies are open and their structure can be defined at will. It can be difficult to define limits of ontologies because of this openness, and the fact that their foundations originate in a branch of philosophy as old as Aristotle, with many ontologists being philosophically trained (Legg, 2007). A short, concise definition of ontologies states that they are a "formal, explicit specification of a shared conceptualization" (Gruber, 1993). Ontologies formally and explicitly describe concepts within a domain using "classes," features, attributes, constrains, and their relations to other concepts. Relationships in ontologies are strongly typed, so any actual form of relationship existing between two concepts can be expressed. Ontologies can be said to be similar to meta-models or class diagrams in software engineering. A "knowledge base" is then produced when instances of previously defined classes are created and connected between themselves, with their properties being filled in with values from an adequate range. The boundary between ontologies and knowledge bases can be a little vague, with many ontologies being accompanied by corresponding knowledge bases (Noy & McGuinness, 2001). Following the previous analogy with software engineering, knowledge bases correspond to actual models or object diagrams. Finally, an ontology language is a formal language that is used to construct an ontology and can be likened to a meta-meta-model.

Ontologies and topic maps are used to describe subjects in subject-based classification, but are also powerful and extendable enough to be used to describe even information objects themselves, that is, to take over the role of a meta-model. This is not strange since meta-models are closely related to ontologies and both are used to analyze and describe relations between concepts. Ontologies can be used to describe all aspects of information resources, including the terms used for subject-based classification and other attributes that can be used for different classification schemes (e.g. functional classification). When resources are annotated using well-structured ontologies, the data contained in them can be easily retrieved, integrated, and compared by both the human users and the machines.

When there are several related ontologies defined in a wider domain of use, or when ontologies need to be interoperable between domains, an upper ontology can be used to improve interoperability. Upper ontologies are domain-independent ontologies that describe general concepts reusable in all knowledge domains. Domain specific ontologies can be derived from upper ontologies. These two levels of ontologies are bridged by the mid-level or utility ontologies that are more concrete than upper ontologies (Semy, Pulvermacher, & Obrst, 2004). However, upper ontologies tend to be very complex, which is a characteristic that works against their widespread implementation. Some authors doubt that these universal ontologies can be used for interoperation in information systems since they are concerned with institutional facts that depend heavily on context and background (Colomb, 2002). Domain ontologies are, for good reasons, often built to be as simple as possible while expressing all necessary concepts and relations. This provides easier understanding, implementation, use, and future changes to existing concepts.

Information Resources Model in E-Government Systems

Governments operate with substantial volumes of documents and other resources and any improvement of efficiency of existing processes can have a large impact on the functioning of government. Better interoperability of various governmental departments and agencies can strongly contribute to improved efficiency by reducing redundancies. This interoperability can be leveraged by using a common framework and vocabulary for data exchange.

A vocabulary for describing resources and their subjects that would be useful in an e-Government context needs to be sufficiently expressive and flexible in order to express all the possible relations between the relevant information objects. Since governments operate in many domains, a decision of which method or approach to use is at least somewhat influenced by the main domain of interest. However, if different units in government, corresponding to different operational domains, implement different solutions for data manipulation, the problem of interoperability would not only remain unsolved, but would become exponentially more complicated for every new solution added to the mix. This is why, even if we analyze and design for one specific domain, the final solution must take into consideration future interoperability with other domains. This chapter concentrates on the interactions between government and citizens and on the appropriate documents that serve to carry information between the government and its environment. The two possible methods that will be analyzed for this domain are faceted taxonomies and ontologies. Other domains where these methods can also be applied will be mentioned briefly.

Faceted Taxonomies in E-Government

The first possible approach is the use of a faceted taxonomy. Standard taxonomies are just a controlled vocabulary with terms ordered hierarchically. They can be applied to a single attribute based on which the subject-based classification of some information resources will be performed. This is, however, not enough for an e-Government context. Using a single attribute to define a subject of a document is a type of a subject-based classification that is suitable for use with scientific and similar papers where the subject is potentially the most important attribute when performing a search. In the e-Government context, many other attributes can have a large importance: the identification of party submitting the document, the time and place of the document becoming a part of the government system and so on. A single taxonomy cannot express this, so the use of a faceted (multidimensional) taxonomy is appropriate.

Faceted taxonomies can have any number of facets, where each facet represents a single, preferably orthogonal, domain of interest that can be said to correspond to the attributes of an information resource (Ranganathan, 1965). This way a faceted taxonomy allows mapping a multidimensional universe along several lines (dimensions). Browsing a faceted taxonomy is performed by the operation of zooming from a view on the entire universe to a view of a specific concept in one of the facets. This process is called a faceted search and the user is presented with a subset of the universe, that is, all the objects that match the selected concept. The user can also start by using a standard search method, have all the results displayed in a faceted taxonomy, and only after that start to narrow the focus by selecting a facet or a specific concept.

Although the facets can be defined freely, the first developer of the principles of a faceted taxonomy has defined five fundamental categories: Time, Space, Energy, Matter and Personality (Ranganathan, 1967), where each category can further be separated into several facets. These categories have the following meaning:

- **Time:** Time attributes in general, as well as concepts like seasons and metrological concepts.
- **Space:** Geographical terms and areas.
- **Energy:** Conceptual, intellectual actions.
- **Matter:** Materials and property.
- **Personality:** Manifestations of personality in a wide range of meanings as well as other concepts that are not attributed to one of the four other categories.

When considering the government documents related to services provided to the citizens, appropriate subjects (terms) need to be defined to describe them. Since these documents are usually labelled by their purpose, this is a simple task. The actual issue is to define appropriate facets and distribute terms among them. A multi-layer approach was applied in (Dadić, Despotović-Zrakić, Barać, Paunović, & Labus, 2012), where facets were distributed into several layers, with the lowest layer containing labels or terms, and labels potentially holding additional sub-labels. The model defined in this paper defines nine facets in six categories: five originally defined by (Ranganathan, 1967), and an additional, "basic," category. These facets are defined by service type (to citizens/to business), document time, request location, status (request, confirmation, etc.), operation performed, immobility, user group, institution in charge (national, regional, local), and topic. This model and two representative documents described by it are shown in Figure 1.

A faceted taxonomy offers several benefits when implemented. First, the model is expandable simply by adding new facets to it. In the context of e-Government, the employees can automatically generate summaries for all documents using their associations with taxonomies. The documents can

Figure 1. Example of the government faceted taxonomy model defined in (Dadić, Despotović-Zrakić, et al., 2012)

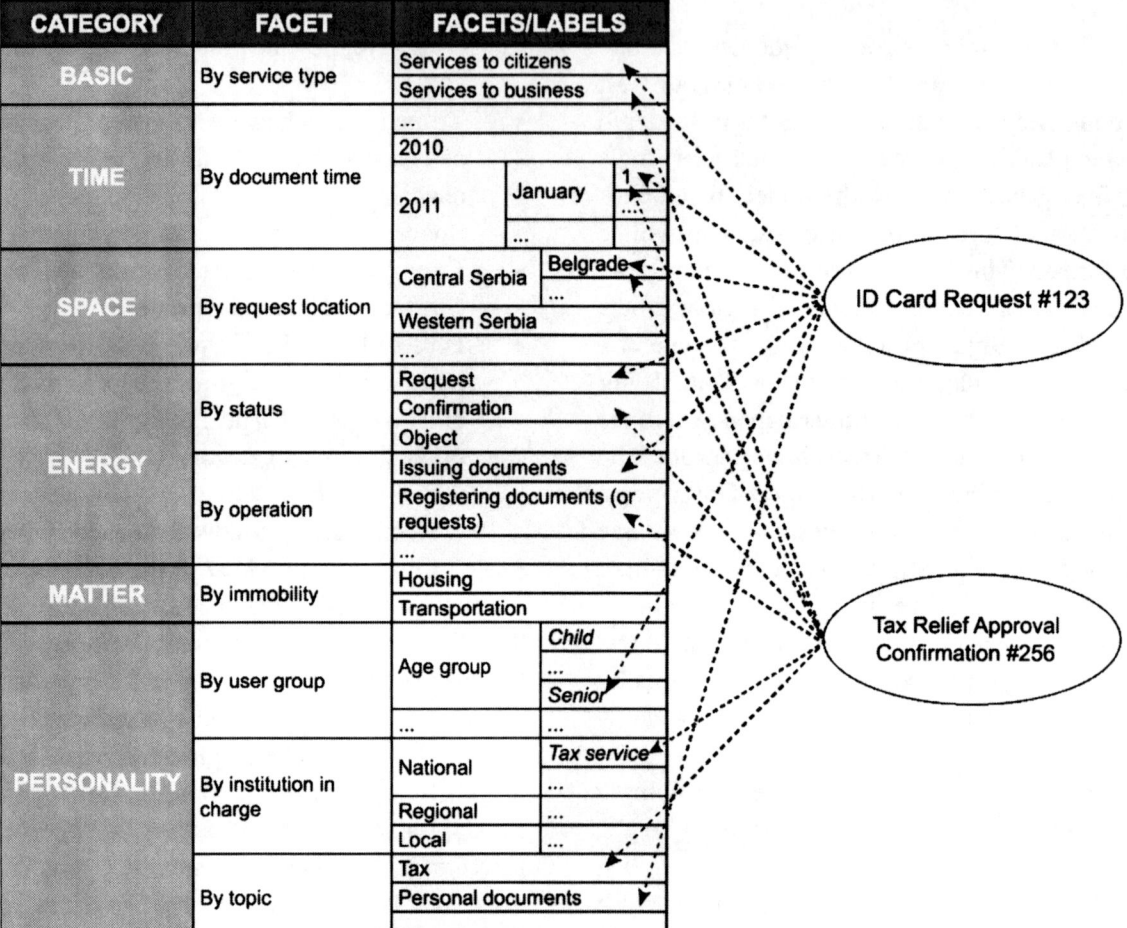

also be searched by using values not contained in them, based on the hierarchical relations between super-terms and sub-terms. The proposed architecture is also easy to implement since it does not require major modifications to the entire system architecture, but only introduces an additional layer that contains the information model (Dadić, Bogdanović, Radenković, Mazinjanjin, & Jovanić, 2012)(Dadić, Labus, Simić, Radenković, & Despotović-Zrakić, 2012).

Since this model applies a faceted taxonomy, each facet is just a simple taxonomy and only relations that can be expressed by it are the hierar-chical sub/super relations between terms. Without advanced types of relations, certain scenarios are difficult to execute. For instance, many documents are related to other documents chronologically – a citizen must first acquire a document, fill it out and submit it, and then perhaps receive another document from the government institution. If government employees wish to follow this connection starting from one of the documents in the chain, they will not be able to do so easily in such a model since the terms in the relevant categories: status, operation, and topic in (Dadić, Despotović-Zrakić, et al., 2012): are not related between themselves.

Another problem in the presented model is that the information about a specific party handling (submitting/receiving) the document is not a part of the model. In the model above, all values are discrete and such information would have to be kept outside of the model, presumably structured to comply with another model better suited for information of such type. Defining a single model where most of the terms would be upgraded to full-fledged concepts with their own attributes and relations would allow execution of such, and more advanced scenarios.

Designing Ontologies

Ontologies allow their creators to define them at will and to make them as powerful as needed. All of the other classification methods can be described using ontologies. For instance, faceted taxonomies can be simply described by defining a type "facet" and creating instances of that type for each existing facet. Top level terms should then be defined as instances of type "term" and a relation "belongs-to-facet" should be defined and applied between them and corresponding facets. Finally, lower level terms should also belong to the type "term" and should be connected to other terms using "broader/narrower" type of relation (Garshol, 2004). If no other relations between concepts in this ontology are allowed, then it is an ontology representing faceted taxonomies. This is only an elementary example, and ontologies are capable of expressing much more complex structures.

If designed properly, ontologies can accurately describe any domain, but caution needs to be exercised. The ontology designer needs to balance the expressiveness of the designed ontology with its complexity. Governments, by nature, regulate the working of all organizations and individuals within a country or a similar political unit. Governments operate and interact with parties in diverse domains so any government information system needs to operate with corresponding information resources or documents. While core functions of a government could be covered by a single ontology, in order to describe concepts appearing in diverse domains, a number of supporting ontologies might be needed. These ontologies could conveniently be selected from an existing framework of an upper, mid, and domain ontologies. If existing domain ontologies are not suitable for the task at hand, generic upper and mid level ontologies can still be used as a general concept/relation vocabulary upon which a new, specific domain ontology will be designed. However, considering the average size of upper ontologies, measured in thousands of concepts, this top-down approach introduces too much complexity for efficient introduction into an e-Government context. A better, tried and tested approach is the opposite, bottom-up approach, where the domain of interest is analyzed and only the concepts of interest are singled out and described by the ontology. This approach inherently provides less interoperability between systems if they are based on differing ontologies, but a sufficient solution can be the use of another method of improving interoperability. For instance, the attributes in the ontology can be based on an existing set of metadata like Dublin Core.

When defining a new ontology, the first step is to define its domain and scope (Noy & McGuinness, 2001). The domain analyzed in this chapter is an e-Government information system that deals with different kinds of documents, mainly those that are produced in citizen-government interactions. In order to properly describe this domain, a wider range of accompanying concepts representing persons, employees, institutions, etc. has to be defined. The ontology defined in this chapter is oriented towards the enhancement of data handling in an e-government context. The same domain can be described using differing ontologies, from different perspectives and using slightly different base concepts. This is an iterative process where a developed ontology is tested in real applications or analyzed with domain experts and then revised accordingly. This ontology represents a first iteration in a wider sense, specifically targeting better data handling and, later on, visualization. Since different governments operate with different organs, services, and documents, only more generic concepts are defined, creating a framework that can be applied to any government. This ontology will, from here on, be referred to as SGOF (Simple Government Ontology Framework).

Specialized tools for ontology and knowledge base development tend to be complex and cumbersome to use. Since this chapter attempts to define a suitable ontological framework or a model rather than a specific ontology, a more general, abstract, and simple approach was used by utilizing UML (Unified Modelling Language) class diagrams. Concepts that exist in these diagrams – classes, attributes, and relations roughly correspond to those in ontology languages like OWL, although they lack a certain amount of semantics in comparison. Still, using UML to construct even full-fledged ontologies is possible and discussed in a number of papers. Many tools exist for modelling in UML; it is widespread in comparison to ontology tools, especially in software engineering and object oriented programming; it also possesses a standard graphical representation. Some papers suggest expanding the standard UML to make it more expressive, often by applying OCL (Object Constraint Language) (Baclawski et al., 2002). It is also possible to use the concept of an UML profile to adapt the basic UML constructs for the purpose of ontology modelling (Gašević, Djuric, & Devedžic, 2009). Transformations from UML to ontology languages can be performed even on standard UML diagrams, without any extensions or profiles, making it possible to use any of the existing, legacy UML diagrams (Xu, Ni, He, Lin, & Yan, 2012).

Simple Government Ontology Framework (SGOF)

The central concept of Simple Government Ontology Framework (SGOF) is the Document concept, representing all the possible types of documents manipulated within an e-Government information system. The Document concept is a generalization of all information resources in the system and should be expanded by new subclasses as new information types are added to the system. In order to ensure some level of interoperability with different systems, Dublin core set of metadata attributes was added to the Document. Specialized types of documents must have their own, specialized attributes and a defined scheme for translation and aggregation of these attributes into Dublin core. This way, when performing a search within the government system, all of the advanced search options can utilize document-specific fields. If resources are exported into an environment that is not familiar with the used ontology, at least some of the semantics will be preserved in the form of digests stored in the Dublin core attributes. Dublin core is suitable for this task since it was intended to be a core element set that can be further augmented with new elements or refined using qualifiers (Weibel, 2005b). It represents a lowest common denominator for resource description and does not intend to replace richer description models but to provide a core set of elements for simple resource description interoperability (Weibel, 2005a).

In SGOF, the Document concept specializes into (not barring the future expansion) the following:

- Interaction document, representing all documents submitted from or issued to external parties, or manipulated internally.
- Law, representing laws.
- Research result, representing information resources from government funded researches, surveys, investigations, and organs that deal with statistics.

These concepts and their relations are shown in figure 2. Since these three types of documents differ greatly, their parts, attributes, and relationships need to be defined separately. A possible comprehensive solution would be to define one or more separate ontologies per document type, where each ontology would explicitly describe the content and the semantics of the document.

Laws could be described using an ontology of legal terms, while the documents containing research results would require a further class hierarchy would need to be defined for different types and areas of research (environment, crime rates, transport, voter turnout, etc.), possibly pointing to a need of several distinct ontologies. For the sake of simplicity, in SGOF the Law documents are defined as a structure composed from hierarchically distributed generic units. This provisionally takes care of their structure, since these units can be stacked both in width and depth as much as needed when creating instances of laws. The lack of semantics in this approach is compensated for by introducing a set of Dublin core attributes into every Law section. A complete solution would undoubtedly be the development of a full-fledged ontology describing general legal concepts and several sub-ontologies describing concepts from domains of human activity to which the law pertains. This is a complex task and since legal ontologies are not directly the topic of this chapter introduction of further complexity in this area was avoided. Still, there are many attempts to introduce ontologies into the legal domain and their review by type and role is given in (Valente, 2005). Due to similar considerations for documents of type "Research result," a flat set of measured instances and their values was selected for the structure of this concept, while relying on Dublin core attributes inherited from the Document to describe the subject of the research. Additionally, "type" attribute was added with a simple controlled vocabulary for describing not the type of the research (which is covered by Dublin core attributes), but the type of the values contained in it (temporal, spatial, etc.). Interaction documents are described in more detail and related to other relevant concepts from the ontology.

The Interaction Documents are further specialized by whether they are created in interactions with the citizens, businesses or other government employees and organs. Under this level are classes of actual government documents, which can number in hundreds, and which define their own specific content and relations. These documents should be grouped into categories by placing them under common super classes, placed in one or more layers between the three abstract types of documents and actual document classes. An example of one actual document class is given in figure 3, with all its relations and attributes, including the inherited ones.

Birth certificate is arbitrarily classified as a "Personal Document," a document that contains personal information about a citizen. Other classes of documents could be "Tax document," "Permit" or any other grouping that makes sense in a specific government context. The specific attributes of this document take their values from the Citizen concept using three connections: the connection between Citizen interaction document and the Citizen for which the document is issued, and two specific connections defined for the documents of type Birth Certificate (mother, father). Every interaction document is issued by a government unit, at a government structure (unless issued using the Internet) and by a government employee, and with one or more specific law sections pertaining to it. Access requirements (inherited from the Interaction Document) should be defined in order to limit the access to citizen personal information. Finally, a Birth certificate inherits Dublin core attributes (Dublin Core Metadata Initiative, 2012) from the topmost concept of its hierarchy: the Document and fills them in accordingly:

- **Contributor:** Government employee that issued the document
- **Coverage:** Government structure where the document was issued
- **Creator:** Government unit that issues the document
- **Date:** Date when the document was issued; should be used "as is" both within the government system and when exporting resources to another system
- **Description:** Similar to subject attribute but padded with predefined text in order to form a human language description.

Figure 2. SGOF hierarchy of document-related concepts

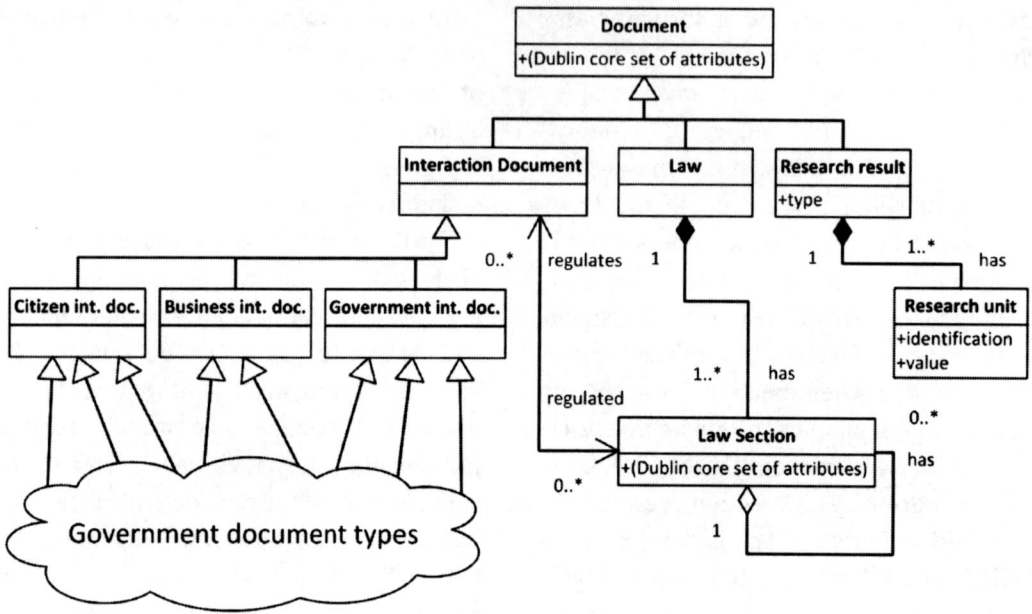

Figure 3. Birth Certificate document, its attributes and relations

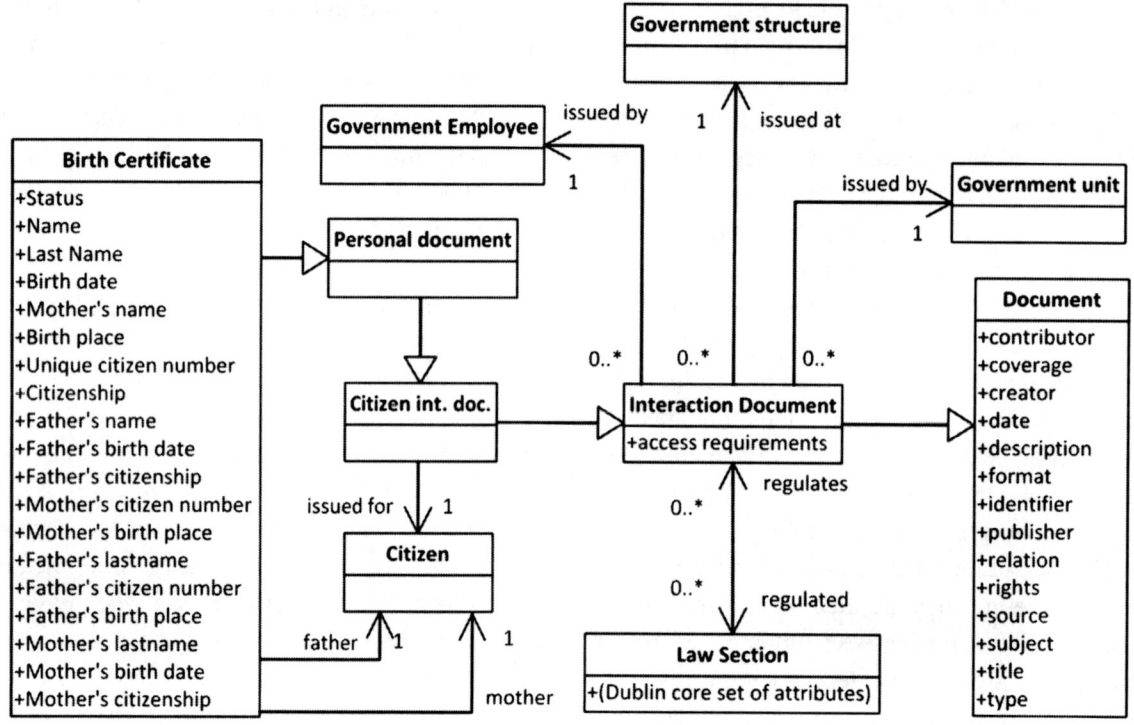

Example: A *birth certificate* (a *Personal Document*) issued on *26.06.2013* by the *Random City Municipality* to citizen John Smith at the *Random street 11, Town hall.*

- **Format:** A format of the information resource itself, using internet media types (MIME) like "pdf," "doc," "png," etc. Likely to be the same for all the documents in the system depending on the way they are handled and digitalized. Should also contain information about whether the document was issued in paper or digitally.
- **Identifier:** Identifies a document; similarly to date attribute, it should be mandatory for all documents in any context (internal/external)
- **Publisher:** Government employee, structure and unit responsible for making the document available outside of the government information system; added only when exporting the document for use in external systems
- **Relation:** Identifiers of other, related documents (if documents are a part of a chain), and laws that pertain to the document in some way other than defining it (which is specified in the Source attribute)
- **Rights:** Especially important because of the privacy concerns related to accessing personal information of citizens. Should contain a description of access requirements normally associated with every document.
- **Source:** Identifier of a law section or sections that define the document
- **Subject:** A combination of several attributes: document type and super types, and government unit that issues the document
- **Title:** Name of the document type combined with either the date or the identifier, depending on the used scheme. Example: "Birth certificate 26.06.2013 #1234"
- **Type:** Dublin core suggests using a specific vocabulary with types like dataset and text;

should be combined with a more specific value created by specifying all super-classes of the document, e.g. "Personal document," "Citizen interaction document," "Interaction document," and "Document" for a document of type Birth certificate.

The remaining concepts are presented in figure 4. A Citizen is represented as a specialization of both the Person class and the Legal personality class and should contain all of the attributes that a government keeps on its citizens. The access to complete citizen information should not be allowed to regular government employees, only to those specific attributes that appear on relevant documents. The Legal personality class also specializes into a Judicial person representing a business or other entity, while the Person class can also be specialized as a Government employee class. This class contains an access rights attribute that specifies to which resources an employee can access, as well as other information like the place they are stationed at, the organizational unit they belong to, and information that can be useful to a government HR department. The organizational units within an government are represented using the Government unit class. Each Government unit can contain other government units within it and has a name and a type. The type should take a value from a controlled vocabulary including appropriate terms at different levels of granularity like "ministry," "office," "court," and so on. Government structure class similarly represents any physical government structure at any level of granularity. It is described by type attribute, for instance "town hall," "office," "counter," geographical location and surface. Finally, a concept of Government service was introduced representing a service that government provides to interested parties. A Government unit provides a Government service, and one or more Government employees are responsible for it. It can require some documents in order to be performed and produce others as a result; it can also have a numerical value for its price or be free. A Government service is further specialized into a Physical service and a virtual service. A Physical service can have a counterpart in the form of a Virtual service and vice versa.

Even though the SGOF was envisioned as a simplified ontology (and therefore a framework for future expansion), it can already provide for some advanced scenarios. All of the available attributes and relations can be used to perform search and browse operations. Documents can be browsed based on the employee that issued the document, location where it was issued, government organ that issued it, related service or law, etc. More specific search queries can also be given, for instance finding all documents issued in some region by a certain class of employees, or all documents that are related to selected set of law sections. Depending on the level of knowledge about the documents in the system, different approaches are possible:

- **No Knowledge about the Structure of The Required Documents:** Documents can still be searched by using keywords that will be compared to all of the document attributes. Even the documents existing outside of the government system will provide comparable performance by searching through Dublin core attributes containing digests of specific document structure.
- **Partial Knowledge about Structure:** Allows users to search for all document types and instances that contain fields of a certain type (e.g. "licence plate number"). Useful when the actual document type is not known.
- **Full Knowledge about Structure:** Documents can be filtered by type and search values defined for every specific attribute of a document.

If the employee has access to citizen database, and if that database is compatible (or fully compliant) to the document-centred ontology, the citizens can be searched by both their attributes and relations to documents issued. Information about government structures, units and employees can be presented to government planners who can use it as a basis for research, efficiency analysis, restructuring and reorganization of government. If enough information about each employee is embedded into the model, human resources department can use the information to supervise employees, their tasks and performance. Research results related to same phenomena could be automatically consolidated and combined with other information that can be extracted from the document sets stored in the system.

Such a system can also be used not only for instance browsing and searching, but to obtain knowledge that is embedded into the ontology. This can be useful for the purpose of training government employees, but can also be presented to the citizens in some form. Citizens could use a public (and limited) access point to browse the document classes, and service, structure, unit, and law classes and instances. They could find out which service requires/provides which document, where they can be obtained, follow a chain of documents and services required for a specific goal, and read relevant laws. If the system is set up to allow user registration and verification of identity, users could also be allowed to review all the documents that were previously issued to them or documents that they submitted to the government. They could also be provided with a shortcut for accessing online services and for submitting electronic documents. Since transparency is an important concept for establishing and maintaining democracy in government, users could also be allowed to browse some of the information about government operations contained in the ontology.

Once an ontology-based system is implemented in a government context, integration with external information sources is only a small step forward.

Figure 4. Concepts describing government units, structures, employees, services, and citizens and other parties

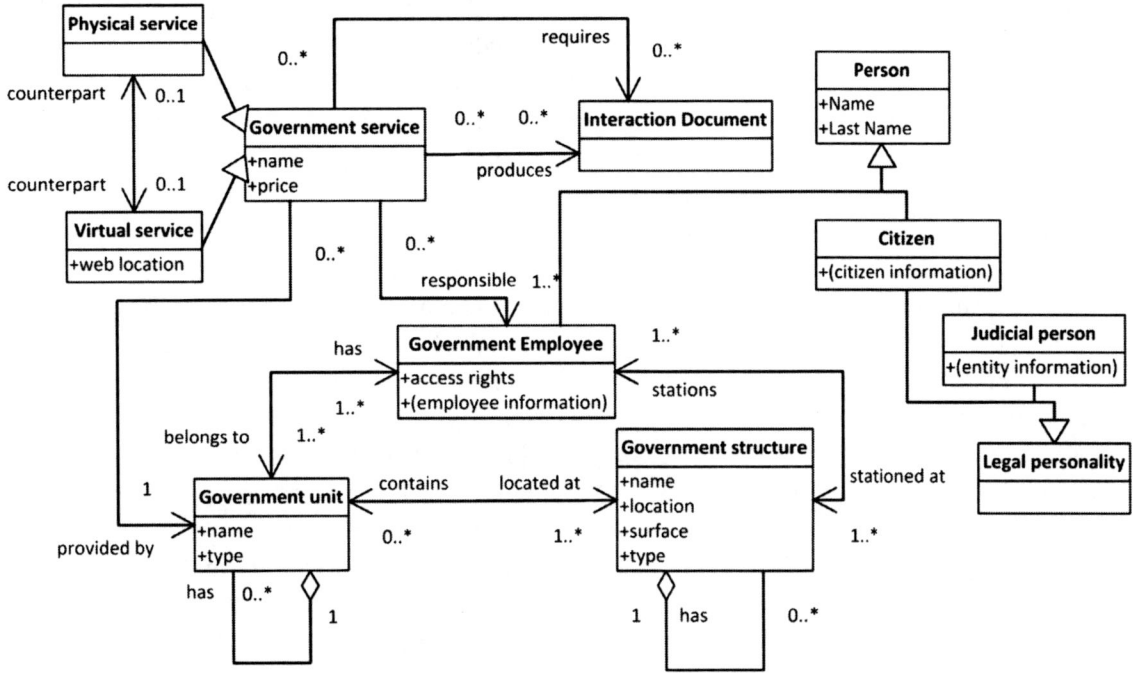

Such integration can allow the e-Government system to utilize the full power of the semantic Web by achieving a synergy between government and external data, and by providing evolvability through the use of standard technologies and interfaces to communicate with any present and future sources. All that is needed is for the ontology and instances to be representable in an ontology language like RDF, and they can be linked to external data (Paunović, Simić, Dadić, Jovanić, & Barać, 2012). This can be utilized to semantically enrich data in the system, e.g., when citizens seek information about marital law, they can be presented with links explaining the concept of marriage, its history, links to churches in the neighbourhood and so on. Another possibility is to allow user search queries to utilize concepts defined outside of the government ontology, based on their relation to existing concepts. For instance, a user can request

all resources that are related to the term "bride," not defined in the SGOF but externally related to the term "marriage" that is likely to be used as an attribute for some documents.

Although there are many possibilities and a lot of potential for expansion of the suggested framework, the main point that needs to be considered is the presentation of search results. Designing a suitable interface for query formulation and result review can improve usability and efficiency of any system, but only to an extent. The standard method of presenting information usually relies on linear or tabular arrangement, but other, visual methods exist. These methods can be applied at many different levels of information review: both individual, collective and global, and can help to more easily comprehend large data sets. Visualization methods are outlined in the following chapter.

DATA VISUALIZATION

Methods and Technologies

Non-technical users are limited in their ability to interpret and exploit linked, semantic data; government employees might inherently understand the domain ontology applied in government context, but the task is not so simple when data is interlinked with external sources and ontologies. The solution is visualization of such data in a coherent and legible manner so that even non-domain and non-technical users can understand its structure and compose queries and discover new information resources (Dadzie & Rowe, 2011). This is especially relevant if government data is made available to citizens that are both diverse in their technical skills (or lack of thereof), and not intimately familiar with this domain.

Data visualization is an interdisciplinary area that uses visual representation of abstract data in order to simplify processes of research and analysis. The goal is to simplify handling of large amounts of complex, distributed, heterogeneous data by relying on human perception. If the problem is defined well enough to be completely automatized, then visualization is not necessary. The role of visualization is then to present data, and test and create hypotheses based on known and partially known datasets. The process of mapping attributes from a dataset into graphic elements is called visual encoding. The main problem is to select a mapping suitable for the dataset at hand, while taking the characteristics of human perception system and purpose/task of visualization in consideration (Munzner, 2009a). (Shneiderman, 1996) gives a type by task taxonomy of information visualizations, describing types of datasets and corresponding possibilities for visualization.

Current approaches for browsing and visualization are outlined in (Dadzie & Rowe, 2011), where they are divided into two types: text based and browsers with visualization options. The importance of both approaches is stated: text approach allows fine grained analysis, while visual browsing allows for the bigger picture to be seen. Visualization can provide insight into the structure of information resources and their associations with external entities, but is not always the best solution for complex analysis. In essence, there is room for both approaches to be used in combination effectively.

Maps are a distinct kind of visualizations. Naturally appropriate for geographical data, but they can also be used for other things, often in combination with distortion and other forms of abstraction with the goal of highlighting data. Some types of map-based visualizations include flow maps featuring stroked lines depicting movement in time and space, choropleth maps that use colour encoding, graduated symbol maps that place symbols over the map, and cartograms where geographic regions are distorted to directly represent a value of a variable (Heer, Bostock, & Ogievetsky, 2010).

Visualization tools that work with semantic data need to be able to generate an overview of the underlying data, and to support filtering and selection of regions of interest that will be visualized in detail. They also must be capable of handling multi-dimensional, hierarchical and networked data, highlighting relationships: an important part of any ontology: within data, and allow data extraction to a format that can be used by third parties (Shneiderman, 1996). Various visualization tools are available in different forms, with different capabilities and roles, but the most interesting are those based on web technologies since they can be integrated into existing web interfaces and require only a browser to be viewed. Visualizations built using such tools can be presented within web pages, portals, and other web-based applications including government informational sites and government portals. One of the most well-known libraries today is the D3 library that stands out by its performance, availability, simplicity, and capability of direct manipulation with presented data (Bostock, Ogievetsky, & Heer, 2011).

Visualizing Data in E-Government Systems

In order to improve data legibility, searchability, browsability, and facilitate easier understanding of government information resources and their relations, three visualization mechanisms were developed and tested on an SGOF-based ontology. Several classes of documents were defined (personal document, tax document, permit, court document, etc.) and a number of instances were created and made searchable/browsable through a simple web interface. D3 library was used in development of visualizations, allowing them to simply be integrated as a part of the web interface.

The process of designing and developing visualization is often done intuitively, after understanding the domain of analysis and the needs of its users. This process can be formalized and represented as four nested levels (Munzner, 2009b). The first level involves understanding the data, tasks, and problems of users in a specific domain. The information system based on SGOF can be useful to many different users, including both the government employees and citizens, and each of them can perform various tasks. Three tasks were selected for visualization:

- Document instances and their internal and external relations.
- Reviewing the structure of the government domain.
- Presenting research results and other information on geographic maps.

The second level requires that all of the domain raw data be mapped to general concepts and types used in information science. Since these visualizations are developed for an existing information system using a SGOF-ontology, the second step is already completed. The SGOF defines concepts and metadata attributes that can directly be used in an information system. The third level is where the actual visual encoding of selected information is performed, and this is described in the following sections. Finally, the fourth step is about the visualization algorithm design, but this step was omitted since it is outside of the scope of this chapter. In short, since the three tasks being visualized include only analyzing single instances and relatively limited sets of concepts, a simple algorithm was developed without constraining it too much with the need for efficiency.

Visualization of Document Instances and Their Internal/External Relations

In order to help government employees when browsing through large numbers of documents, a simple visualization of the current selected document instance and its relations was developed. The main point that influenced the development of this visualization was the uniformity of interaction documents that figure in the government context. The government information system could contain thousands or tens of thousands of records about issued birth certificates, with all of them having the same structure and relations at the class level, and only the attribute values differing from instance to instance. This is not conductive to visualization: displaying a mass of documents and their values can only increase confusion in comparison to the standard presentation methods like pagination and sorting of results by value. The visualization that was designed was therefore concentrated on a single instance, with the goal of allowing a quick overview of information contained in the document and quick navigation to some of the related concepts. This visualization is shown in figure 5.

Since the role of the visualization is to augment search processes, it was integrated into the existing system using the split-screen presentation. The visualization is flexible and can be turned on or off at will. The scenario is as follows: the users types in a search query and then browses a list of the results. When he clicks (or performs a mouse-over, depending on the system setting) on a document, the visualization is shown on the

Figure 5. Split screen visualization of document instances

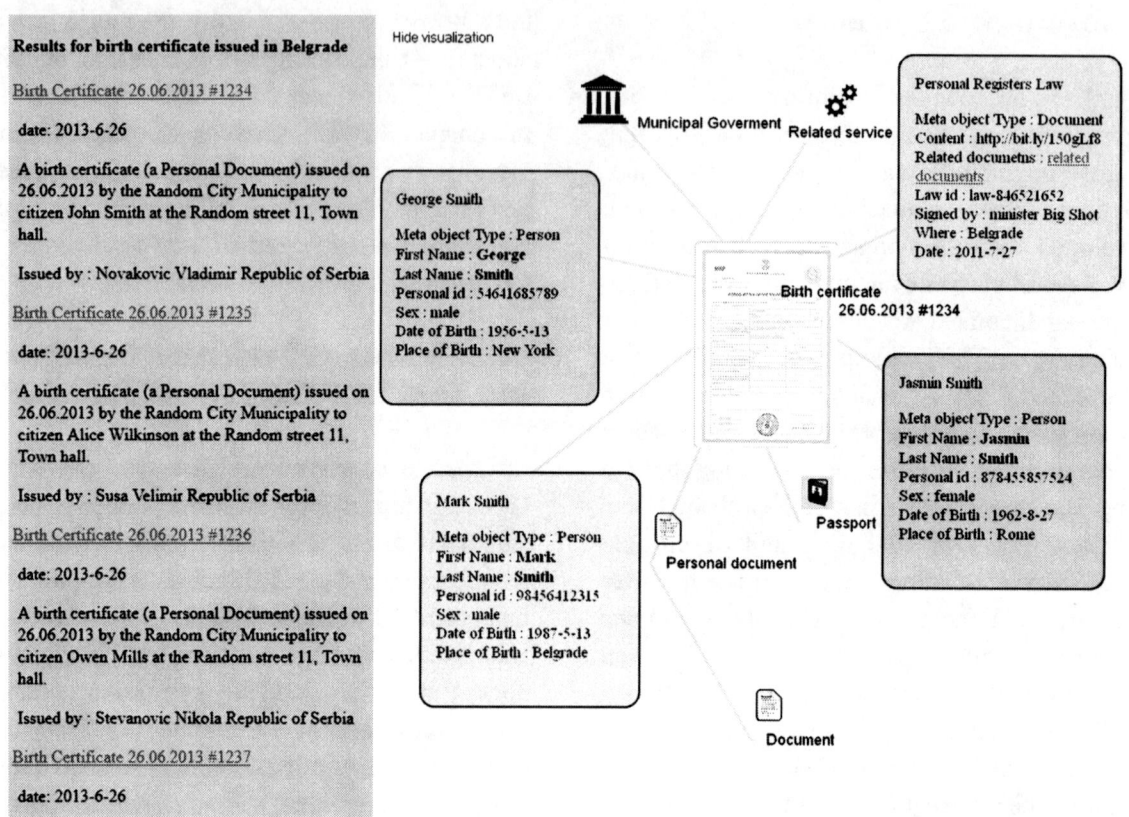

right side of the screen. The visualization displays some of the general concepts of the government ontology and some of the specific instances of said concepts related to the document in focus. The focus of the visualization can be shifted from the document by clicking on any of the entities shown in the visualization. For instance, in Figure 5, the law section concept also displays a part of the said section on screen, with the entire section (and law) displayable by clicking on it. Clicking on the government entity that issued the document or on the related services transfers the user to the conceptual level and a somewhat different presentation described in the following section. Instances that are related to the document (e.g. persons in Figure 5) also display most of their content on screen and can be either displayed in full or used to construct a new search query. Finally, if the document in focus was used to perform some

government service, this relation is shown in the visualization and can also be followed-up.

Visualization of Government Domain Structure

The citizens cannot access information about document instances (except possibly for laws and publicly available research results), but can still utilize the SGOF ontology by exploring the concepts and their relations in order to gain insight into some aspect of government functioning. This can also be useful for the government employees as a way of transferring knowledge about the domain to trainees, or as a starting point for browsing of instances by first selecting a starting concept. The visualization is shown in figure 6.

The user can start the visual browsing by entering a search term in a provided search field or

Figure 6. Visualization of birth certificate document type

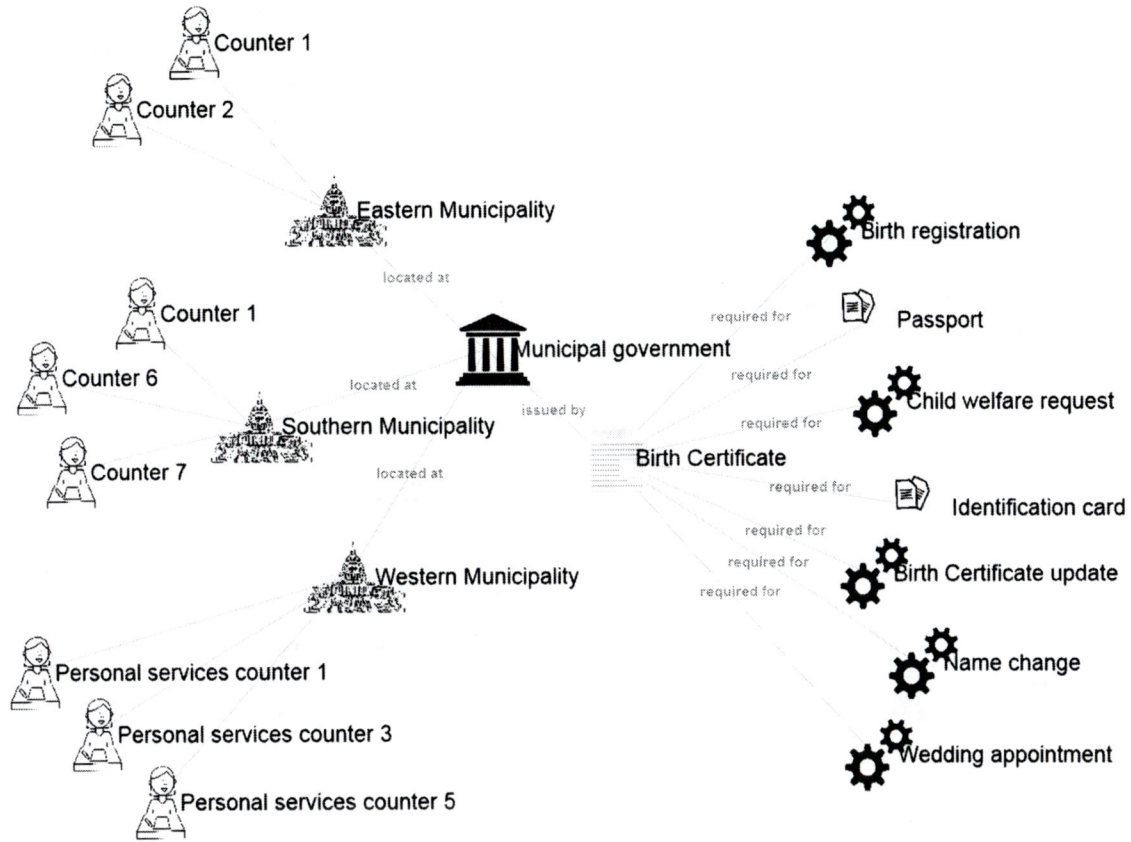

by selecting one of the categories corresponding to the concepts at the top of the hierarchy in the ontology (i.e. document, government unit, government structure, etc.). He can also start directly from the visual perspective where he is shown a high-level visual overview of top-level concepts, roughly corresponding to the actual diagrams of the ontology shown in Figures 2 and 4. When a concept is selected, the focus shifts and the presentation is changed to show appropriate attributes. The previous figure shows the Birth Certificate concept and its relations. Both the concepts and their relationships are named and the user can get additional information about them displayed by performing a mouse-over. For the Birth Certificate example, performing a mouse-over on it offers a link leading to the listing of municipalities that offer this service online, while doing the same operation on presented municipalities offers a

direct link to the municipality's online service. Services that require a Birth Certificate in order to be performed are shown, connected with the "required for" relation, as well as documents like "Passport" that actually represent an abstraction of the corresponding service (i.e. "Issuing a passport"). The user can also select the super class of the document and view all of the documents in that class, or select the related law section and read it. He can also select the government unit (Municipal government) or one of the government structures to get more information about them.

This way of browsing government information not only gives an overview of the domain, but also provides quick access to the actual pages and services. Any addition to the government ontology will also immediately be reflected in its visualization, thereby avoiding discrepancies and stale data.

Presenting Research Results and Aggregate Document Information on Geographic Maps

The research document concept of the SGOF was not developed in detail, but even with basic semantic annotation of the collected data, several possibilities for visualization exist. The type attribute of the research document defines its visualization type; out of several values of the type attribute, the visualization was developed for the "spatial" type where "identification" assumes the form of X and Y coordinates (latitude and longitude) and "value" attribute is a measured value of some phenomenon on said coordinates. The same visualization can be applied to aggregate document information obtained from the instances in the government information storage. For instance, information about issued construction permits can be extracted and represented on the map using different shapes and colours for different structure types. A simple example is shown in Figure 7.

This type of visualisation can be useful to decision-makers in the government, but can also be presented to the general public, where it is especially useful for making scientific and statistic data easily understandable. If the data is semantically annotated, no manual action is needed for the presentation: the visualization can be generated automatically using the underlying data. This, however, might require additional options for visualization tuning like changing colours, shapes, sizes of objects, zoom level or rotation of the map and so on.

DISCUSSION

The ontology framework presented in this chapter makes an important step towards standardization of services and government operation. It can be beneficial to several groups of users, including both the citizens and several categories of government employees:

Figure 7. Visualization of average salary by municipality in the city of Belgrade

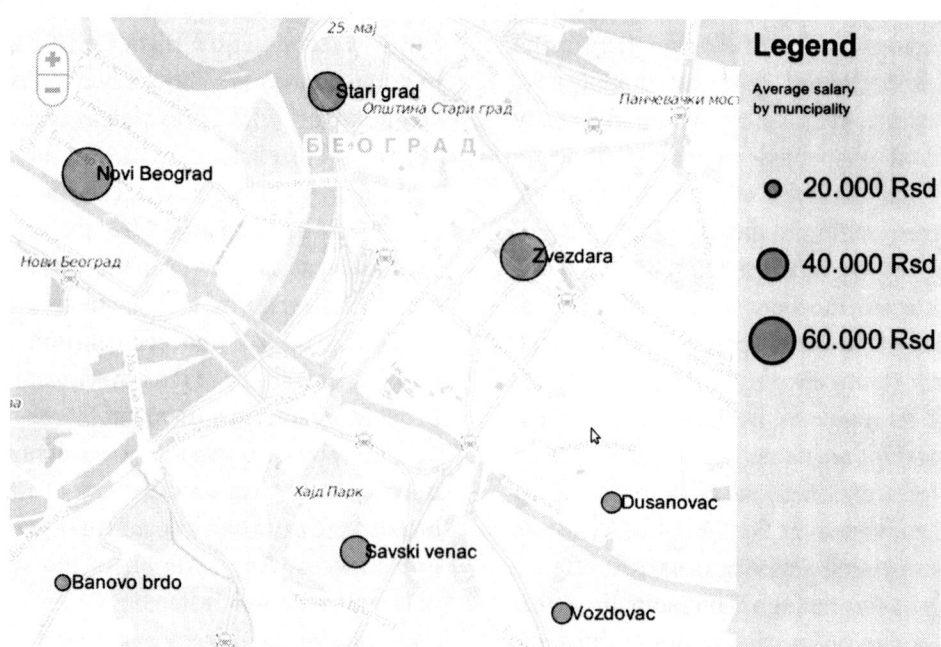

- New employees: knowledge embedded into the ontology allows easier transfer of knowledge.
- Employees that directly work with documents and other information resources.
- Human resources department.
- Decision making unit, organ, department, or top management.
- Citizens interested in available government services and needed documents.
- Businesses.

The presented framework also leaves many possibilities and points of expansion to be explored, allowing additional use-cases to be added for each of the user groups.

As described in this chapter, the processes of browsing and searching for documents can be enhanced by employing ontology-defined concepts, attributes, and relations between information resources as a basis for classification and narrowing of the results. Further improvements to SGOF could be made by applying a knowledge-based approach for organization of retrieved documents. An existing ontology and knowledge model could be further utilized by combining it with an additional query model that defines a number of general user query categories. Each query corresponds to a certain desired search outcome and defines a specific, natural grouping of results based on the types in the ontology (Pratt et al., 1999). When a user queries the system, his query is compared to the query types defined in the query model, an appropriate type is selected, and the result set is adjusted according to the rule defined in the query model. This approach allows users to input queries resembling natural language, and can be easy to implement on top of an existing ontology if user queries are not too diverse.

One of the main goals of introducing ontologies is to achieve interoperability and integration of different environments. SGOF and other government ontologies that describe concepts like government services can be used to form a basis for an integrated e-Government environment. Online services can be automatically discovered, orchestrated, and provided to citizens based on their goals. This requires that all services, their inputs and outputs be semantically annotated. The SGOF defines these concepts, but mainly on the document level; for a more flexible system a higher level of granularity is needed, with individual document fields figuring as concepts in the ontology with their own relations and value constraints. Such a sub-ontology can then directly be utilized for service composition and generation of web forms for data input (Stadlhofer, Salhofer, & Tretter, 2009). If all services needed to achieve a goal are available online, the chain of document (data) submission/receiving can be simplified by removing redundancy (e.g. inputting name on every document in the chain) and in-between steps can be abstracted. When e-Government services are available online to citizens, or internally to government employees, safety becomes an important concern. Digital signatures can be applied in order to achieve authentication, non-repudiation, and integrity of digitally signed electronic documents. Management of such documents can be performed from a centralized point, implementation-wise (Đokić, Labus, Jevremović, Stokić, & Milić, 2012). A comprehensive solution would also require introduction of more detailed security concepts in the ontology, including such as Roles, Rights, and Requirements.

If all services and documents are modelled in detail, and all their relevant concepts are integrated into the ontology or added as sub-ontologies, a problem of administrative responsibility and change management becomes relevant. Different documents will contain different concepts that might require introduction of new ontologies. For instance, a construction permit will introduce concepts of different types of structures from the legal standpoint. A government ontology should be complete and cover all of the relevant, general concepts, but also allow incremental development. Specific government entities should then be given authority to add, edit, and remove concepts and instances from the ontology and the government information system. Each government entity would be charged with regulating those concepts that it deals with in reality. As long as the core set of concepts remains unchanged, and is used in a standard way by all of the government entities developing their own concepts and sub-ontologies, interoperability can be achieved (Vassilakis & Lepouras, 2006). However, a problem can appear when the entity in charge changes concepts in the ontology and appropriate changes need to propagate to the actual service implementation. Performing this process manually can cause discrepancies between the administrative and technical side. Automatization of service configuration and reconfiguration can eliminate such errors (Apostolou, Stojanovic, Lobo, Miró, & Papadakis, 2005) by further expanding the service describing sub-ontology in such a way that all aspects of its service functioning is described in it.

Limitations and problems of SGOF mostly stem from its simplicity and general orientation towards a single aspect of government (government-citizen interaction). The theme of this chapter was to lay down the grounds for future consideration and expansion, to discuss the possible outcomes of applying certain technologies, and to present some visual prototypes based on visualization technologies. For this purpose, the topic of access rights and security was mostly ignored, but should be considered in future. This topic is relevant for a generic framework like SGOF and security features should be embedded in order to provide a safe environment for future expansion. Access rights, requirements, and policies should be defined for most or all elements of the ontology. For instance, a government employee could inherit access rights defined by his institution, his position in hierarchy, geographical location, or any combination of relevant factors. Every information resource and attribute should also have corresponding, fine-grained access requirements. Although security is extremely important when dealing with citizen and other sensitive information, constraints should be placed appropriately in order not to limit the effectiveness of the system. Access requirements and rights are also likely to change depending on the external factors, and an interesting direction for future research would be to consider some algorithms and methods for their dynamic adjustments.

The visualizations used in complement with SGOF were mostly simple, using graphs and simple shapes displayed on a map to present information. Still, they offer a number of advantages in their use: easy overview, discovery and knowledge acquisition of the structure of the government domain, a more understandable way of viewing document instance information, interactivity and so on. Simplicity in design was selected for practical purposes, since cluttering the screen with elements that serve no purpose other than aesthetic can have the opposite effect and impact intelligibility. More complex methods of visualization are often of limited use and tied in to a specific domain or task; some of more exotic methods can be found in (Heer et al., 2010). It is presumed that the government ontology is a growing, evolving definition, so existing visualizations should also be adjusted and improved accordingly. If many additional concepts are added to the ontology, or if instances (which are inherently numerous) are visualized in some form, additional functionalities should be added to the visualization. For instance, the user can be given a zoom option, with concepts appearing at different resolutions of zoom according to their importance and with a "mini-map" showing the user's global position. Selective filtering of concepts, relations, instances, and attributes should also be provided.

CONCLUSION

When every area of human activity is augmented using information and communication systems, problems of interoperability between systems and coping with the complexity within a system become important. Ontologies establish a controlled, but easily expandable vocabulary that can be used as a basis for achieving interoperability, and add rich semantics helping both the machines and the human beings in understanding the domain of interest. Visualization can especially help with the latter task, by transferring the cognitive load to a more visual and intuitively understandable domain. In order to build visualization mechanisms, data must be understood by the machines, so using an existing ontology with already semantically annotated data can simplify this task. Although businesses are the ones to push the borders and experiment with new approaches in handling and representing the data, governments face the same or even more pronounced problems, and possibilities of implementing similar mechanisms should be explored.

Several methods of classification were explored in this chapter, with ontologies as the most powerful solution that was further explored. A simple government ontology framework (SGOF) was designed to describe mainly government documents, while leaving enough room for any specific government to specialize defined concepts according to their own system. A basic implementation was developed in order to demonstrate the functioning of the system and to serve as a platform for developing visualizations. Three different visualizations were developed, mainly concentrating on exploring the structure of the government domain. Since the SGOF was envisioned as a framework, both the ontology and the visualizations are flexible enough and hopefully can be of use to practitioners in the government context.

ACKNOWLEDGMENT

The authors are thankful to the Ministry of Education, Science and Technological Development of Republic of Serbia for the financial support grant, number 174031.

REFERENCES

Apostolou, D., Stojanovic, L., Lobo, T. P., Miró, J. C., & Papadakis, A. (2005). Configuring E-Government Services Using Ontologies. In M. Funabashi, & A. Grzech (Eds.), *Challenges of Expanding Internet: E-Commerce* (Vol. 189, pp. 141–155). Springer, US: E-Business, and E-Government. doi:10.1007/0-387-29773-1_10

Baclawski, K., Kokar, M. K., Kogut, P. A., Hart, L., Smith, J., & Holmes, W. S. I. et al. (2002). Extending the Unified Modeling Language for Ontology Development. *Software & Systems Modeling*, *1*(2), 142–156. doi:10.1007/s10270-002-0008-4

Bostock, M., Ogievetsky, V., & Heer, J. (2011). D^3: Data-Driven Documents. *IEEE Transactions on Visualization and Computer Graphics*, *17*(12), 2301–2309. doi:10.1109/TVCG.2011.185 PMID:22034350

Burnett, K., Ng, K. B., & Park, S. (1999). A Comparison of the Two Traditions of Metadata Development. *Journal of the American Society for Information Science American Society for Information Science*, *50*(13), 1209–1217. doi:10.1002/(SICI)1097-4571(1999)50:13<1209::AID-ASI6>3.0.CO;2-Y

Colomb, R. M. (2002). *Use of Upper Ontologies for Interoperation of Information Systems: A Tutorial*. Padova, Italy: Academic Press.

Dadić, J., Bogdanović, Z., Radenković, M., Mazinjanjin, Đ., & Jovanić, B. (2012). Developing a Multifaceted Model for Scaffolding Information Management in E-Government Systems. *Metalurgia International*, *17*(12), 140–146.

Dadić, J., Despotović-Zrakić, M., Barać, D., Paunović, L., & Labus, A. (2012). Managing E-Government Information Resources Using Faceted Taxonomy. In M. Gasco (Ed.), *12th European Conference on eGovernment (ECEG 2012)* (pp. 169–175). Barcelona: Academic Publishing International Limited.

Dadić, J., Labus, A., Simić, K., Radenković, B., & Despotović-Zrakić, M. (2012). A Model for Structuring Information Resources in E-Government. *Innovative Issues and Approaches in Social Sciences*, *5*(2), 104–117. doi:10.12959/issn.1855-0541.IIASS-2012-no2-art07

Dadzie, A.-S., & Rowe, M. (2011). Approaches to Visualising Linked Data: A Survey. *Semantic Web*, *2*(2), 89–124. doi: doi:10.3233/SW-2011-0037

Đokić, D., Labus, A., Jevremović, S., Stokić, A., & Milić, A. (2012). Portal for the Management of Digitally Signed Electronic Documents. *Metalurgia International*, *17*(9), 120–128.

Dublin Core Metadata Initiative. (2012). *Dublin Core Metadata Element Set, Version 1.1*. Retrieved June 22, 2013, from http://dublincore.org/documents/2012/06/14/dces/

Garshol, L. M. (2004). Metadata? Thesauri? Taxonomies? Topic Maps! Making Sense of it all. *Journal of Information Science*, *30*(4), 378–391. doi:10.1177/0165551504045856

Gašević, D., Djuric, D., & Devedžic, V. (2009). The Ontology UML Profile. In *Model Driven Engineering and Ontology Development* (2nd ed., pp. 235–243). Berlin: Springer. doi:10.1007/978-3-642-00282-3_9

Gruber, T. R. (1993). A Translation Approach to Portable Ontology Specifications. *Knowledge Acquisition*, *5*(2), 199–220. doi:10.1006/knac.1993.1008

Heer, J., Bostock, M., & Ogievetsky, V. (2010). A Tour Through the Visualization Zoo. *Communications of the ACM*, *53*(6), 59–67. doi:10.1145/1743546.1743567

Interoperability Solutions for European Public Administrations. (2010). *European Interoperability Framework for European public services, Annex 2* (pp. 1–40). Retrieved from http://ec.europa.eu/isa/documents/isa_annex_ii_eif_en.pdf

Klijn, E., & Koppenjan, J. F. M. (2000). Politicians and Interactive Decision Making: Institutional Spoilsports or Playmakers. *Public Administration*, *78*(2), 365–387. doi:10.1111/1467-9299.00210

Landsbergen, D. Jr, & Wolken, G. Jr. (2001). Realizing the Promise: Government Information Systems and the Fourth Generation of Information Technology. *Public Administration Review*, *61*(2), 206–220. doi:10.1111/0033-3352.00023

Legg, C. (2007). Ontologies on the Semantic Web. *Annual Review of Information Science & Technology*, *41*(1), 407–451. doi:10.1002/aris.2007.1440410116

Munzner, T. (2009a). Visualization. In *Fundamentals of Computer Graphics* (pp. 675–720). Academic Press.

Munzner, T. (2009b). A Nested Model for Visualization Design and Validation. *IEEE Transactions on Visualization and Computer Graphics*, *15*(6), 921–928. doi:10.1109/TVCG.2009.111 PMID:19834155

Noy, N. F., & McGuinness, D. L. (2001). *Ontology Development 101 : A Guide to Creating Your First Ontology*. Academic Press.

Paunović, L., Simić, K., Dadić, J., Jovanić, B., & Barać, D. (2012). The Impact of Applying the Concept of the Semantic Web In E-Government. *Innovative Issues and Approaches in Social Sciences*, *5*(2), 161–179. doi:10.12959/issn.1855-0541.IIASS-2012-no2-art11

Pratt, W., Hearst, M. A., & Fagan, L. M. (1999). A Knowledge-Based Approach to Organizing Retrieved Documents. In *Proceedings of the sixteenth national conference on Artificial intelligence and the eleventh Innovative applications of artificial intelligence conference innovative applications of artificial intelligence* (pp. 80–85). AAAI.

Ranganathan, S. R. (1965). *The Colon Classification*. New Brunswick, NJ: Rutgers University Press.

Ranganathan, S. R. (1967). *Prolegomena to Library Classification* (3rd ed.). New York: Asia Publishing House.

Sacco, G. M. (2000). Dynamic Taxonomies: A Model for Large Information Bases. *IEEE Transactions on Knowledge and Data Engineering, 12*(3), 468–479. doi:10.1109/69.846296

Scholl, H. J., & Klischewski, R. (2007). E-Government Integration and Interoperability: Framing the Research Agenda. *International Journal of Public Administration, 30*(8-9), 889–920. doi:10.1080/01900690701402668

Semy, S. K., Pulvermacher, M. K., & Obrst, L. J. (2004). *Toward the Use of an Upper Ontology for U. S. Government and U. S. Military Domains: An Evaluation*. Retrieved from http://www.dtic.mil/cgi-bin/GetTRDoc?AD=ADA459575

Shneiderman, B. (1996). The Eyes Have It: A Task by Data Type Taxonomy for Information Visualizations. In *Proceedings of IEEE Symposium on Visual Languages* (pp. 336–343). IEEE.

Stadlhofer, B., Salhofer, P., & Tretter, G. (2009). Ontology Driven E-Government. *2009 Fourth International Conference on Systems, 7*(4), 251–255. doi:10.1109/ICONS.2009.20

Uddin, M. N., & Janecek, P. (2007). The Implementation of Faceted Classification in Web Site Searching and Browsing. *Online Information Review, 31*(2), 218–233. doi:10.1108/14684520710747248

Valente, A. (2005). Types and Roles of Legal Ontologies. In V. R. Benjamins, P. Casanovas, J. Breuker, & A. Gangemi (Eds.), *Law and the Semantic Web* (Vol. 3369, pp. 65–76). Springer. doi:10.1007/978-3-540-32253-5_5

Vassilakis, C., & Lepouras, G. (2006). Ontology for e-Government Public Services. In *Encyclopedia of E-Commerce* (pp. 865–870). E-Government, and Mobile Commerce.

Weibel, S. (2005a). The Dublin Core: A Simple Content Description Model for Electronic Resources. *Bulletin of the American Society for Information Science and Technology, 24*(1), 9–11. doi:10.1002/bult.70

Weibel, S. (2005b). The State of the Dublin Core Metadata Initiative: April 1999. *Bulletin of the American Society for Information Science and Technology, 25*(5), 18–22. doi:10.1002/bult.127

Xu, Z., Ni, Y., He, W., Lin, L., & Yan, Q. (2012). Automatic Extraction of OWL Ontologies From UML Class Diagrams: a Semantics-Preserving Approach. *World Wide Web (Bussum), 15*(5-6), 517–545. doi:10.1007/s11280-011-0147-z

ADDITIONAL READING

Amar, R. A., & Stasko, J. T. (2005). Knowledge precepts for design and evaluation of information visualizations. *IEEE Transactions on Visualization and Computer Graphics, 11*(4), 432–442. doi:10.1109/TVCG.2005.63 PMID:16138553

Barać, D., Đorđe, M., & Despotović, M. (2009). Risk Management in Developing E-Government Web Portal. Global instability reflections (pp. 465–485).

Chen, C., Yeh, J., & Sie, S. (2005). Government Ontology and Thesaurus Construction: A Taiwanese Experience. In E. Fox, E. Neuhold, P. Premsmit, & V. Wuwongse (Eds.), *Digital Libraries: Implementing Strategies and Sharing Experiences* (Vol. 3815, pp. 263–272). Springer Berlin Heidelberg. doi:10.1007/11599517_30

Chiotti, O. (2006). A Process for Building a Domain Ontology: an Experience in Developing a Government Budgetary Ontology, *72*(c).

Corcho, O., Fernández-López, M., Gómez-Pérez, A., & López-Cima, A. (2005). Building Legal Ontologies with METHONTOLOGY and WebODE. In V. R. Benjamins, P. Casanovas, J. Breuker, & A. Gangemi (Eds.), *Law and the Semantic Web* (Vol. 3369, pp. 142–157). Springer Berlin Heidelberg. doi:10.1007/978-3-540-32253-5_9

Fonou Dombeu, J. V., & Huisman, M. (2011). Combining Ontology Development Methodologies and Semantic Web Platforms for E-government Domain Ontology Development. *International journal of Web & Semantic Technology, 2*(2), 12–25. doi:10.5121/ijwest.2011.2202

Gómez-Pérez, A., Ortiz-Rodríguez, F., & Villazón-Terrazas, B. (2006). Legal Ontologies for the Spanish e-Government. In R. Marín, E. Onaindía, A. Bugarín, & J. Santos (Eds.), *Current Topics in Artificial Intelligence* (Vol. 4177, pp. 301–310). Springer Berlin Heidelberg. doi:10.1007/11881216_32

Grandi, F., Mandreoli, F., Martoglia, R., Ronchetti, E., Scalas, M. R., Tiberio, P., et al. (2006). Semantic Web Techniques for Personalization of eGovernment Services. In J. Roddick, V. R. Benjamins, S. Si-said Cherfi, R. Chiang, C. Claramunt, R. Elmasri, F. Grandi, et al. (Eds.), Advances in Conceptual Modeling: Theory and Practice (Vol. 4231, pp. 435–444). Springer Berlin Heidelberg. doi: doi:10.1007/11908883_51

Hellmann, S., Stadler, C., Lehmann, J., & Auer, S. (2009). DBpedia Live Extraction. In R. Meersman, T. Dillon, & H. Pillar (Eds.), *On the Move to Meaningful Internet Systems: OTM 2009* (pp. 1209–1223). Springer Berlin Heidelberg. doi:10.1007/978-3-642-05151-7_33

Hinkelmann, K., Thönssen, B., & Wolff, D. (2010). Ontologies for E-government. In R. Poli, M. Healy, & A. Kameas (Eds.), *Theory and Applications of Ontology: Computer Applications* (pp. 429–462). Springer Netherlands. doi:10.1007/978-90-481-8847-5_19

Mazinjanin, Đ., Bogdanović, Z., & Despotović, M. (2009). Communication Methods in the City Government G2C Service. *Journal For Management Theory And Practice, 53*, 27–33.

Milojković, J., Janković, S., Kostić, M., Despotović-Zrakić, M., & Bogdanović, Z. (2012). Building Public Key Infrastructure for E-Government. *Metalurgia International, 17*(11), 227–234.

Pepper, S. (2000). *The TAO of Topic Maps* (pp. 1–21). Retrieved from http://www.ontopia.net/topicmaps/materials/tao.html

Sarantis, D., & Askounis, D. (2010). Knowledge Exploitation via Ontology Development in e-Government Project Management. *International Journal of Digital Society, 1*(4), 246–255.

Schandl, B., Haslhofer, B., Bürger, T., Langegger, A., & Halb, W. (2011). Linked Data and multimedia: the state of affairs. *Multimedia Tools and Applications, 59*(2), 523–556. doi:10.1007/s11042-011-0762-9

Smith, T., Noble, S., Avenell, D., & Lally, G. (2009). DataViz: Improving Data Visualisation for the Public Sector (pp. 1–30). London.

Stolpnik, A. (2009). *Visual Hints for Semantic Graph Exploration*. Tel-Aviv University, October 2009.

Stowers, G. (2013). *The Use of Data Visualization in Government* (pp. 1–49). Retrieved from http://www.businessofgovernment.org/sites/default/files/The Use of Visualization in Government.pdf

Visser, U., Stuckenschmidt, H., Schuster, G., Neumann, H., & H, S. (2001). Ontology-Based Integration of Information — A Survey of Existing Approaches. *IJCAI--01 Workshop: Ontologies and Information Sharing* (pp. 108–117).

KEY TERMS AND DEFINITIONS

Controlled Vocabulary: A set of terms that represent allowed values of a property.

Domain Ontology: An ontology that describes all of the relevant concepts in a single domain of interest.

Interoperability: A property of organizations and, more commonly, technical systems, that allows them to work together to achieve common goals.

Knowledge Base: A set of instances of concepts defined in an ontology, connected with instances of relations and with properties having values from a specific range.

Meta-Model: A model that defines data structures and relations describing data and resources in some domain.

Metadata: Data that describes other data; can be used by data users, managers, software agents and other entities to provide various services including data management, browsing, searching, restructuring, analysis, distribution, aggregation, and adaptation.

Ontology: A formal, explicit model describing concepts in the real world using classes, relations, and attributes.

Ontology Language: The language used to encode the ontology and the knowledge base.

Semantic Web: Movement that promotes standard data formats and semantic annotation of web pages in order to achieve interoperability and creating, sharing, integrating, and reusing of data on a global level using a new class of powerful semantic applications.

Subject-Based Classification: Using subjects (terms) to describe an object (resource).

Topic Maps: An ontology framework for information retrieval that defines constructs specifically for subject (topic) description.

Upper Ontology: An ontology that defines only high-level, abstract or general concepts that can be reused in a multitude of more specific ontologies.

Visualization: Visual representation of data, usually generated to facilitate better understanding of data or some aspect of it.

Chapter 3
Connected Services Delivery Framework:
Towards Interoperable Government

Mohammed Al-Husban
Southampton Solent University, UK

Carl Adams
University of Portsmouth, UK

ABSTRACT

Efficient public service delivery is a primary task of public administration within any governance model. The main theme of modern governance implies an integrated, effective, and citizen-centric practices of government and administration as a prerequisite for a long-term positive development of the economy. Electronic public service delivery via e-government portal has become a convenient means for the customers—citizens and businesses—to fulfil their requirements. However, the quality of service delivery is heavily based on the level of integration of the services between different partners in the back office. Service integration requires good governance among partners in agencies in various departments and sometimes at different government levels. This chapter provides an interoperability integration framework that connects closely coordinated services based on Service-Oriented Architecture, Enterprise Service Bus, and Web services. The proposed framework is presented as an attempt to align the organizational structures and processes of different government departments while reducing implementation and ownership costs. The framework is applied to a realistic case example of integrating three different public services, namely applying for a Tourism Agency License, applying for a Vocational License, and applying for No Criminal Record Certificate, in a highly interoperable manner and a high level of adaptability to existing government policies and priorities.

INTRODUCTION

Governments around the world are taking serious steps to improve collaboration and integration across their departments. Conventionally, governments have been planned and organised with vertical structures, aligned to deliver a wide range of services to citizen such as health and education services, and to businesses such as investment facilities and legal regulations (Noreng, 1980). This structural separation provides efficiency, clear lines of accountability and concentration

DOI: 10.4018/978-1-4666-6082-3.ch003

of specific range of related service (Hyde, 2008). However, vertical governmental structures are not well equipped to deal with many recent public delivery issues, especially those require cross portfolio such as social security programs, unemployment insurance, food assistance and healthcare (Stegarescu, 2006).

Responding to complex service delivery problems requires governments to modernise their ICT infrastructure, and leverage this infrastructure within the public sector in order to better share information, internally and externally, and to deliver integrated services. Moreover, global trends such as rising citizen expectations, budgetary constraints and global competition for investment, have driven governments to review their service delivery. Therefore, a necessity has arisen, as ICT for service delivery is being revisited to improve integration and connectedness. Accordingly, a form of services and processes integration in relation to service delivery has emerged, and is recently found not only in e-government agendas but also at national and sub-national modernization initiatives. This reform is referred to as "Integrated Public Governance" that it may be the successor to the New Public Management (Kenneth, 2009).

Connected service delivery is defined as the process of integrating public services to a convenient, seamless and single point of access portal, through which public services can be accessed, utilized and completed. The end user of this portal, refers to as Customer of Public Administration in many literature, could be a citizen, business or other public administrations who may seek to utilize a service (Klischewski, 2007).

This chapter investigates the connected public service delivery through the integration of back office processes. It provides an interoperability integration framework that connects closely coordinated services based on Service Oriented Architecture (SOA), in an attempt to align the organizational structures and processes of different government departments. This alignment will help to achieve better IT utilization and return on investment in e-government initiatives. As we believe that this integration is not only necessary but also critical at this stage of e-government implantation in developing countries.

The chapter builds on insights and learning from best practices in large corporate integration projects, using SOA and ESB (Enterprise Service Bus) technologies, and applies this to e-government service integration and provision. The integration framework suggested in this chapter draws together proven techniques and integration frameworks from several proceeding architectures and design styles, with new universal open standards and integration technologies that have the potential to provide a technically achievable framework for modern e-government implementations.

In the scale and scope of e-government implementation, public services are not only implemented using a multitude of technologies and platforms, but are also by a wide range of people practices, application codes and interactions between stakeholders and the system. Therefore, a realistic case example of integrating three public services based on the suggested integration framework is presented in this chapter.

CONNECTED SERVICE DELIVERY: DEFINITION

Connected service delivery in the context of this chapter is broadly defined as the provision of an integrated cluster of public services, joint up and connected in ways that suit customer requirements, sourced from a range of partner organisations and service providers. The services being connected may be separate service areas from within the government agency, or alternatively, the partners may include other government agencies, private business and voluntary sector.

This connected collaborative approach may be accomplished by merging structures, sharing budgets, combining in joint teams or sharing information between distinct departments, or developing a joint customer interface such as a website or portal (Great Britain, Cabinet Office, 2000). Connected services are those involving more than one agency that are coordinated and integrated around the needs of the individual citizen.

Service integration occurs horizontally and/ or vertically:

- **Horizontal Service Integration (Intra-Governmental Integration and Collaboration):** Integrating some or all of specific set of services within one level of government (Dean & Boutilier, 2011). It can also refer to the integration of services in government agencies with different functionality that has some relation in common to the clients. This integration model focuses on the economies of scale derived from collaborative approaches to procurement, the efficiencies and knowledge management advantages associated with developing common IT infrastructure, and merged front counter services with the associated benefits of consolidating real estate (Dean & Boutilier, 2011).
- **Vertical Service Integration (Inter-Governmental Integration):** A service integration in which two or more levels of government in the same function collaborate to integrate service delivery. The vertical integration addresses the integration between different levels of government but in the same functional areas, for instance, the integration of local level business license application being linked to state and government level to obtain an employer identification number (Chopra, 2005).

Connected service delivery is also the use of modern information and communication technologies to improve the efficiency and effectiveness of government services provision. Integrating public services into a convenient, seamless and single point of access portal will aid government initiatives to achieve the followings:

- Improve and enhance public service provision.
- Improve public participation and involvement to ensure that citizens have a direct voice in public decisions.
- Improve government transparency, accountability and participation.

Connected service delivery provides the common framework and guideline in the implementation of the following e-Government interactions:

- Government to Citizen (G2C).
- Government to Business (G2B).
- Government to Employee (G2E).
- Government to Government (G2G).

CONNECTED SERVICE DEDLIVERY: KEY DRIVERS

Governments in a number of modern democracies are undertaking ambitious reforms to transform their public services and are moving toward the concept of connected service delivery. Two key drivers are pushing towards integrating the delivery of citizen oriented services across the disparate level of government. The first key driver is the citizen-centric approach towards public service delivery, as governments such as Australia, Canada and UK have recently experienced better eservices uptake level and more efficient internal services integration (Reding, 2006). The second driver is the current implementation of collaborative networks by e-government initiatives to collaborate with the private sector, which has dramatically improved the efficiency of public transactions across organisational boundaries.

The vision of modern e-government projects has been to accomplish rapid and flexible integration of systems across all government departments in order to achieve the following:

- Reduce implementation and ownership costs.
- Improve public service provision.
- Improve government transparency, accountability and participation.
- Achieve better IT utilization and return on investment.

E-government projects have succeeded when they have been planned and designed around the needs and requirements of business and citizen, and that can only be realized by the wide back office integration between government departments. This integration will minimize duplication and redundancy across government entities. In an example of an individual is setting up a new Tourism Agency, and needs to provide various documents to satisfy and comply with legal and professional rules, including Criminal Records, Vocational License, Social Security Debts, etc. Such information is already available in several government entities and should be accessed automatically. In practice, this example implies service integration and alignment so that individual government departments share information and avoid the need for the client to provide information that is already held by different government entities.

CONNECTED SERVICE DELIVERY: CHALLENGES

Providing more public services with less public spending is an ongoing challenge for all modern governments (Curristine et al.., 2007). However, the structural and organisational dependencies between government agencies in developing countries form a network structure, in which a multitude of interdependent actors exist (Bruijn et al.., 2000). Therefore, the implementation of connected service delivery presents a major challenge for government organizations. It requires government departments to adapt their business processes to the service delivery chain, but often the organizational goals of these departments are not in line with the goals of the chain as a whole.

The challenges of connected service delivery can be categorised into two major categories: Business challenges that include:

- Funding issues.
- Organisational culture, resistance to change and structural dependencies of government silos.
- The explicit absence of common legal framework to define integration responsibilities and to support electronic service delivery.
- The lack of evidence base of identifying integration opportunities or assessing integration costs and benefits.
- Financial problems in the current pooled budget of the conventionally integrated public services.

Technical challenges that include:

- Over-engineering at the technical level is considered the most realistic technical problematic in the domain of interoperability in the developing countries (CSTransform, 2011). Much of the technical standards in many electronic government interoperability framework eGIF are at a level of details which is exhausting and rather unnecessary.

The following issues can be listed as the most common challenges in public service integration and delivery in developing countries:

- Lack of service centricity in back end implementation of e-government systems, which led to data duplication and service redundancy.
- Lack of technical integration between systems within and across government agencies, which led to reduce productivity in shared public services and potential revenue leaks and losses

- The absence of common technical framework that provides a common workflow to integrate desperate business applications.
- Security and privacy of exchanged documents over government networks.
- Paper based forms remain the primary data collection methods for most government in developing countries.

SOA, WEB SERVICES, AND ESB AS INTEGRATION FRAMEWORK

The ultimate objective of connected service delivery is to enable cross integration between closely coordinated services, which if technically and organisationally achieved, it will help to integrate the process of public service provision for best use by the end user, as well as to avoid duplication and redundancy in government agencies.

Using SOA (service oriented architecture), ESB (enterprise service bus) and web services to connect and integrate service delivery will enable multiple government constituents and service providers to reuse developed assets, share data and provide a seamless panel of highly integrated services. Implementing a flexible SOA solution for governing, integrating, deploying, securing, and managing services, irrespective of the platforms on which they were created, will increasing the effectiveness, efficiency, and quality of service. Moreover, an SOA flexible solution will help to overcome the current organisational resistance which is presented as one of the most challenging barriers hindering service connectedness (Curristine et al.., 2007).

Service-Oriented Architecture (SOA) is a set of design principles, technologies and practices for managing and communicating enterprise digital services to support dynamic, flexible business processes. As an architectural approach, SOA is inherently platform, technology, and protocol neutral and independent (Papazoglou et al.., 2007). In recent years, SOA has emerged as the leading integration and architecture framework in modern complex and heterogeneous computing environment (Colan, 2004).

Web services have been utilised to develop SOA applications. They can be implemented to support the integration of application and systems of different levels of e-government aimed at both public individuals and private businesses. Web services provide the SOA with a set of features that support the overall architecture, reusability of business components and loosely coupled building blocks, for instance provide services to end-user applications and other services through heterogeneous networks, these features make SOA the best architecture match for public service integration (Putnik et al.., 2007).

Enterprise Service Bus (ESB) is software architecture for middleware that provides fundamental services for more complex architectures. It is a set of rules and principles for integrating multiple systems and applications together over a bus topology infrastructure. ESB architecture enables the connection of multiple systems and applications that run on parallel fashion over different platform regardless the programming languages and programming models (Berthold et al.., 2013). ESB provides common communication and integration services. ESB uses industry standards for most of the services provided; it facilitates cross-platform interoperability and has become the logical choice for organisations looking to implement SOA.

Service-Oriented Architecture (SOA)

Service-Oriented Architecture is a development of distributed computing based on the request and reply design paradigm for synchronous and asynchronous application (Kodali, 2005). SOA offers an extensible but flexible and composable approach to reusing existing applications and services, and extends in scale and scope to constructing new ones (Hutchison et al., 2005).

High degree of Interoperability and distribution is a key for public service integration, and that is due to the numerous entities they imply (Sellami et al., 2007). A service-oriented architecture and web services is a way forward to solve a substantial number of public service integration issues. In SOA context, software components and applications are modelled as services, whereas the main focus in application design will be placed on combining slightly connected services to construct larger applications. Therefore, service provider must offer a service description explaining the required procedure to invoke the offered service, which will allow information to be exchanged in an interoperable manner (Kodali, 2005).

In SOA context, applications and functions are modularized and presented as services for consumer applications in this case service requestor. The most momentous factor of these applications and services is they are inherently loosely coupled. As defined through separation of concerns, the service interface is independent of the implementation, as service developers and system integrators can develop new applications by composing one or more services without knowing the service's underlying technicalities. Accordingly, an existing application or a service can be implemented either in .NET or J2EE, and the service consumer or requestor can be on a different language.

The key actors of SOA include services, dynamic discovery, and messages. A service is a retrievable routine that is made available over a network. A service exposes an interface contract, which defines the behaviour of the service, and the messages it accepts and returns. The term service is often used interchangeably with the term provider, which specifically denotes the entity that provides the service.

- **The Service Consumer:** Any entity or individual request a service.
- **The Service Provider:** The entity which creates the service and makes it available to other entities.
- **The Service Registry:** The directory in which all services are stored in a searchable and semantic manner.

Service register or directory stores service interfaces where they are categorized based on different services offered. Service consumers can look up a particular service by dynamically querying for services based on various categorization features. This process is referred to as the dynamic discovery of services. Service consumers or clients consume services through messages. Because interface contracts are platform- and language-independent, messages are typically

Figure 1. SOA actors

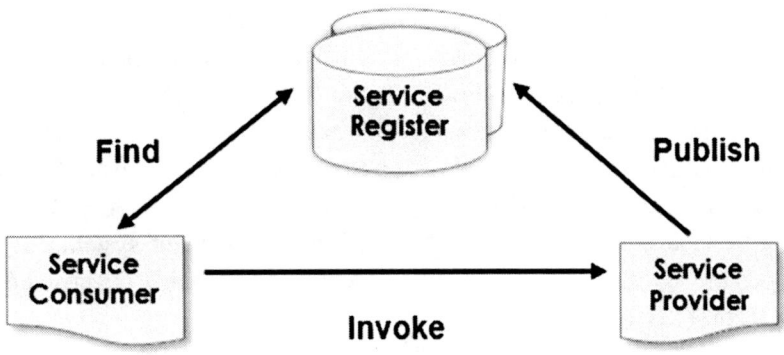

constructed using XML documents that conform to an XML schema.

In order to implement SOA, two main technical components of the SOA infrastructure need to be constructed. The first component is the SOA backbone structure, which can be achieved by the implementation of Enterprise Service Bus. ESB provides robust communication, intelligent routing, and sophisticated translation and transformation of services. ESB as the middle ware will be where point to point interaction between services and application is established. The second component is web services. This technology provides standards for message exchange and service description. Web services utilize the XML-based Simple Object Access Protocol (SOAP) over HTTP for communication between service providers and consumers. Services are exposed as interfaces defined by Web Service Definition Language (WSDL), whose semantics are defined in XML. Universal Description, Discovery and Integration (UDDI), a language-independent protocol, is used to interact with registries and look for services. All of these features make Web services an excellent choice for developing SOA applications.

E-Government applications and services require both high level of interoperability and security. The interoperability is a prerequisite in the services integration process due to the huge number of heterogeneous participants in such applications. Considering the privacy and confidentiality of the exchanged information between governmental agencies and citizens, it is significant to underpin the security demand. A service oriented approach presents a solution for those requirements. It facilitates the interconnection between diverse heterogeneous administrations through well-defined standards (Sellami et al., 2007). Web services are commonly used as an implementation for service oriented approach. They offer interoperable entities and propose standards to secure them. This feature has pushed forward for further e-government web services development.

Service Oriented Delivery

A Service Oriented Delivery (SOD) based solution needs to rapidly transform existing applications, data, and content into web services using a completely non-intrusive approach that requires no changes to the existing applications. An SOA based solution will reduce the dependency on back-end applications and the necessity to write code every time there is a change in policy, and introduces new software that promotes the direct collaboration of citizens and government departments irrespective of the delivery model.

Realising a connected service delivery based on SOA entails a substantial organisational shift, from public services structured, planned and designed around the fragmentation of public administration towards a citizen centric government, whereas public services are designed and structured around customer needs and demands.

Current successful practices of citizen centric governance have emphasized better outcomes and performance over processes (Roy, 2008). Citizen-centric governance emerged as a result of a reform movement that evolved during the 80s and 90s, well known as New Public Management, which placed customer service at the core of public mission, instituted business inspired management flexibility, and used market and competitive forces in the design and delivery of services (International Council for IT in Government Administration, 2006).

Citizen centric government is all about turning the focus of government around citizen needs, looking at the service delivery though the eyes of the citizens (so that the needs of the citizens come first) rather than operational or other imperatives of the government systems. It is designed and developed to enable government:

- To advance service delivery levels.
- To increase the uptake of online services, thereby improve sustainability and encourage investment in e-government services.
- To improve citizen satisfaction with government services

Shifting towards citizen centric government involves two primary elements. Firstly, at the connected service delivery portal, service provision is based on customers' context and situations (OECD, 2009). Therefore, customers should be able to access these services in a well-structured and clear manner meeting their perspective and needs. This view must not conflict with the different public authorities or private service providers that actually deliver the public service. Secondly, the specific responsibilities and fragmentation of public administration in respect to service production and service delivery must not be hampered. Consequently, a smooth integration of two distinct requirements is required: an external, citizen requirements; and internal, public administration requirements.

Citizens' Requirements

Citizen requirements can be defined as what services, applications, interactions and facilities citizen require from e-government application. According to Sellami and Jmaiel (2007), the most commonly recognized citizen requirements are given below:

Eservices and information accessibility: A greater public access online services and information is becoming increasingly a citizen demand. Citizens expect more responsive, integrated and efficient services from their governments; they do not want to deal with multiple providers. Web services are a promising technology in the context of SOA, and it can be used to provide citizens and involved government agencies (Service providers) with eservices. The presentation layer of the suggested architecture offers citizens

ubiquities accessibility to the information they need. Additionally, to avoid the routing problem in the proposed ESB, an orchestrator process is technically required in order to offer the citizen good quality e-services in replacements of the complicated administrative services. Orchestration engine will aid to coordinate and orchestrate shared, composite and integrated services which require multiple service providers and yet multi –procedural steps, including forms and approvals, to complete.

Multiple access channels: This requirement is to facilitate customer self-service including face to face and telephone but also the use of technologies such as the Internet, kiosks or DTV. Multi-channel accessibility will improve access to services including making sure that access is available for anyone who wants it (social inclusion). This is met by separating the presentation layer from the application layer in the proposed architecture. This technical separation of concerns between the application layer and the presentation layer will not allow multiple accesses, but it will enable the presentation layer to be more technologically independent platform, as the presentation layer can run and implement several forms and protocols (WAP and Web server), without interrupting the systematic procedure at application layer.

Authentication facility: Citizen wants to be sure of what government or private business department that they are doing business with. An emergent need has arisen for the necessary assurance level for identification and authentication of persons and parties using e-government services. SSL3 protocol along with digital certificate use is proposed in the model to authenticate and identify citizens. Authentication process will also help to comply with Personal Data Protection acts, as well as will improve data confidentiality, as information sent by the citizen has to remain confidential. The data exchanged between the presentation layer and the application layer will have to be encrypted with the public key contained in the digital certificate of the concerned

Administration Requirements

Administration requirements can be defined as internal government requirements which need to be achieved by the proposed connected service delivery model. The most common administration requirements areas are as follows:

Authentication at Administration Level and Between Government Constituents: Government agencies as service providers should be authenticated before retrieve, call or request the service, this will help to orchestrate and route services in the middle ware (ESB) and help to store service retrieval index for statistical purposes, and this index can also be used to measure service uptake and consumption rate by the service providers and consumers. SOAP will be the primary communication protocol between connected service providers over the SOA; authentication can be assured through a digital signature in SAML assertions contained in the exchanged SOAP messages. The signature will be executed in accordance with the eGIF guidelines set by the e-government initiatives in an attempt of not interfering with the interoperability constraint of the proposed architecture. During the signature validation for this SAML assertion, the authenticator (in this case a Service Provider Authenticator) will try to find a Validation Alias element with the value of its Key attribute. This alias references a certificate in the Key Store that will be used to check the signature validity. Depending on the privacy, security and confidentiality of the closely connected service, a certificate of each service provider involved in the completion of that service will be required. On the other hand, citizen authentication is equally vital for the government agencies, service providers and consumers. The government agencies must be able to authenticate a citizen interacting and requesting a service. The proposed model authenticates users by using (Security Assertion Mark-up Language) SAML as an XML standard for exchanging authentication and authorization data between security domains. SAML has the

feature like platform independent and is mainly applied to Single Sign-On (SSO) (Yoon Jae Kim, 2009).

Interoperability in Terms of Service Delivery: Interoperability is an ultimate goal of service oriented design paradigm. It is an attempt to make all cross organisational systems and applications work together in an interoperable manner. As governments have been planned and organised with vertical structures, each government department aligned to deliver particular services. Therefore, heterogeneity in the technologies and the platforms has emerged. This heterogeneity and lack of universal standards can be overcome by using the service oriented approach, which will allow the co-operation between heterogeneous systems due to its universal independent platform, whereas different applications with different source codes can interconnect in services manner, each service must conform to the web service based on service-level agreement listed in eGIF guidelines set by the e-government initiatives.

Data Integrity and Confidentiality: Exchanged SOAP messages over the ESB will have an attached digital signature which makes it possible to the governmental agency to verify certificate for data validation purposes. Digital signature will also be extended to the SOAP message in accordance to the WS-Security specifications based on SAML open standard data format. When digital signatures are enabled, the authenticator will look at the Validating Alias table for an entry that matches the value of the Key attribute with the host name of the Issuer of the SAML assertion.

Public Service Integration: Governments are looking into integrating most of their interactions of G2G, G2C, G2E and G2B into a convenient, seamless and single point of access portal, through which public services can be accessed, utilized and completed. A service orientation approach to integrate services is technically effective and realistically achievable due to undependability nature of SOA of any vendor, product or technology. Moreover, web services deployment over

SOA will reinforce reusability of public service to improve the internal constancy of shared services, increase service quality through multiple testing cycles by different service consumers and reduce the number and internal complexity of services. Connecting public service based on service oriented architecture will enable government agencies to create services by wrapping the existing applications and reuse them to assemble new and composite applications.

INTEGRATION FRAMEWORK FOR INTEROPERTABILITY OF CONNECTED SERVICE DELIVERY (IFICSD)

The multifaceted nature of the government organisational and technical structures resulted in increasing the complexity of the service delivery. However, as connected service delivery requires government departments to cooperate and collaborate across organizational borders to provide connected services, a cross interoperable system has to be introduced. Integration framework for interoperability of connected service delivery is an efficiently achieved philosophy that provides an inclusive answer to all problems stem from shared services, and all other cut-cross services that require two or more governmental departments to cooperate efficiently and effectively to provide.

Integration framework for interoperability of connected service delivery (IFICSD) will help connect all actors involved in service delivery in an integrated environment. The framework provides an effective means of communication between all of those actors to share data, information and services in a constituent and consistent manner. IFICS is formed as an interconnected structure that offers an integrated environment for all actors to communicate together in a seamless manner, through using Enterprise service bus technology, an integrated medium for all involved actors to be plugged onto the system, without having to

deal with migration technical issues and legacy problems. As a result, government agencies are expected to publish their existing services in a form of web services into the ESB and make them available to other systems to use, share and utilize using a unified web service description framework based on SOA.

Considering the requirements from citizen and government agencies, a multi-tier architecture should be introduced in order to logically separate presentation, application processing, and data management functions in e-government applications, which would enable various existing e-government applications to be integrated with new ones in extensible fashion.

IFICS is architecturally split into four layers, namely: the client layer, the presentation layer, the application layer and the data layer. The framework presumes a full implementation of a secured government network though which connectivity and document exchange amongst government departments is provided.

Client layer is where users access the application, and considered the main access channel for citizens, businesses and government agencies to access government services and information through a single access consolidated browser based interface. Access channels will enable clients (a citizen or a business) to connect through the browser, it will interact with the application layer via the presentation layer .The communication with the presentation layer will be established using the HTTP protocol secured by SSL. The need of HTTPS protocol (HTTP over SSL) stems from current security threats circling the Internet, and the sensitivity of data being exchanged over the network, which will exponentially increase user trust on online government services. On the other hand, the use of digital certificate will enable clients to use HTTPS to verify the identity of the connected site.

Government agencies will directly interact with the application layer when successfully authenticated. As a service consumer, government agencies will be connected to service discovery engine, which will enable government agencies to invoke and request governmental services through the application layer. The communication with the application layer will be established through the implementation and utilization of SOAP over HTTP. Using HTTP will enable the bypass of the conventional firewalls in distributed applications (Sellami et al., 2007), and will highly assist to maintain a higher degree of confidentiality and integrity of the exchanged messages over the ESB, as SOAP messages will be digitally signed according to the WS-Security specifications and conformed to eGIF.

The presentation layer acts as an intermediary tier between the client layer and the application layer. It manages the interfaces for the clients interacting with the e-government application. The presentation layer contains web server to handle requests from clients connected via web browsers, as well as a WAP server for mobile clients. The technical and architectural separation between the presentation layer and the application layer provides a better accessibility via different channels such as web browsers and mobile clients without changing the application implementation. This separation of concerns will provide better data integrity over the national network. The communications protocol with the application layer will be using SOAP over HTTP.

The application layer is the core of the integration framework for interoperability of connected service delivery (IFICSD), and it consists of the following components and elements as illustrated in Figure 2.

Enterprise Service Bus: ESB is used in order to replace the point to point approach with a single, centralized place to integrate systems, and it does so in a service-oriented manner. The ESB will play a vital role in the structure as it will be used to discover, bundle and deliver agencies eservices. The bus provides message delivery services based on standards such as SOAP, HTTP. The ESB is utilized for high-throughput, guaranteed message

delivery to a variety of service providers and consumers. It enables the use of multiple protocols and performs transformation and routing of service requests. The ESB enables services to interact with each other based on the quality of service requirements of the individual transactions, and will exhibit the following capabilities:

- **Communication:** The ESB will support routing of messages and addressing of massaging style as request/response between service requestor and service provider, and publish/subscribe between service providers and service repository (El Haddad, 2009). SOAP will be implanted as a transport protocol within the ESB in order to make all services widely available. This will enable location transparency and service substitution, which will eventually enable the decoupling of the consumer view of services from their implementation through the service panel.
- **Interaction:** ESB will provide an interface definition format and associated messaging model (such as WSDL and SOAP) to allow the decoupling of technical aspects of service interactions.
- **Management and Autonomic:** ESB will provide a consistent administration model across a potentially distributed infrastructure, including control over naming, routing, addressing, and transformation capabilities, which will lead to efficient services management in the government (Balani, 2010).

Service Repository (SR): A centralized database structure where all available government services are stored and published. The service repository integrates information about a service from multiple sources and stores it in a centralized database. Service information may include design artefacts, deployment topologies, service code repository and service monitoring stats. The

Figure 2. Integration framework for interoperability of connected service delivery (IFICSD)

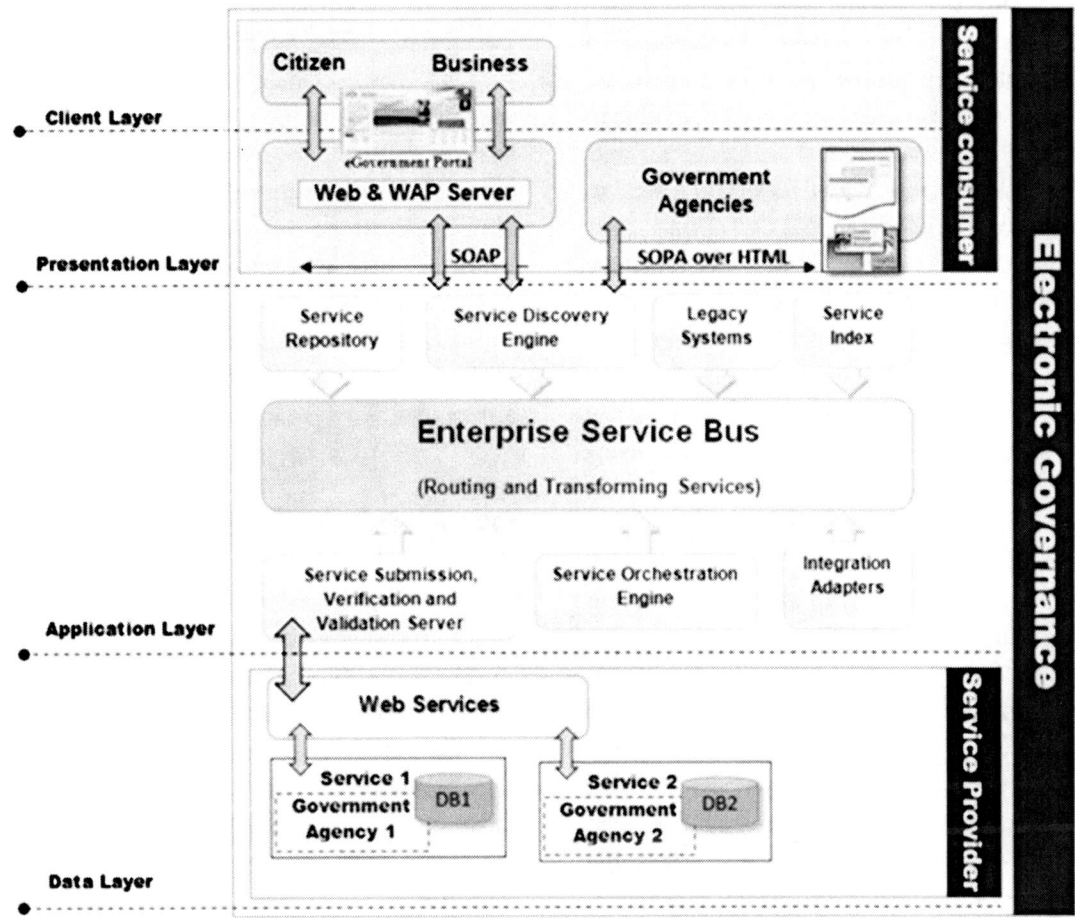

service repository will consist of services registry that contains service descriptions, service contracts and service policies that describe a service like determining a service end-point address at runtime based on the service unique name or ID.

Service Submission, Verification and Validation System (SSVVS): A system that enables service providers to publish and submit their web services through the IFICS in the service repository, as government entities wish to plug their services must comply with a predefined eGIF, such system will ensure a high level of compliance from the service providers.

Service Discovery Engine (SDE): This engine is plugged into the ESB for the purpose of finding a suitable service for a given task or applications. SDE are containers that host the business logic

and processing rules of the service components. SDE process the message information received from the Service Infrastructure. SDE will match the needs of the service requestor/consumer with the appropriate service provider in the service repository, in the basis of, when a service is invoked; SDE will request a list of services from service repository and reply with a list of services that fit the requirements of the requestor. The clean functional separation between service repository and service discovery engine will create efficiency, clear lines of security and functionality of the increasing number of web services, and help to control the complexity associated with newly emerged composite web services from combining two or more other web services.

Integration Adapters: Are connectors and adaptors for all required technologies, message formats and protocols available. Depending on the ESB solution a government is implementing, sometimes not all clients and services are able to understand the protocols and messages formats the ESB uses natively. As such there is a need to be able to bridge between ESB end points.

Service Orchestration Engine: Is a software engine built on top of the message-oriented middle backbone of the ESB. These orchestration engines are capable of orchestrating long-running business transactions through stored state. It primarily serves as a coordinator and manager of conversations among web services. Such orchestration can be simple logic such as single two ways conversation or a nonlinear complex and multistep government transaction with exception handling and compensation logic.

Service Index: An artificial intelligent agent to monitor service contract adherence and ensures correct service delivery. It will play a key role in support service consumer choosing a service with appropriate metrics. Service index will also be used for the service prioritisation process in iterative pattern, led by transaction criteria and stakeholders perceptions. It will provide an impartial view

of service phasing, planning, consumption, and collect data in order to identify key services for portal personalization methodology.

Data Layer: This allows government departments to maintain their data model and backend system. As such, government departments in the service provision side will be expected to publish their applications and services in a form of web services. That can be achieved by plugging directly onto the Service Submission Verification and Validation Server to ensure the best compliance with a predefined eGIF, and the service-level agreement.

OVERVIEW OF THE CASE STUDY

The recent reforms in Jordanian government's structure and organization, which was accompanied by the technological transformation through the adoption of e-government systems, have led to the emergent of a new set of services that can be simply described as closely coordinated. While these services are inherently connected, they are highly dependent on each other as shown in Figure 3.

Figure 3. Public service dependency

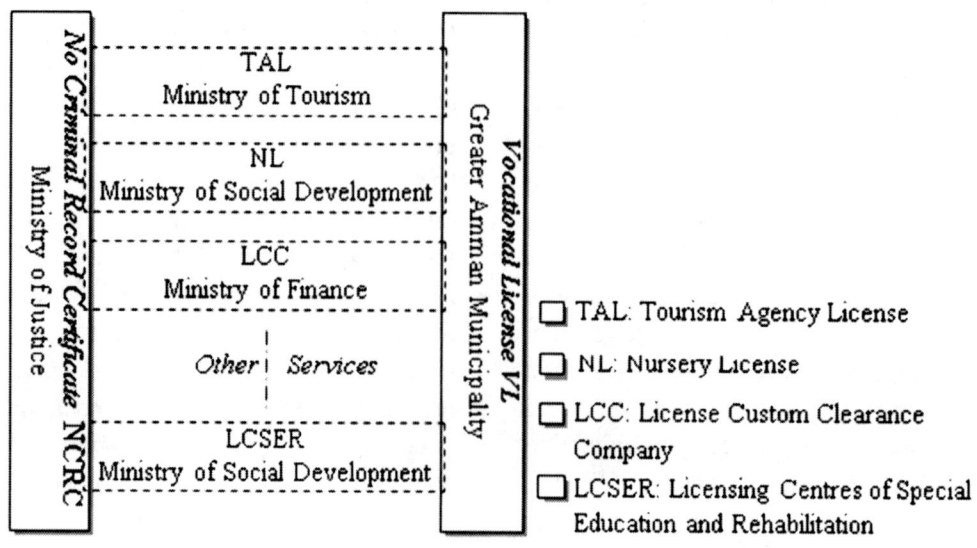

Online licencing services have been flagged up as the most demanded services in developing countries (CSTransform, 2011). Therefore, Jordanian e-government program immediately responded to the call and encouraged governmental constituents and service providers to offer online licensing services. As a result, a form of services and processes integration in relation to service delivery has emerged, explicitly in services such as Vocational Licensing (VL) and No Criminal Record Certificate (NCRC) and other closely coordinated services which cannot be completed without fulfilling other dependent services.

From legal and organizational perspectives, in some scenarios these services are prerequisite to each other and the service cannot be fulfilled without validating the preconditioned service. An example of these closely coordinated services is the licensing services in Jordan, as most of the licensing services require either applying for a Vocational Licence (VL) and a Non-Criminal Record Certificate (NCRC).

VL Vocational license is a document that formally enables companies to operate within the Municipality of Greater Amman. This basic information characterizes the business itself and serves as a reference for physical inspections of the Municipality of Greater Amman representatives. Anytime the business information changes, it requires official update of the vocational license. A vocational license must be renewed annually (Jordanian e-Government Program, 2006).

NCRC No Criminal Record Certificate is a certificate that is requested from Jordanians or non-Jordanians residing in Jordan (in this case the certificate indicates no-criminal record in Jordan only) in order to apply for a vacancy in public sector or in some private companies. The certificates are also necessary for obtaining a license for opening a particular type of business such as Internet cafe. There are no regulations regarding the validity period of the certificate. Generally, certificates that are not older than 3 months are accepted (Jordanian e-Government Program, 2006).

These services are provided by different agencies and can be access and utilized from several locations and different websites, and yet required for several professions licensing such as Child Care (Nursery) and Licensing Tourism and Travel Agencies (TAL) as depicted in the figure 3. The service that requires the completion of other service or services will be named in this chapter as dependent service. Jordanian government provides links to these disparate systems from its e-government portal. Citizens will eventually get forwarded to different portals to complete the requested services, which is in nature similar to the conventional approach of requesting the citizen to physically visit two or more governmental departments to complete a service.

In the domain of dependant services such as TAL, the service is subject to validate by two preconditions, which are obtaining a Vocational License and presenting valid a Non-Criminal Record Certificate as illustrated in Figure 3. Consequently, citizen now is required to visit three different portals, with three different login details, to fill in three similar online forms in order to get a service completed.

The first portal is provided by a service provider which is Ministry of Tourism and Antiquity (MoTA), the second service (NCRC) is offered by the Ministry of Justice and the third portal is provided by the Greater Amman Municipality. Figure 3 illustrates the current technical overlap between the two services (VL and NCRC) and the dependant service such as TAL, NL, LCC and LCSER.

A significant number of government services require the completion of two other major services (VL and NCRC) in order to be fulfilled and delivered to client. The vast majority of these government services is likely to share a common procedural and information flow, but in the end, they all need to be validated legally by verifying the eligibility of the applicant through NCRC, and professionally through the process of issuing Vocational License VL. This has contributed to the emergence of new aspects of e-government interoperability, namely as organisational interoperability, semantic interoperability and technical interoperability.

Connecting these services has the effect of changing the relationship between the components that provide or manage them. In some cases, these components will need to be more closely coordinated, and this may be achieved by a combination of new technological mechanisms and organizational/ administrative measures, or even by reorganizing the structure of government. If the responsibility for delivering a service is split between several departments, this is likely to result in higher costs as well as bigger margin for error and delay.

The Service Background

A travel agency is a private retailer that provides tourism related services to the public on behalf of suppliers such as airlines, car rentals, cruise lines, hotels, railways, and package tours Ref. In order to start a tourism business in Jordan, a license of travel and Tourism Company and office must be obtained from the Ministry of Tourism and Antiquates (MoTA), this license allows registrant to carry on the following activities:

- Sells, resells, or offers to sell travel tickets.
- Sells or arranges travel with accommodation.
- Makes or offers to make travel arrangements.
- Advertises any of the previously mentioned activities.

The Travel Agency Licensing (TAL) service was chosen because it doesn't only require the competent authority but the implication of at least three governmental administrations, which are called dependant services in this chapter. The service is basically applying for a license of Travel and Tourism Company and Office and review the application. Traditionally, the TAL service life cycle

involves a chain of interconnected procedures, each of which takes time and effort to be fulfilled. However, in line with the e-government national agenda, the Ministry of Tourism and Antiquities has carried out several technical improvements in terms of the services provided, as a result, an optimised version of the ministry's website has recently been launched as part of an outsourced contract with imagine technology.

According to the TAL Life Cycle Figure 4, when a service consumer begins with the procedure, an initial application should be submitted at the MoTA, stating the business operations, plans and all other related details, which will enable the MoTA to determine the eligibility, exemption and accordingly provide the permissions required in order to obtain the License. At this stage, service requestor is required to provide a No Criminal Record Certificate (NCRC), a Vocational License (VL) which should be acquired previously, and a commercial name registration with phrase travel and tourism, and commercial record certificate with capital of no less than 50,000 JD for travel and tourism purposes, and trademark, if any, all issued by Ministry of Industry and Trade, issue date not to exceed 3 months from application date.

Consequently, the MoTA will issue an initial approval _Valid for 30 days_ for the service requestor to carry on in the business establishment process. In the following step, the service requestor is expected to present a bank guarantee and employees contract for five employees at least, and then the MoTA will carry out an inspection to verify all the stated specification. In the next stage, a JTTAA (Jordanian Travel and Tourism Agencies Association) membership must be submitted along with the licensing application. At the last stage, a certain amount of fees must be paid to the MoTA and the approval is to be issued within two to four weeks.

The explicit overlap between TAL service and NCRC and VL services underpins a necessity to integrate the three different services without having to eliminate the organisational boundaries

Figure 4. TAL service life cycle

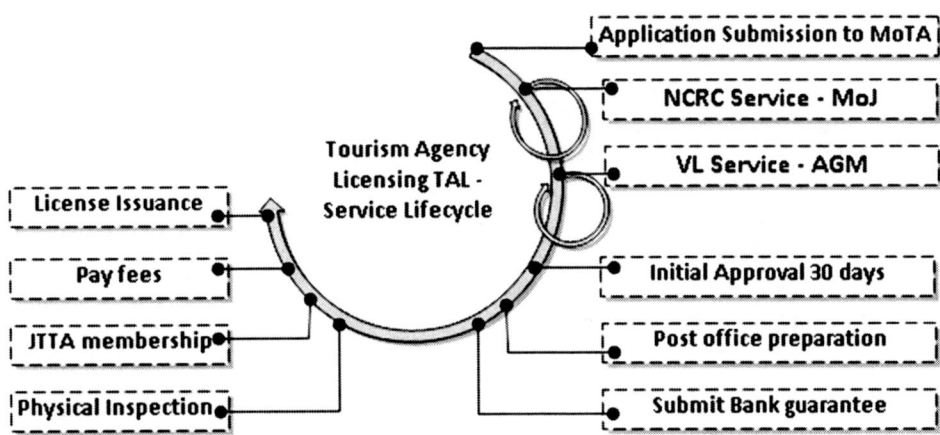

between the government entities that provide them. As part of the e-government implementation in Jordan, all ministries and government agencies were encouraged to move their presence onto the web, by having a website that concludes basic information and some valid application forms that might be required by citizens. Accordingly, some of the services that can be labelled as the most commonly used services have moved online, like NCRC and VL, in an attempt to ease access to these services and provide constant delivery of common services. Nevertheless, public service consumer still has to visit three different websites and verify his personal details several times in order to complete a service.

The Online Service Architecture

Applying for Tourism and Travel Agency License requires the customer to follow a set of procedures predefined by the competent authority, in the case the Ministry of Tourism and Antiquities (MoTA). Traditionally, if a citizen wants to apply for a Vocational License for Tourism Agency Licensing, the current procedure requires completing at least three main forms, provided by three different authorities, which are the ministry of Tourism (MoTA), the Ministry of Justice (MOJ) for the purpose of No Criminal Record affirmation, and

Greater Amman Municipality (GAM) for vocational licensing purposes, as depicted in Figure 5.

The service consumer is actually dealing in this case with a composite service that consists of two other dependant services, which are No Criminal Record Certificate (NCRC) provided by MOJ and Vocational License (VL) provided by GAM, without fulfilling these dependant services, ministry of tourism in Jordan cannot issue a travel agency license. All necessary forms are available on the Internet in an online format, as it can be seen in Figure 5. Typically, the procedure begins with MoTA, as an online registration is essentially required to log in the system, therefore, the citizen has to visit the competent department physically, and that usually through calling the ministry, make an appointment, and eventually, the process ends with acquiring an account, which will allow the citizen to access the online form.

In the following step, citizen has to apply for a Non-Criminal Record Certificate, and that requires the citizen to follow the same conventional procedure, except in this case, a signing up process at the portal is provided, so citizen can fulfil this service online. The portal provides a means of communication with the potential citizen to apply for a new NCRC, or review and edit an existing application. In order to complete the service, citizen has to apply to Vocational License

Figure 5. TAL, NCRC and VL services online presence

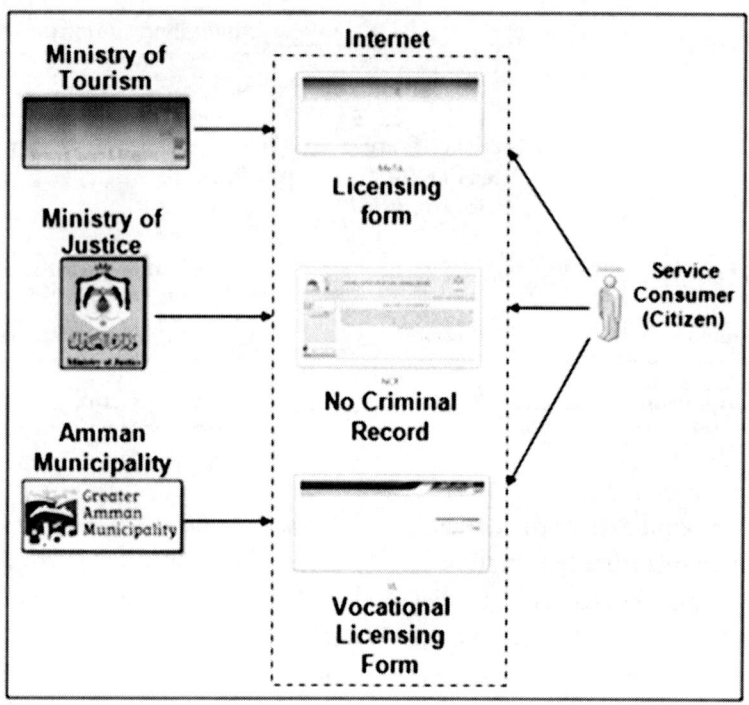

online, this service is provided by GAM, and it provides several features to the end users ranging from applying for a new license, edit an existing license, and verify the status of a license. However, registration process for a new account has to be done physically thorough the GAM.

Despite the fact that all forms are available online, the process involves a set of procedures that are no different than the traditional procedure, which requires citizen to be presented in each governmental department and interact directly and physically with the service provider, as well as the same data is being entered multiple times which will increase the possibility of negative data redundancy and duplication. As a result, the process of applying for TAL is time consuming, labour intensive and costly from government oriented view, and multifaceted, multi procedural steps and long process from citizen oriented view.

Moreover, the current state of TAL clearly indicates the lack of coordination between government services and governmental entities. It points out an ambiguity and inefficiency in procedures performed due to electronic flow of documents, high costs related to managing services across the government and external entities and the extensive time that is needed to complete the service.

Integration Framework for TAL, VL, and NCRC

Integrating public services based on a service oriented approach needs to rapidly transform existing government applications, data, and content into web services using a completely non-intrusive approach that requires no fundamental changes to the existing applications.

Service Provider Side

In order to implement TAL, VL and NCRC in the proposed integration framework, common goals and shared purpose by public services organisations need to be developed, particularly in front

Table 1. Current online service providers

Service	Vendor	Platform	Link
NCRC	Optimiza	JSPX	https://ncrc.moj.gov.jo/NCRCApplication/faces/com/moj/ncrc/issuingncrc/customer.jspx
VL	Optimiza	ASPX	http://www.ammancity.gov.jo/ar/eservices/login.asp
TAL	ESKADENI	ASP.NET	http://professions.mota.gov.jo/

line mode, in this case, Ministry of Justice (MOJ), Greater Amman Municipality (GAM) and Ministry of Tourism (MOT), which will lead to respond effectively to the increasingly complex demands of increasingly sophisticated consumers of public services with multiple needs and life challenges. The technical vertical structure of the mentioned ministries poses a major interoperability challenge, as backend and legacy systems of service providers are developed, maintained and administrated in distinctive environments. Therefore, service providers are expected to cooperate and collaborate by addressing a cross-cutting policy, strategic and operational demands of public services.

As TAL is a shared and closely coordinated service, service providers involved in TAL service are expected to transform their shared and existing applications and online systems to asynchronous web services, with a standard message-based interface using the same interoperable language SOAP. Those applications can be agreed upon within the government department depending on citizen demands, degree of digitalisation and level of dependency by other services. Web services need to be conformed to a predefined legal, technical and organisational exercise intends to resolve and prevent issues arising from incompatible content of different computer systems, namely electronic Government Interoperability Framework eGIF. EGovernment initiatives should produce a set of architectural guidelines to serve as an initial reference to all architectural standards to be distributed and used by various eservices implementation projects. In case of Jordan, the compliance to the Jordanian eGIF release 1.0 needs to be met before service providers can publish their web services.

Involved service providers in TAL service should publish their web services, namely TAL, VL, and NCRC onto the ESB, the Service Submission, Verification and Validation Server will validate and verify the submission and confirms compliance with eGIF. The technical separation between the data layer and application layer will help to maintain a high level of security, as well as allowing government departments to restrict direct access to their internal systems, which would reduce cross organisational resistance to minimum levels.

Service Consumer Side

Service consumer, citizen, business or government department will connect to e-government portal via the provided access channels. Government services will be offered and presented on the portal based on predetermined agreement between e-government initiative program and the service provider. Service consumer is expected to login to the web service of issuing a tourism agency licence through the application interface via service discovery engine. Once authenticated successfully based on TAL WS-Authorization, service consumer will be prompted to fill in the electronic application form with the necessary information that meet the requirements of tourism agency licensing service of MoTA.

As demonstrated in figure 4, the next procedure in the service life cycle is to present a valid NCRC and VL, as these service should ideally be published as web services, the service orchestration server should be notified when a shared service is invoked, which is in this case TAL. At this point of the application, TAL service will verify a valid NCRC and VL based on its WS-Policy, and accordingly, invoke NCRC web service and VL web service under the monitoring of the service orchestration server. Once web services are invoked, the service index will record the process for future use. This invocation will generate four different scenarios:

1. **New Service Consumer:** In this scenario, TAL service consumer has neither obtained NCRC nor VL. Once TAL web service is invoked and the initial electronic form is filled, TAL will invoke NCRC web service and exchange the required data for the application using XML messages. This will lead the additional fees of a new NCRC application to be added to TAL application. As the consumer is new, TAL web service will also invoke VL and exchange the necessary data, due to the nature of VL application based on VL WS-Context, further details may be required, as such, service consumer will be notified to fill in the required information, and a new application fee will be also added to the total amount. Considering the time needed to TAL, NCRC and VL, consumer will be issued a TAL ID, NCRC ID and VL ID through email or SMS, and pay the total application fees through the e-payment gateway shared service. See figure 6.

2. **Returned Service Consumer, NCRC Valid:** In this Scenario, TAL service consumer has previously obtained NCRC but has not yet applied for VL. Once TAL web service is invoked, TAL will invoke NCRC web service and verify the validity of the application, if successfully granted by a valid NCRC ID, TAL web service will invoke VL, in this case VL is not valid, TAL web service will exchange the necessary data, due to the nature of VL application based on VL WS-Context, further details may be required, as such, service consumer will be notified to

fill in the required information, and a new application fee will be also added to the total amount. In the end, TAL consumer will be issued TAL ID and VL ID through email or SMS.

3. **Returned Service Consumer, VL Valid:** In this Scenario, TAL service consumer has previously obtained VL but has not yet applied for NCRC. Once TAL web service is invoked, TAL will invoke NCRC web service and verify the validity of the application, as NCRC is not obtained, TAL will exchange required data for the application using XML messages. This will lead the additional fees of a new NCRC application to be added to TAL application. In the end of the application, consumer will be issued NCRC ID and TAL ID.

4. **Returned Service Consumer, NCRC and VL Valid:** As illustrated in figure 6, in this Scenario, TAL service consumer has previously obtained NCRC and VL. Once TAL web service is invoked, TAL will invoke NCRC web service and validate the application, if successfully granted by a valid NCRC ID, TAL web service will invoke VL web service and validate the VL ID. As both IDs have been previously obtained, consumer will pay the application fees through the e-payment gateway shared service. TAL will issue an ID and initial license approval for 30 days.

Once service consumer is issued TAL ID and the application will be forwarded to the licensing committee in MOTA to make a decision based on the information provided, if further details are required, service consumer can always check the status of the application and supply the information as needed. If the application has been approved, service consumer will be send a notification email or SMS.

In the light of the TAL service requirement, the case study of TAL service concludes that

Figure 6. Web services invocation scenarios

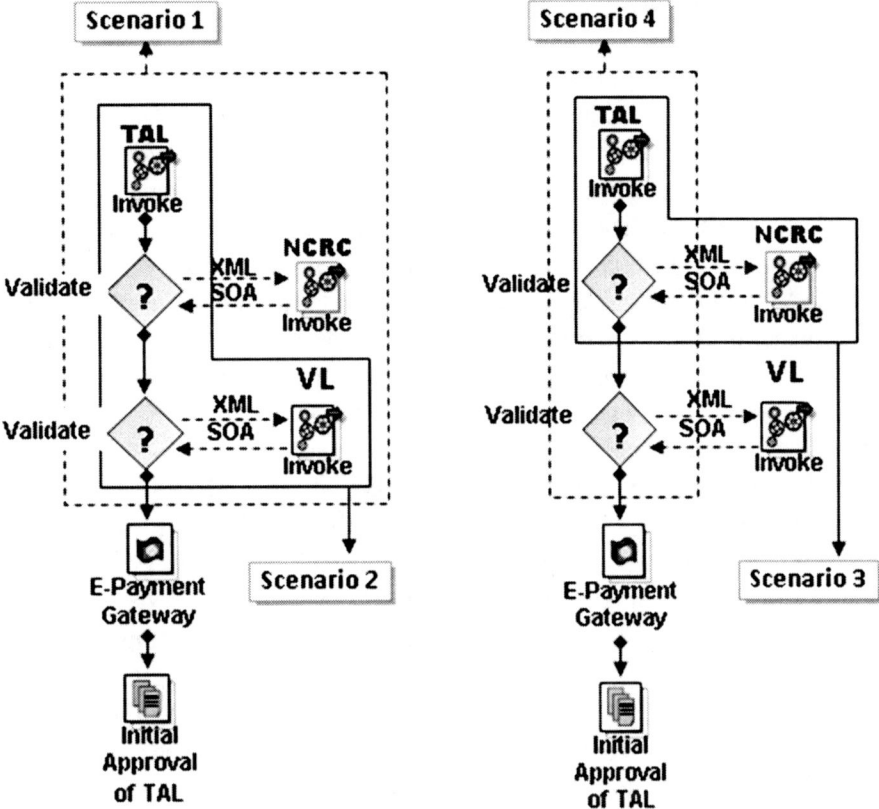

the development of e-government applications and public services in form of web services will potentially offer several advantages which can be described as follows:

- Web services will allow government to integrate disparate applications and data formats with relative ease. TAL, NCRC and VL Web services will be highly interoperable and fully integrated on both service provider and consumers sides. Service provider can maintain their internal legacy systems and data integrity while making it possible for service consumers to use their web service. Using WSDL definition, service consumer can effectively exchange data with the service, in a way that neither part needs to know how the other is implemented or in what format its underlying data is stored.

- Web services usability will help reduce data redundancy and multiple form interaction. When TAL, VL and NCRC services are implemented as web services on the ESB, service consumer doesn't have to be redirected to different applications and websites, as service can be completed through the same application interface, which will not only reduce data redundancy but also reduce the time needed to complete the service.
- The successful implementation of TAL, VL and NCRC will lead to reduce the load of procedural workflow required to complete the service, as well as reduce the workload of employees in different government departments.

PROPOSED E-GOVERNMENT INTEROPERTABILITY FRAMEWORK

E-Government Initiatives should issue an architectural guideline for government entities wishing to implement and integrate their web services into the e-government designated ESB. The E-service Architecture Guideline serves as an initial reference to all architectural standards to be distributed and used by various eservices implementation projects. In order to guarantee an adequate level of service integration, procurement, deployment, all government entities are required to comply with this architectural guideline. The table 2 sums up the most significant rules which government entities are required to comply with, as well as the standards whereas TAL service has to be planned, designed and structured around.

WS – Addressing

WS-Addressing provides transport-neutral mechanisms to address Web services and messages. Specifically, this specification defines XML elements to identify Web service endpoints and to secure end-to-end endpoint identification in messages. This specification enables messaging systems to support message transmission through networks that include processing nodes such as endpoint managers, firewalls, and gateways in a transport-neutral manner (W3C, 2011).

WS – Policy

The Web Services Policy Framework (WS-Policy) provides a general purpose model and corresponding syntax to describe the policies of a Web Service. WS-Policy defines a base set of constructs that can be used and extended by other Web services specifications to describe a broad range of service requirements and capabilities (W3C, 2011).

WS-Coordination

This specification describes an extensible framework for providing protocols that coordinate the actions of distributed applications. Such coordination protocols are used to support a number of applications, including those that need to reach consistent agreement on the outcome of distributed activities (IBM,2009).

WS-AtomicTransaction

This specification provides the definition of the atomic transaction coordination type that is to be used with the extensible coordination framework

Table 2. Proposed eGIF

No.	The Rule	Phase
1	Comply with UDDIUDDI standards	eService Publish
2	Understand the Guaranteed delivery feature	eService Publish
3	WS – Addressing	eService Design
4	WS – Policy	eService Design
5	WS – Coordination	eService Design
6	WS – Atomic Transaction	eService Design
7	WS – Business Activity	eService Design
8	WS – Brokered Notification	eService Design
9	WS – Topics	eService Design
10	SLASLA (Service Level Agreement) contracts	eService Deployment

described in the WS-Coordination specification. The specification defines three specific agreement coordination protocols for the atomic transaction coordination type: completion, volatile two-phase commit, and durable two-phase commit (IBM, 2009).

WS-BusinessActivity

This specification provides the definition of the business activity coordination type that is to be used with the extensible coordination framework described in the WS-Coordination specification. The specification defines two specific agreement coordination protocols for the business activity coordination type: Business Agreement With and Business Agreement with Coordinator Completion (IBM,2009).

WS – BrokeredNotification

WS-BrokeredNotification defines the Web services interface for the Notification Broker. A NotificationBroker is an intermediary which, among other things, allows publication of messages from entities that are not themselves service providers. It includes standard message exchanges to be implemented by NotificationBroker service providers along with operational requirements expected of service providers and requestors that participate in brokered notifications. This work relies upon WS-BaseNotification and WS-Topics, as well as the "Publish-Subscribe Notification for Web Services" document (IBM,2009).

Service-Level Agreement

A service-level agreement (SLA) is a part of a service contract where the level of service is formally defined. In practice, the term SLA is sometimes used to refer to the contracted delivery time (of the service) or performance. As an example, Internet service providers will commonly include service level agreements within the terms of their contracts with customers to define the level(s)

of service being sold in plain language terms. In this case the SLA will typically have a technical definition in terms of mean time between failures (MTBF), mean time to repair or mean time to recovery (MTTR); various data rates and similar measurable details.

CONCLUSION

The prerequisite for effective information and data sharing between government departments is technical interoperability. This Technical level interoperability deals with the technical aspects of interconnecting government systems and standardisation of data format and data exchange protocols, and the extensive use of web services, all these elements can serve purposefully towards the concept of Service Oriented Architecture SOA.

This chapter investigates the connected public service delivery through the integration of back office processes. It provides an interoperability integration framework that connects closely coordinated services based on Service Oriented Architecture, in an attempt to align the organizational structures and processes of different government departments (Service Providers). As we believe that this integration is not only necessary but also crucial at this stage of e-government implantation in developing countries. The framework suggested in this chapter provides a high level of semantic integration through the deployment of Service Oriented Architecture (SOA), Enterprise Service Bus (ESB) and Web services, which will help to integrate the process of public service provision for best use by the end user, as well as to avoid duplication and redundancy in government agencies.

Governments and other public institutions around the world are under increasing pressure to provide practical evidence in support of policies, decisions and practice on how to integrate services rather than necessity to integration. Consequently, this chapter is in a particular importance for government officials working on e-government implementation projects, as it provides rich insights into how to integrate public service using evidence-base case study of existing government services, as well as contribute to an understanding of good practice and what works well. The chapter also examines the underlying systemic difficulties in public service provision, and doesn't suggest combining or joining up three deferent service providers or eliminating the structural or organisational borders between them, but alternatively, it presents a manoeuvre to integrate the current eservices across organisational border.

This chapter provides an efficient use case example of integrating some of the currently existing online services in the Jordanian e-government platform. A set of eservices in the Jordanian e-government portal have been investigated, namely Vocational Licensing (VL) and No Criminal Record Certificate (NCRC) and Licensing services are presented as a realistic case study to implement the suggested integration framework.

The chapter presents an opportunity for researchers to assess the needs to improve public service provision and evaluate the effectiveness of different models. Researchers can also help in identifying areas where savings can be made or where services could be delivered more effectively.

REFERENCES

Australian State Services Authority. (2007). *Joined up government - A review of national and international experiences*. Author.

Balani, N. (2010). *ESB over Service Oriented Architecture frameworks*. Retrieved November 2013, from http://naveenbalani.com/index.php/2010/05/esb-frameworks/

Berthold, M., Jürgen, K., Normann, H., Schmeidel, K., Schmutz, G., Trops, M., & Winterberg, T. (2013). *Enterprise Service Bus*. Part of the Industrial SOA article series.

Chopra, K. R. (2005). *Ecosystems and Human Well-Being: Policy Responses: Findings of the Responses Working Group*. Island Press.

Colan, M. (2004, April 21). *Service-oriented architecture expands the vision of web services – part 1*. IBM Corporation.

CSTransform. (2011). *E-Government Interoperability: A comparative analysis of 30 countries*. London: CS Transform Limited.

Curristine, T., Lonti, Z., & Joumard, I. (2007). Improving Public Sector Efficiency: Challenges and Opportunities. *OECD Journal on Budgeting*, 7(1), 4–9. doi:10.1787/budget-v7-art6-en

de Bruijn, J. A., & Heuvelhof, E. F. (2000). *Networks and Decision Making*. Utrecht, The Netherlands: Lemma.

Dean, T., & Boutilier, M. (2011). *Joint service delivery in federal countries: Research report prepared for the forum of federation*. The Global Network on Federalism.

Di Maio, A. (2004). *Move 'Joined-Up Government' From Theory to Reality*. Gartner Industry Research, ID Number: G00123844.

El Haddad, J. (2009). Service-Oriented Architecture and Web Services. In *State of the art: languages for services interface description and for services composition*. Paris: INRIA Paris-Rocquencourt.

Gauld, R., Gray, A., & McComb, S. (2009). How responsive is E-Government? Evidence from Australia and New Zealand. *Government Information Quarterly*, 26(1), 69–74. doi:10.1016/j.giq.2008.02.002

Hentrich, C., & Zdun, U. (2012). *Process-Driven SOA: Patterns for Aligning Business and IT*. Taylor & Francis. Retrieved November 2013, from http://books.google.co.uk/books?id=HRHk6nG5Z3QC

Hutchison, B., Johnson, K., & Schmidt, M. (2005). *Increasing IT flexibility with IBM WebSphere ESB software*. BM Software Group.

Hyde, J. (2008). How to make the rhetoric of joined-up government really work. *Australia and New Zealand Health Policy*. doi:10.1186/1743-8462-5-22 PMID:18983680

IBM. (2009). *Developer Works Digital Library*. Retrieved November 2013, from www.ibm.com/developerworks/library/ws-coor/

International Council for IT in Government Administration (ICA). (2006). *Executive Summary. Country Reports*. Author.

Jordanian e-Government Program. (2006). *Jordan e-Governmen Strategy*. Retrieved November 2013, from www.thieswittig.eu%2Fdocs%2FMPC_Strategies%2FJordan%2FJordan_e-Governmen-Strategy.pdf&ei=_AaJUp_0IpGthQe3kYDoCw&usg=AFQjCNF6vGV2J9PB9F6vzAeozWhoiswKbw&sig2=AOwpOup_MMQhZ7TqQlZKeQ

Kenneth, K. (2009). Moving towards integrated public governance: improving service delivery through community engagement. *International Review of Administrative Sciences*, 75(2), 239–254. doi:10.1177/0020852309104174

Kim. (2009). *Access Control Service Oriented Architecture Security*. Retrieved November 2013, from http://www.cs.wustl.edu/~jain/cse571-09/ftp/soa/

Klischewski, R. (2007). E-Government Integration and Interoperability: Framing the Research Agenda. *International Journal of Public Administration*.

Kodali, R. (2005). *An introduction to SOA, What is service-oriented architecture?* Retrieved November 2013, from http://www.javaworld.com/javaworld/jw-06-2005/jw-0613-soa.html

Layne, K., & Lee, J. (2001). Developing Fully Functional E-government: A four-stage model. *Government Information Quarterly*, *18*(2), 122–136. doi:10.1016/S0740-624X(01)00066-1

Marche, S., & McNiven, D. (2003). *E-Government and E-Governance: The Future Isn't What It Used To Be*. Canadian Journal of Administrative Sciences / Revue Canadienne des Sciences de l'Administration.

Noreng, O. (1980). *The oil industry and government strategy in the North Sea*. Retrieved November 2013, from http://books.google.co.uk/books?id=jZIOAAAAQAAJ&printsec=frontcover#v=onepage&q&f=false

OECD. (2009). *e-Government Studies Rethinking e-Government Services User-Centred Approaches: User-Centred Approaches*. OECD Publishing. Retrieved November 2013, from http://books.google.co.uk/books?id=Y2f1N2HVtLsC

Papazoglou, M., & Willem-Jan, H. (2007). *Service oriented architectures: approaches, technologies and research issues, 16*(3), 389-415.

Putnik, G., & Cunha, M. M. (2007). *Knowledge and Technology Management in Virtual Organizations: Issues, Trends, Opportunities and Solutions*. Retrieved November 2013, from http://books.google.co.uk/books?id=nXn5BgqYdcUC

Reddick, C., & Turner, M. (2012). Channel choice and public service delivery in Canada: Comparing e-government to traditional service delivery. *Government Information Quarterly*, 29(1), 1–11. doi:10.1016/j.giq.2011.03.005

Roy, J., & Langford, J. (2008). *Integrating Service Delivery across Levels of Government: Case Studies of Canada and Other Countries*. IBM Centre for the Business of Government.

Sellami, M., & Jmaiel, M. (2007). *A Secured Service-Oriented Architecture for E-government in Tunisia*. ReDCAD research unit. 8-9.

Stegarescu, D. (2006). *Decentralised government in an integrating world: quantitative studies for OECD countries*. Springer Science & Business.

W3C. (2004). *Web Services Addressing (WS-Addressing)*. Retrieved from http://www.w3.org/Submission/ws-addressing/

ADDITIONAL READING

Bakvis, H. & Juillet, L. (2004). The Horizontal Challenge: Line Departments, Central Agencies, and Leadership.Ottawa: Canada school of Public service.

Bellamy, C., & Raab, C. (2005). Joined up Government and Privacy in the United Kingdom: Managing Tensions Between data Protection and social Policy. Part II. *Public Administration*, 83.

Castellano, M. (2005). An e-Government Cooperative Framework for Government Agencies. 38th Hawaii International Conference on System Sciences. Hawaii, IEEE.

Donnelley, R. (2007). Transforming Public Services: The Next Phase of Reform Progress Report. (Produced for the Scottish executive). Edinburgh: Scottish executive.

Dutton, W., & Peitu, M. (2007). *Reconfiguring Government-Public Engagements: Enhancing the Communicative Power of Citizens*. Oxford Internet Institute.

Entwistle, T. & Martin, s. (2005). From Competition to Collaboration in Public service delivery: A New Agenda for Research. *Public Administration*, 83.

Goldsmith, S., & Eggers, W. D. (2004). *Governing by Networks - The New Shape of the Public Sector*. Washington, DC: Brookings Institution Press.

Maleyeff, J. (2007). *Improving Service Delivery in Government with Lean Six Sigma*. Washington, DC: IBM Center for the Business of Government.

Pardo, T., & Dadayan, L. (2006). *Service New Bruns-wick. Public ROI—Advancing Return on Investment series*. Center for Technology in Government.

Raymond, R. (2007). Enterprise integration in e-government. Transforming Government: People, Process and Policy, Vol. 1 Is: 1.

Reding, V. (2006). E-Government Developments, E-Government for All Europeans. Australia's E-Government Strategy – New Service Agenda, IOS press: 29-41.

Reed, B. (2004). *Accountability in a shared services world*. Canberra: Government of Australia.

Soliman, K. S. (2003). E-government: a strategic operations management framework for service delivery. E-Government, Bradford. UK: Emerald Group Publishing Limited: 13-25.

Stolba, N., Tjoa, A. M., Mueck, T., & Banek, M. Federated Data Warehouse Approach to Support the National and International Interoperability of Healthcare Information Systems. Austria: Federal Ministry for education, science and Culture.

Wing, L. (2005). Barriers to e-government integration. Journal of Enterprise Information Management, Vol. 18 Is: 5.

KEY TERMS AND DEFINITIONS

Administration Requirements: Internal government requirements which need to be achieved by the proposed connected service delivery model.

Citizen Requirements: What services, applications, interactions and facilities citizen require from e-government application.

Connected Service Delivery: The process of integrating public services to a convenient, seamless and single point of access portal, through which public services can be accessed, utilized and completed.

Enterprise Service Bus (ESB): A software architecture for middleware that provides fundamental services for more complex architectures. It is a set of rules and principles for integrating multiple systems and applications together over a bus topology infrastructure.

Horizontal Service Integration: The integration of services in government agencies with different functionality that has some relation in common to the clients. This integration model focuses on the economies of scale derived from collaborative approaches to procurement and the efficiencies and knowledge management advantages associated with developing common IT infrastructure.

HTTP (Hypertext Transfer Protocol): A set of rules for transferring files of different type over the World Wide Web.

SAML: (Security Assertion Mark-up Language): An XML-based open standard data format for exchanging authentication and authorization data between a service consumer and a service provider.

Service-Level Agreement (SLA): is a part of a service contract where the level of service is formally defined. In practice, the term SLA is sometimes used to refer to the contracted delivery time (of the service) or performance.

Service Orchestration Engine: is a software engine built on top of the message-oriented of the ESB. It is capable of orchestrating long-running business transactions through stored state, and it primarily serves as a coordinator and manager of conversations among web services.

Service-Oriented Architecture (SOA): A set of design principles, technologies and practices for managing and communicating enterprise digital services to support dynamic, flexible business processes.

SOAP (Simple Object Access Protocol): A network protocol for information exchange in distributed environments. It uses XML over HTTP as a mechanism to communicate between different computer programs. SOAP has gained status as a common messaging protocol in Web services and SOA projects.

SSL (Secure Sockets Layer): A commonly-used protocol for managing the security of a message transmission on the Internet.

The Service Consumer: Any entity or individual request a service in SOA environment.

The Service Provider: The entity which creates the service and makes it available to other entities in SOA Environment.

The Service Registry: The directory in which all services are stored in a searchable and semantic manner in SOA environment.

UDDI (Universal Description, Discovery and Integration): An XML-based registry of web services that are available for use in an SOA environment.

Vertical Service Integration: The integration of services in which two or more levels of government in the same function collaborate on integrates service delivery initiatives.

Web Service: A software system designed to support interoperable machine-to-machine interaction over a network. It has an interface described in a machine-processable format. This technology provides standards for message exchange and service description. Web services utilize the XML-based Simple Object Access Protocol (SOAP) over HTTP for communication between service providers and consumers.

WSDL (Web Service Definition Language): An XML-based interface description language that is used for describing the functionality offered by a web service.

Section 2

Web 2.0 Technologies and E–Participation for Next Generation E–Government

Chapter 4
Web 2.0, ICT Infrastructure, and Training Provision for E–Government Readiness in Nigeria

Oghogho Ikponmwosa
Landmark University, Nigeria

ABSTRACT

This chapter presents a discussion on e-readiness, Web 2.0, social media, mobile/wireless technologies, and other Information and Communication Technologies (ICTs) that can help to facilitate the attainment and sustenance of an e-ready environment necessary to enhance e-governance in Nigeria. The chapter aims to clearly articulate the necessary steps to be taken to provide all stakeholders with a blueprint of areas and factors on which to focus. An assessment of how e-ready the Nigerian government and its citizens currently are and the requirements necessary for further steps to be taken (such as policies, programmes, and processes to be put in place, infrastructures to be acquired, and training provisions to equip Nigerian citizens and government officials with the capacity to benefit from and sustain the use of acquired e-technologies) are also presented. Specific ways by which Nigeria can harness the various emerging technologies (social media, Web 2.0, and mobile/wireless technologies) are highlighted. If employed appropriately, these technologies can help to provide improved processes, increased efficiency, improved transparency, and citizen's effective participation and involvement in governance to further improve the lives of Nigerian citizens.

INTRODUCTION

In today's world, we cannot overemphasize the need to pass on relevant information (in a timely and accurate manner irrespective of distance) from those who have it to those who need it. This could be between members of legislative, judicial and executive arms of government, government to government, government and their citizens, patients and their doctors, business executives and their labour force, parents and their children, religious leaders and their followers, salespersons

DOI: 10.4018/978-1-4666-6082-3.ch004

and buyers, students and their teachers, etc. The numerous e-application platforms have the potential to readily and efficiently facilitate the process of information transfer among these groups if underlying technologies are properly harnessed and employed. It is therefore necessary that nations take pragmatic steps towards achieving an e-ready feat if they wish to be well positioned to benefit from the e-platform and relevant emerging technologies including the social media and mobile technologies.

According to Oghogho and Ezomo (2013), the emergence of several e-application platforms such as e-governance, e-business, e-science and engineering, e-health, e-learning, e-agricultural, e-procurement, e-banking, e-justice, etc. has changed the way communication activities is being carried out around the word. These digital platforms use digital technologies to deliver several e-services by making it easy and possible for those who have information to pass them on to those who need them. The result is that decisions are made quickly and efficiently based on up-to-date information, business deals are quickly concluded, learning from home is easier, etc. Every aspect of our modern day life has been influenced by several emerging e-technological applications. Improved productivity and efficiency can now be delivered by using these e-platforms as tools or means to achieve the desired end. The emergence and growth of the various e-technologies and their applications is leading the world towards a global society were physical land boundaries are no longer sufficient to separate people who are from different societies with different cultures, ideologies and beliefs.

Sharma and Vaisla (2011) did a review of e-governance applications in public health care (a vital responsibility of the government) for rural areas of Uttarakhand province in India through ICT applications at National levels. They presented facts on tele-medicine, tele-referal services and health information dissemination using videoconferencing, mobile phones and other ICTs. Oghogho and Ezomo (2013) presented specific examples of the use of ICTs in health, governance, business and education stating their potential to enhance national development when properly harnessed.

Newer emerging e-technologies are now presenting the world with better ways of harnessing ICTs. The World Wide Web (WWW) has entered a new era called Web 2.0 with features such as social media or networks, blogs, wikis, etc. which allow people to effectively connect, create and share live streaming or recorded audio, video, text, pictures, etc. information or knowledge. The emergence of several mobile/wireless technologies is presently at the fore front in providing and enhancing the enabling environment for harnessing Web 2.0 features. Today we have several widespread emerging mobile and wireless devices such as Laptops, Notebooks, Ipods, Ipads, Iphones, Palmtops, PDAs, etc. equipped with numerous features and specifications necessary for accessing the WWW for information transfer (Carmel & Gabriela, 2011). These enhanced features range from touch and sliding screen, video capturing, recording, editing and sharing, bluetooth, 3G and 4G capabilities, radio FM and AM, Music player/ MP3, e-mail applications, HTML browsers, Internet connections, etc. These enhanced features enable users to access (download or view online) other's information and send (upload or comment online) their own information thus enhancing the information sharing and the diffusion process.

As our world becomes increasingly digital or electronic, all organs of government cannot continue to resist the adoption and sustained use of electronic technologies to deliver government information and services to citizens. This is evident from the experiences of countries like Republic of Korea, Seychelles and Kenya where, e-governance has made governance processes and service delivery faster, cheaper, more effective, transparent and accountable (UNDESA, 2012; Nigeria Intel, 2013). E-governance refers to the provision of public services to the citizens via electronic platforms (especially the worldwide

Web or the Internet) while enabling proactive and active participation of citizens in governance, consultation, knowledge policy, service delivery, etc. Hong and Nadler (2012) in their investigation of which candidates the public discusses online in an election campaign, found out (from their analysis) that on the average, a 10% increase in the number of traditional media mentions for a politician is associated with a 4 to 6% increase in the number of Twitter mentions. Governments must therefore take a step ahead of those being governed in harnessing and creating a favourable e-ready environment in order to keep being in charge.

A common feature of e-governance is the automation or computerization of existing paper-based procedures to enhance access to, and delivery of government services to the citizens (Akunyili, 2010). According to Rosa et al. (2013), e-justice is presently being deployed in many countries all over the world despite the risk factors associated with this deployment. Any country's institutions of governance exert primordial influences on the economic growth and prosperity, societal stability and wellbeing of the citizens. The beliefs and confidence of the people on the fairness and justness of their governing institutions will to a large extent, determine the long term stability of such societies and their governments. The Author believes that the e-platform (enhanced by the use of social media, web2.0 and other mobile/wireless technologies) will enable governments to secure this confidence from the citizens because the citizens will be able to easily access relevant government information and can readily participate in governance. Achieving an e-ready feat is therefore one of the numerous vital steps for any government that wants to have economic prosperity, societal stability and the wellbeing of their citizens.

This chapter discusses the concept of e-readiness, ICTs, Web2.0 and Mobile/Wireless technologies as well as Nigeria's state of e-readiness. It also discusses some necessary steps to be taken

to achieve an *e-ready feat* in Nigeria. Some ways by which Nigeria can attain and benefit from the e-ready feat, through harnessing the various emerging e-technologies (social media, Web 2.0 and mobile/wireless technologies) for enhancement, better efficiency, transparency, citizen's participation or involvement in governance and improvement on the lives of the citizens, are also highlighted.

THE CONCEPT OF E-READINESS

E-readiness is about acquiring and putting the necessary technologies and infrastructures in place, pursuing policies that empower the citizens to have the economic capacity to own ICT equipment (mobile phones, computers, scanners, etc.), training the citizens to become acquainted with and able to use such ICTs and creating the enabling environment so that the information diffusion process using these technologies will be sustained. When any Nation becomes e-ready, harnessing and processing information (so as to make timely, accurate and well informed decisions) using several innovative digital application platforms (e-government, e-business, telemedicine or e-health, e-learning, e-democracy, e-agriculture, e-procurement, e-banking, etc.) to improve the lives of citizens, accelerate the growth of business organizations and to improve transparency and efficiency of governance will be readily facilitated.

Information and Communication Technology (ICT) is a broad term that covers the various technical means by which information is processed and transferred from a sender to a receiver. The term ICT emerged from the convergence of information technology (IT) and telecommunications technology. According to Oghogho and Ezomo (2013) ICT is an umbrella term that includes any communication device or application, encompassing: radio, television, mobile and fixed phones, computer and network hardware and software, satellite systems, surveillance systems, and so on, (as well as the various services and applications associated with them, such as videoconferencing, social media, distance learning, etc.) necessary for the delivery of information in the form of audio, data, video, image etc. from Point A to Point B.

The Internet and other established and emerging e-technologies that will facilitate processes of e-governance must be in place and the citizens must have access to them and able to use such technologies before any nation can be said to be e-ready. Apart from the general Internet (which is a mega network made up of smaller networks such as Wireless Local area Networks (WLANs or Wifi), Wide Area Networks (WAN), WIMAX, Long Term Evolution (LTE) networks), the other technologies involved in the e-process include: web2.0 and mobile/wireless technologies, broadband technologies, fixed line telephones and fax machines, surveillance systems, television (analogue and digital), FM and AM radios, tracking systems etc. Before a nation is said to be e-ready, there has to be a willingness on the part of the government to both use and make these technologies available for e-governance processes, ensure they are accepted for usage by the generality of the citizens who must also be trained and equipped with the capacity and economic power to access and use them.

E-READINESS, WEB 2.0 AND MOBILE/WIRELESS TECHNOLOGIES

Achieving an e-ready feat is indeed a huge task for any developing nation. However the use of emerging e-technologies and their applications such as social media, Web2.0 and mobile/wireless technologies will readily enhance and facilitate the provision of an e-ready environment for e-governance in any nation. We cannot really talk about having an e-ready environment if these technologies are not already being used by the citizens on a large scale. According to Wright et al. (2009), all Web 2.0 systems must be delivered

on the Web, allow each user who can access the system the ability to participate in the discussion and development of contents as well as impose only a minimal amount of authority or editorial control. Web 2.0 (sometimes considered more as a movement or philosophy than a precise technology) facilitates online communities, free and open sharing, interactivity and collaboration (Wright et al., 2009). Web2.0 and other emerging mobile technologies have been found to be very effective in enhancing information creation and sharing in education (Nulgultham, 2012, Smith & Peck, 2010, Holotescu & Grosseck, 2011, Baltaci-Goktalay & Ozdilek, 2010, Daud & Zakaria, 2011), in health (Wright et el., 2009) and in governance (Bonsón, et al., 2012, Fedotova, et al., 2012).

Howe (2006) lists four general types of processes that reflect ways of interaction within Web 2.0 as: for sharing contributed content by user ("where you make it"), for large sets of contributed content by user ("where you name it"), for the development of content collections by the community of users ("where you work on it") and for finding objects, trends and overviews of contributions ("where you find it"). Using Social media, Web 2.0 and mobile/wireless technologies allow users to share and access information or knowledge anywhere and anytime irrespective of location. There is flexibility, a wider audience is reached and personal messages are received entirely while creating a long-lasting interaction for multiple purposes. According to Wright et al. (2009), Web 2.0 applications share some common principles and policies, namely: (i) Using the Web as an application and content deployment platform, (ii) Leveraging the Web as a participatory and not merely as a publishing platform, (iii) Providing valuable content in addition to simply offering useful tools, (iv) Treating users as co-developers, and lastly (v) Supporting syndication of services and content, as opposed to central control.

Web 2.0 is still under debate, because a lot of concepts have been causing confusion and ambiguity about the term when compared with

Web 1.0. There is the argument that Web 2.0 is not technically different from Web 1.0 since they both use the same protocol (Hypertext control protocol (HTTP)) and the same make up format (Hypertext Makeup Language (HTML) (Wright, 2009). Several authors have differentiated Web 2.0 from Web 1.0 in the following ways (Sandoval-Almazan, 2011):

- Web 2.0 facilitates flexible design, creative reuse and updating;
- Offers the user an enriched and interactive interface;
- Facilitates collaboration for creating and modifying content;
- Allows new applications to be created by reusing and combining data and sources;
- Establishes social networks between people who have the same interests; and
- Supports cooperation in gathering collective intelligence.

Social media, Web 2.0 and mobile/wireless technologies are vital among other requirements to enhancing the achievement of an e-ready feat necessary for a successful implementation of e-governance in any country because they provide an innovative platform that applies Internet technology in a way that users are enabled to be more creative, more involved, and readily able to relate with each other anywhere, anytime and on the move. They provide a more active, interactive, participatory and collaborative environment suitable for e-governance anywhere anytime. Social networks, social bookmaking, Instant messaging, Wikis, Internet telephony using VoIP, Audio or video conferencing (NetMeeting) and blogs can be easily created to discuss governance issues to harness solutions, ideas and innovations on how to proceed with plans, goals and objectives in various aspects of governance. This means important government decisions can still be made without delays while the chief executive is away on any official trip. He can append his signature

electronically and give instructions as well as receive feedback on them in real time using social media. Also citizens are more readily able to participate in the governance process as they can easily have access to them and are enabled to participate through several social media, Web2.0 and mobile/wireless technologies now available at quite affordable prices in the market.

Although, Web 2.0 is an attractive technology, it also comes with some limitations (Holotescu, 2011) e.g.:

- Most of the mobile functions are underexploited by a large percentage of the community of users.
- There is difficulty in content development, hence; only few members actually make useful contributions.
- Technology might not function for the purpose or aim for which it was targeted or it may be difficult to respond to the learning needs of the user community
- Additional training is always required for it to be successfully deployed to achieve its aim.

Despite these challenges, Social media, Web 2.0 and mobile/wireless technologies have generally been affirmed as having the potential to provide any nation with the tools to create and sustain the desired e-ready environment necessary for enhancing e-governance.

E-READINESS IN NIGERIA

Use of social media, Web2.0 and mobile/wireless technologies will not have a significant impact in enhancing e-governance in Nigeria if the "backbone" necessary to provide an e-ready environment is not put in place. In Nigeria today, several governance processes have been digitized showing the will of Nigerian governments to pursue an e-ready feat. However, can we say that

Nigeria is e-ready, considering the fact that the Federal Government and several states of the federation are yet to make a firm commitment towards providing governance majorly through the e-platform? Provision of telecommunications infrastructure, easy and low cost access to the various e-technologies and training provision for Nigerians to enable them use them are the major hurdles to overcome. Many Nigerians do not have access to the necessary e-technologies; lack the economic power to acquire them and the skills to utilize them. Web 2.0 technologies are not widely deployed in Nigeria. The number of blogs, wikis, Social networks, social bookmaking, Instant messaging, Internet telephony using VoIP, Audio or video conferencing (NetMeeting), etc. created and managed by Nigerians are grossly infinitesimal compared with the rest part of the world.

According to Internetworldstatistics.com (2013), Nigeria leads in the total number of people using the Internet in Africa as 28.9% of the total population of African Internet users are in Nigeria. Nigeria however has an Internet penetration rate of 28% (which is lower than that of Morocco (51%) and Tunisia (39.1%) as at June 2012. Table1 shows the Internet penetration rate of Nigeria and selected countries of the world as at June 2012 while Table 2 shows Nigeria's e-government development index placed side by side with that of selected countries of the world.

Table1, that represents data extracted from Internetworldstats.com (2013), clearly shows that Nigeria's Internet penetration (percentage population) is still very low when compared with the selected nations of the world. According to a report on the global e-discussion organized by the World Bank Institute's Business, Competitiveness, and Development Team and the Research and Innovation for Organizations and Societies (RiOS) Institute, developed nations on the average reported more than 300 secure Internet servers per 1 million people but developing nations reported less than 2 (Braund, et al., 2006).

From Table 2, that represents data extracted from United Nations E-government Survey (2012), it is clear that Nigeria and the other African countries listed are very far behind in e-government development when compared with the rest part of the world. The same survey gave Nigeria's Telecommunication infrastructure index and its components as follows: Index value (0.1270), Estimated Internet users per 100 users (28.43),

Table1. Internet usage of selected countries

S/N	Country	Continent	Population	Internet Users	Internet Penetration (% Population)
1	Nigeria	Africa	170,123,740	48,366,179	28.4
2	Morocco	Africa	32,309,239	16,477,712	51
3	Tunisia	Africa	10,732,900	4,196,564	39.1
4	Argentina	The Americas	42,192,494	28,000,000	66.4
5	Brazil	The Americas	193,946,886	88,494,756	45.6
6	United States	The Americas	313,847,465	245,203,319	78.1
7	Japan	Asia	127,368,088	101,228,736	79.5
8	South Korea	Asia	48,860,500	40,329,660	82.5
9	Philippines	Asia	103,775,002	33,600,000	32.4
10	France	Europe	65,630,692	52,228,905	79.6
11	Germany	Europe	81,305,856	67,483,860	83
12	United Kingdom	Europe	63,047,162	52,731,209	83.6

Table 2. E-Government development Index for selected Countries in the world

S/N	Country	Continent	Rank	Index Value	Online service Component	Telecommunication Infrastructure Component	Human Capital Component
1	Nigeria	Africa	162	0.2676	0.2222	0.1270	0.4535
2	Morocco	Africa	120	0.4209	0.5425	0.2772	0.4430
3	Tunisia	Africa	103	0.4833	0.4771	0.2886	0.6841
4	Seychelles	Africa	84	0.5192	0.3333	0.4037	0.8204
5	Argentina	The Americas	56	0.6228	0.5294	0.4352	0.9038
6	Brazil	The Americas	59	0.6167	0.6732	0.3568	0.8202
7	United States	The Americas	5	0.8687	1.0000	0.6860	0.9202
8	Japan	Asia	18	0.8019	0.8627	0.6460	0.8969
9	Republic of South Korea	Asia	1	0.9283	1.0000	0.8356	0.9494
10	Philippines	Asia	88	0.5130	0.4967	0.2082	0.8341
11	France	Europe	6	0.8635	0.8758	0.7902	0.9244
12	Germany	Europe	17	0.8079	0.7516	0.7750	0.8971
13	United Kingdom	Europe	3	0.8960	0.9739	0.8135	0.9007

Main fixed phone lines per 100 inhabitants (0.66), Mobile subscribers per 100 inhabitants (55.10), Fixed Internet subscriptions per 100 inhabitants (0.12) and Fixed broad band per 100 inhabitants (0.06). These values are far lower than that of many countries in the world.

Despite these short falls, Nigerian Internet users grew between the year 2000 and 2012 with 24,083.1% which is far higher than that of many countries in Africa and in the rest part of the world (Internetworldstatistics.com, 2013). This growth was a direct result of the deregulation of the telecommunications sector in 1992 and issuance of licences to mobile operators. The Minister of Communication Technology, Mrs Omobola Johnson, recently decried low rate of adoption of ICTs by Nigerians and added that 0.9 per cent of households owned a PC, 3.6 per cent had access to one and 3.1 per cent accessed the Internet (NAN 2, 2013). According to her, Nigeria has an average Internet download speed of 1.38Mbps and most Nigerians are excluded from the growth and development that can be aided by information and communication technology (Aregbesola, 2012).

She said the Federal Government through its TIAP project had established connectivity in 17 Universities and their Teaching hospitals using fibre optics technology (NAN 1, 2013). Also, about 204 Institutions have joined the project, (with 74 institutions in 2012 alone) since the beginning of the programme implementation.

In another programme called the School Access Project (SAP), (which provides classmate PCs, with e-learning content and accessories, solar power solutions, high speed Internet connectivity and wireless network deployment to government public schools), over 823 schools have benefited between 2010 and 2013 (NAN 1, 2013). Another initiative (Students Computer Ownership Scheme) which allows students to purchase laptops using a low interest facility with a monthly repayment plan of less than $32 has also been flagged off by the ministry with several Federal and State Universities participating in the scheme.

In April 2013, an IT Incubation Centre was launched in Lagos by the Federal Ministry of Communication Technology with another to be launched in July in Cross-River (NAN 2, 2013).

This will no doubt follow the Chinese model of creating Technology parks which will catalyse the ICT industry by helping Nigerian government and entrepreneurs to create successful processes and businesses supported by ICTs.

After the Nigerian Telecommunications sector was de-regulated in 1992, the sector did not experience a boom until the Federal government under the leadership of President Olusegun Obasanjo took the bold step to auction GSM licenses in 2001. This led to a massive growth in the sector from less than 500,000 fixed telephone lines to about 110 million mobile cellular active subscribers by the end of 2012 (BuddeCom, 2013). The growth of mobile users in Nigeria had an accelerated pace due to lower prices resulting from competition between the mobile service providers and a growing demand for mobile broadband services. Today, Nigeria, with a market penetration of about 70% as at early 2013, is Africa's largest mobile market (BuddeCom, 2013).

The rapid growth of the mobile users came with its own problems: network congestion and poor quality of service which is a big setback to the e-ready feat. The influx, diffusion and usage of cheap and affordable 3G and 4G mobile phones from China, (having numerous functions including Web access capabilities), has the potential to aid and make the e-governance process in Nigeria a success. However, the challenge of overcoming network congestion and providing sufficient bandwidth must be tackled to create an enabling environment for the e-technologies to become acceptable to the citizens who will have to use them in the e-governance processes.

Large parts of Africa gained access to international fibre bandwidth for the first time via submarine cables in 2009 and 2010 (Oghogho and Ezomo, 2013, Akunyili, 2010). In order to complement the fiber connectivity and provide more bandwidth for the nation, Nigeria has successfully launched NigConSat-1R satellite on 19th December 2011 (Aregbesola, 2012). NigComSat-1R is a hybrid satellite for broadcast telecommunications and navigational services, with footprints in over 35 African countries, parts of Europe and parts of Asia. Apart from Sat-3, which is jointly owned by some African countries, Main One and Globacom are the first indigenous companies to bring in submarine fiber optic cables to Africa from Europe. Additional base stations are also being added to existing mobile base stations (making a total number of over 20,000 cells) to support the ever increasing demand for bandwidth (BuddeCom, 2013).

Although, Web 2.0 technologies have been identified as having the potential to facilitate and enhance an e-ready feat for e-governance, they have not been wide diffused and used in Nigeria for governance. Mobile and wireless technologies have however gained some grounds in the country. Taleb and Sohrabi (2012) in their study of student's viewpoint about the educational use of mobile phones to support their learning process affirmed the fact that e-inclusion is made more possible through use of mobile phones for information downloads or for learning purposes. These steps need to be intensified and pursued to have these technologies fully accepted, affordable and diffused so that a large percentage of the citizens can afford, access and use them in the e-governance process. In line with this, the government of Nigeria in 2006 established a public corporation known as Galaxy Backbone to provide the technological platform for e-governance, and is working on a comprehensive broadband policy and vision document which will provide broadband definition, performance indicators, incentives for investment, macroeconomic targets, deployment guidelines and citizens charter (Akunyili, 2010). Galaxy Backbone is presently supporting technology services to Federal Ministries, Departments and Agencies (MDAs), building ICT infrastructure to boost the sophistication and effectiveness of the Nigerian government to tackle security challenges and improve connectivity to modern technology in the country (Adeniyi, 2012). The technology company in the year 2012 received

One hundred million dollars ($100million) for an ICT infrastructure rollout of public priority projects underway (Adeniyi, 2012).

Radio and Television services (both digital and analogue) are also presently serving as a platform for participation of Nigerian citizens in governance as there are regular programmes hosted by the government, the Civil society, the media houses, etc. which provide government information to the public and make provision for them to respond and say their views. These types of forums have helped to shape the opinion of the Citizens by making them aware of relevant government information and able to participate in the overall decision being made. Digital satellite television (DSTV) has made transfer of local and international information to many rural communities (where there are no Internet access and availability of mobile and fixed telephone networks) possible. DSTV however has the challenge of high purchase and installation cost and monthly charges as well as the unavailability of electrical power in several rural communities coupled with the epileptic supply in areas where electric power is available. The use of Radio and television services to enable Nigerian Citizens participate in governance was clearly demonstrated in the 2011 elections where live coverage of the events, and the use of other e-technologies such as social media, Web 2.0, mobile phones, electronic voting machines, the Internet, etc. made the process more transparent with easy monitoring of results and increased confidence of the electorates in the process, hence the international observers generally agreed that the elections were free and fair (Oghogho and Ezomo, 2013).

Despite some progress made so far, the author believes Nigeria is still not e-ready and cannot fully deploy e-governance without facing severe bottleneck challenges. The basis of evaluation of e-governance according to United Nation Public Administration Network Global e-Governance Readiness Index are: the state of e-government readiness based website assessment, telecom-munications infrastructure and human development as well as the state of e-participation of the citizens (UN e-government survey, 2012). Nigeria was ranked 150th in the world with an E-governance development index of 0.2687 in 2010 but dropped to 162nd position with an E-governance development index of 0.2676 in 2012. No African country made the first 83 among the world ranking. Seychelles which ranked first in Africa with an E-governance development index of 0.5192 ranked 84th in the world.

Nigeria's Internet penetration (percentage population) has to be improved from 28.4% to at least 70%. Also our available bandwidth is far from being sufficient considering the population of the country assuming everyone were to possess the capacity to access the Internet so as to participate in the e-governance process using social media and Web 2.0 technologies. Broad band access to the Internet is very low (Aregbesola, 2012). The present cost of bandwidth is also very high considering the low per capita (PPP) income of the average Nigerian. Most Nigerians cannot afford to spend their income on bandwidth to access the Internet when they have not taken care of their other basic needs like food, clothing and shelter. According to Adenike and Oyesoji, (2010), the high level of illiteracy, poverty and low socio-economic status coupled with high rate of paternal and maternal deprivation of student academic needs, which was necessitated by poor socio-economic situation of the country has thrown many parents into untold financial problems such as poverty, lack of money to purchase necessary textbooks and working materials for their kids.

Also, we do not presently have an efficient and reliable central electronic data base from which government information are stored and retrieved. Other important electronic systems such as GPS systems, Closed circuit television (CCTV) systems in our roads and public buildings, National digital ID cards, etc. are not yet fully deployed, diffused and in widespread use across the country. Electronic banking, e-payments and other e-transaction processes are still at their elementary stages of development. The steps taken by the Nigerian government towards having a cashless society has not yet yielded much results due to numerous issues ranging from unwillingness to accept the change, lack of trust of Nigerians in the process and the unavailability of the required facilities, equipment, skilled manpower, logistics, etc. needed to make it work.

The power supply available to Nigerian citizens is epileptic and insufficient. This is a challenge because the e-technologies need power supply to function. Other back up power sources such as generators increase the overall cost of using these e-technologies which can be a source of discouragement in their usage. The awareness of Nigerian citizens of the benefits of e-governance and their confidence that their input into the overall government processes will count is still very low. A large percentage of Nigerians do not presently possess the economic power to own ICT infrastructures and the skills to use.

The use of social media, Web 2.0 and mobile/wireless technologies in governance is still at its infancy in Nigeria even though these technologies have been identified as having the potential to provide the desired e-ready environment necessary to enhance e-governance in any nation. According to Sandoval-Almazan et al. (2011), communication that arises from the interaction between government and citizens can take place in different ways:

- Information stage where government sites only display information on the activities of public administration.
- Interaction stage which deploy applications that allow interactions between governments and citizens such as forms for asking questions and making enquiries, forums, or automated applications such as virtual public servers.
- The transaction stage which focuses on interchange of services and application processes with a well defined cycle and in many cases the payment of fees. The transaction stage is an extension of the interaction stage with a focus on e-commerce.
- Integration Stage which make reference to the capability of the site to present itself as a single window for providing services to the citizens and transparently making known which agency or agencies are in charge of delivering the services or information.
- Participation Stage which offer citizens the ability to socialize and in this way obtain full interaction. At this stage, communication is most extensive, taking place between government and citizen, between dependencies, between citizens and providing feedback.

Social media, Web 2.0 and mobile/wireless technologies have the potential to enhance the interaction, transaction, integration and participation stages. These technologies share certain characteristics, such as the generation and classification of information and content in a collective manner, the integration of communities, and the production and consumption of socially distributed knowledge (Sandoval-Almazan et al., 2011).

Using Social media and Web 2.0 tools and applications on electronic government sites is just the first step because there is the need to put in place strategies and clear approaches detailing what these tools and applications are expected to achieve. According to Sandoval-Almazan et al. (2011), Government 2.0 has great potential to transform and improve relations between governments, citizens, companies and other interest groups, but these tools must be combined with a clear vision and effective strategies if their effects are to be valuable and meaningful to governments and citizens alike, as well as to society as a whole.

There is still much to do if Nigeria must begin and sustain the e-governance process and reap its numerous benefits.

FACILITATING THE E-READY PROCESS

The author highlights four stages that will help Nigeria facilitate and sustain the e-ready feat as:

- E-awareness and sustainable e-policy formulation.
- ICT infrastructure acquisition.
- Development of e-governance processes and procedures.
- Training provisions for Nigerian Citizens.

E-Awareness and Sustainable E-Policy Formulation

Availability of sufficient bandwidth to facilitate fast and easy access to the Internet at affordable costs (a feat that can be achieved by acquisition of more ICT infrastructure) has already been identified in this paper as one of the major hurdles to be overcome to facilitate the e-ready feat in Nigeria. However, before proceeding with infrastructure acquisition, there is the need to embark on an enlightenment campaign alongside the acquisition of ICT infrastructure which will be followed by a training program that outlines the benefits of e-governance so as to sensitize all Nigerians on the need to embark on the e-governance project. This is necessary because awareness of the numerous benefits of e-governance and the role of the citizens in the process will be vital to its success. Nigerian Citizens, who are the consumers, need to appreciate the need to invest their hard earned resources in acquiring access to the Internet to use social media, Web 2.0 tools and applications and other e-technologies because of the huge returns they will get from them which comes in the form of knowledge empowerment, enhanced productivity in business, easy communications with friends and family, involvement or participation in e-governance, etc.

The first step to take to achieve an e-ready feat is to begin creating the awareness of the various e-platforms, the various e-technologies (e.g. Web 2.0) and their potential benefits when properly harnessed. People are always resistant to change and may perceive the efforts to have a paradigm shift towards e-governance by the government as a waste of their time and resources if they do not fully appreciate the benefits it will present to them. This awareness campaign should identify and consider barriers to ICT adoption and diffusion (concerns about the privacy of data or security issues, the need for face-to-face interaction in some transactions and processes, present non usage of the various e-technologies by the citizens, finding support staff for the various e-platforms, cost of implementing e-governance, the challenge of making the necessary organizational changes, the inadequacy of legal protection for Internet purchase, overcoming challenges in the use of social media and Web2.0 technologies in governance, etc.) and make strategic efforts during the campaign to point out pragmatic steps to be taken to overcome or at least tackle a large percentage of these issues. This is necessary because the citizens may still refuse to accept the change even after seeing the benefits, if they do not see and appreciate the commitment of the government to tackle such challenges. The media (both print and electronic) will play a vital role in this campaign process.

The awareness campaign should begin with the setting up of a unit in the Ministry of Information and Communications saddled with the responsibility to develop the e-awareness message, collaborate with all stakeholders and reach Nigerians with the message using various media. It is now ever more essential that governments exploit all possible delivery channels in order to reach out to as many people as possible, no matter how poor, illiterate or isolated (United Nations E-Government Survey 2012: Executive Summary). The e-awareness message should be developed in audio, video and graphic (picture) formats which will be transmitted to the Citizens using various print and electronic media. Today mobile phones have reached about 80% of the populace (Iroko, 2012). Mobile/wireless technologies should also be used to diffuse the awareness message because the wide campaign and continuous expansion of the telecommunications industry in Nigeria has provided wider coverage even in several rural communities (Onwuemele, 2011), hence they can easily be used to deliver relevant, innovative and useful information to Nigerians at much reduced costs.

Social media and Web 2.0 platforms such as online forums (Social network Sites), wikis, blogs, etc. should also be used to diffuse the awareness message because most Nigerians who already are using the Internet (especially the youths) are registered in and visit one social network, blog or wiki forum regularly. Traditional and Religious Institutions should also be involved in the information diffusion process because the people have respect for them and will easily accept the message and perceive it to be of value if they are told about it directly by these Institutions. The e-awareness message should also be entrenched in our education curriculum at all levels so that the up-coming generations are informed of its potential as they grow up into adulthood. With the diffusion and acceptance of the e-governance message, the huge investment in ICT infrastructure will be easily harnessed and put to use so that it reflects quickly in National development.

An e-policy that provides a legal regulatory framework for the e-governance processes should be developed and signed into law. The regulatory e-policy is necessary because in a World Bank study in 2006, inadequate legal protection and concerns about privacy issues were seen to be among the top obstacles of ICT diffusion and usage (Braund et al., 2006). The Citizens need to be assured that their information is safe and kept confidential as much as possible and will not get into wrong hands. In the regulatory legal frame work, the various e-crimes and the punishments should be captured. Judicial independence, independent regulatory authorities, fair administration of justice, personal security and private property protection, and protection of financial assets,

should be emphasized. Transparency in financial transactions in government budget, central bank, the treasury, public enterprises, civil service and the public sector, pricing systems, exchange and trade regimes, banking systems, etc. should be contained in the regulatory legal e-policy. The regulatory e-policy laws have to be implemented in a transparent, consistent and effective manner to guarantee sustainable development both on the long and short run. The e-policy should also capture ways of harnessing the potential of using social media, Web 2.0 and wireless/mobile technologies in enhancing e-governance.

ICT Infrastructure Acquisition

Having a solid ICT infrastructure backbone is vital to having a successful e-governance plan in Nigeria. According to Obiozo, W.E (2013), the African continent has a lot to worry about, especially on the invention, adoption, development, training, availability and application of ICT resources and tools in different sectors of the African economy. Many African nations are lagging behind in ICT development and adoption in their urban areas, not to mention the rural community settings. Investing massively in Telecommunication infrastructure is the bedrock of achieving a solid ICT infrastructure backbone for Nigeria. Telecommunications infrastructure covers the transmission media and processes, the access facilities, the distribution loops, the network interconnectivity, network interoperability, the trunk backbone, etc.

The Nigerian government has made some progress in this direction through the deregulation of the sector but more steps should be taken to make the service provided by the various telecommunication companies become more reliable and efficient by improving the quality of service delivery. This should be done through a combined public and private initiative. ICT growth does not only depend on the large scale deployment of telecommunications infrastructure but also on the availability of mobile phone, use of social media and Web 2.0 technologies, fixed telephones, computers and their networks (wireless and wired), printers, scanners, video recorders, surveillance devices (e.g. CCTV, GPS systems, etc.), television, radio, digital cameras, fax machines, etc. The amount of ICT usage by the Citizens to access government information and participate in governance is also very vital and relevant to the success of the e-governance process.

It is necessary that Nigeria develops a plan to acquire necessary ICT infrastructures in phases as we drive towards the e-ready feat. This requires huge investments but following a laid down plan will help the nation to progress steadily. The annual budgets should reflect the commitment of government to pursue the plan and their implementation should be closely monitored.

The plan should focus more on providing equipment and technologies that provide access to the Internet (in both rural and urban communities), reliable and efficient National and regional data banks and servers, etc. A centralized approach which allows different government institutions to be interlinked should be deployed except in cases where this is unreasonable due to factors such as security issues. Infrastructure acquisition should be pursued in such a way that social equity and inclusion are enhanced by eliminating institutional barriers to citizen inclusion. There should be an equitable distribution of Citizen's opportunities for participation in governance through ICTs; else the aim of achieving an all-inclusive national development using e-governance will be defeated as the digital divide between the citizens of the same nation will become even greater. Web 2.0 and Mobile/Wireless technologies should also be given some priority because they have become the most rapidly adapted technologies to provide e-services both in Urban and rural communities with the potential to readily facilitate e-inclusion and participation in governance.

Appropriate policies to encourage private participation in the infrastructure acquisition process should also be developed as government alone cannot bear the overall burden of providing the ICT infrastructures necessary for achieving the e-ready feat in a reasonable time. Electric power infrastructure must be improved to provide a sustainable platform for the large scale rollout of telecommunications infrastructure. Costs of providing telecommunication services have been increased due to unavailability of stable power supply in the country which has led to installation of generators by the service providers, a situation which reduces the economic power of the citizens to afford and use ICTs.

Nigeria can also incorporate the provision of free access to e-governance services by providing free public Internet services where the Citizens can access government information and can participate in governance. However procedures have to be put in place to avoid the wrong usage of such infrastructures for purposes other than that for which they were established.

Development of E-Governance Processes and Procedures

Governance involves several processes and procedures which vary from very simple to very complex. Some of these processes require face to face communication while others require paper work, telephone calls, emails, etc. E-governance aims to minimise (if not completely eliminate) the use of face to face communication and paper work by developing e-solutions that efficiently and reliably replaces them. There is therefore the need for appropriate e-governance processes and procedures to be developed so as to facilitate a smooth transition from using normal processes to e-processes in governance. In line with this, Galaxy Backbone has already been contracted by the Nigerian government to help in developing technology service solutions to Federal Ministries Departments and Agencies (MDAs). The question however is whether they have the capacity, technical expertise and manpower as a single company to perform this task alone without the government also bringing in established giants in the e-service solution business?

There is also the need for all government (public) and private Institutions to have well equipped ICT units having highly skilled technical personnel saddled with the responsibility to move the establishments towards e-readiness by developing innovative e-processes and procedural solutions that will replace the existing processes and procedures. Equivalent e-processes and procedures to replace normal paper and face to face communications should be developed and tested and subsequently fully deployed. An effective approach to developing these e-solutions is to consider the numerous telecommunications infrastructures available, the possibility for the citizens to possess and use them as well as the present governance structure and the various procedures used in governance transactions. Integrating the large scale use of social media and Web 2.0 technologies in the e-governance processes should be given priority.

E-solutions which will focus to provide a platform for equality in the participation of the Citizens will have to be developed for health, education, administration, finance, democracy, business, agriculture, banking, etc. The solutions should as much as possible provide the possibility of participation using a wide variety of channels (equipment and technologies) so that Citizens who have limited access to some equipment and technologies can still participate in governance without being marginalized. Special focus should be given to creating processes and procedures that allow the use of social media, Web 2.0 and mobile/wireless technologies in participating in governance. Citizens have diverse needs and demands for services; therefore it is no longer sustainable for governments to utilize one preferred way of service provision over the other, hence it is now more essential that governments exploit all possible delivery channels in order to reach out to as many people as possible, no matter how poor, illiterate or isolated (United Nations E-Government Survey 2012: Executive Summary). The e-solutions should also be designed not considering only the supplier (government) perspective but also the consumer (Citizen) demands and needs.

All government Institutions must establish a sustained online presence with at least basic services so as to build the Citizen's trust in e-governance. They should have official social media pages and may even create their websites where citizens can register and interact with government officials and other citizens either using tweets or instant messages. This is necessary because Citizens will become frustrated and may develop a negative attitude towards e-governance if they experience delays or are unable to access government services whenever they desire. According to Al-Hujran et al. (2011) perceived usefulness, perceived ease of use and attitude are significant indicators of citizens' intention to use state government services online.

The e-solutions should also provide a structurally integrated and united purpose government providing collaboration between Nigeria and other nations, between Federal, State and Local governments and between Public Institutions, Private sector and Civil societies.

Training Provision for Nigerian Citizens

Apart from the inadequate infrastructure available, the differences in skills and the lack of large scale Citizen's capacity to access information is presently playing a major role in enlarging the digital divide between the developed world and the developing nations. Provision of relevant training to Citizens without any bias or discrimination is therefore a necessary tool towards achieving a successful e-governance process. The Nigerian people are the most important resource available, hence empowering the Citizens to possess the capacity to use ICTs while eliminating barriers to this process should be a matter of national priority.

ICT training should be entrenched into our education curriculum at all levels. We need necessary training to provide skilled technical personnel, who are able to develop, deploy, maintain and sustain ICT infrastructures all over the country (both in rural and urban communities). There is also the need to provide the training platform for Citizens to be able to use the basic ICTs infrastructures and equipment to access the Internet. There will also be the need to provide training on the various e-solutions and applications developed through which e-governance processes and procedures will be carried out. Programmes and strategies that encourage the sustained use of ICTs as well as strengthen the capacities of the Public, Civil and Private organizations for effective exploitation of ICTs to address their needs should be designed and implemented. They should be focused on strengthening the competitiveness of stakeholders in the Public and private sectors as well as empower them to enable optimal contribution to building a better and more sustainable society. Appropriate capacity building programmes are designed to get the most out of people and organizations while contributing towards establishing a more inclusive and prosperous knowledge based society (Jidaw, 2013).

There should be customized training programmes, including those based on globally recognized and accepted certification standards. The training should create a platform to encourage female participation so as to overcome ICT training diffusion inequality due to the present trend of gender bias in the country. Training provisions should evolve from a long-term vision and an IT management programme which should be aligned with the overall e-governance strategy and technical integration of ICT systems.

Free ICT Training centres run by government, Non-Governmental Organizations (NGOs) and other Private organizations should also be established to aid the training of Nigerians. One way to do this is to set up ICT Academy Research Centres (ICTARC) to boost Nigeria's ICT technical capacity so as to facilitate the replacement of expatriate expertise with highly skilled Nigerians. Training of Nigerians in ICT capacity should not be limited to usage but also in creating and adding to already existing technologies hence; nanotechnologies, microelectronics and embedded systems technologies design and development

training centres which form a bed rock for ICT infrastructure design should be pursued. This will be an expensive venture in the short run but on the long run, it will help Nigeria to develop the capacity to create ICT solutions that suites our needs.

EXAMPLES OF WAYS NIGERIA CAN BENEFIT FROM THE E-READY FEAT

There are several ways that Nigeria can benefit from the e-ready feat if properly pursued. Some specific examples are cited as follows:

Core Governance Processes

ICTs are making several government functions easier, cheaper, faster and more reliable. VoIP and intercom technologies are helping to cut down communication costs in several private and government institutions (Oghogho, et al., 2012). ICTs are being used in facilitating operation of the Judicial, Executive and Legislative arms of government, acquisition of National data bank, Personnel verification schemes, Pension schemes, Salary and wage payments, Tax collection and clearance, Training, regulatory functions, security provision, Administration, etc. Transactions and money transfers can easily be traced due to e-banking procedure now being used in governance, a situation that can help government and the civil society to easily trace and deal with corrupt practices.

Electoral Processes

The use of ICTs in electoral processes makes the entire process more transparent and inclusive. This was clearly demonstrated in the just concluded 2011 elections in Nigeria where more Nigerians where able to trust the process and decided to participate by casting their votes. However, ICTs should be more involved in both the planning and execution stages for the next election so as to correct the short comings experienced in 2011. Social media and Web 2.0 technologies can be deployed before, during and after elections to both monitor and harness the feelings of the public on such matters. President Goodluck Jonathan used his Facebook page to gather support from Nigerians while campaigning for the 2011 elections. The youths were glad to be able to talk directly with their President. During elections, social media especially tweets can be used to identify and monitor the progress of elections in different parts of the country.

Education

The use of ICTs in education has become common place because it has facilitated easy and cheaper access to all forms of information both within the same institution of learning and their outside linkages. Several education processes ranging from payment of fees, submission of assignments, downloading of notes, search for information online, receiving lectures online, etc. are now possible using ICTs. Social media and Web 2.0 technologies are now at the forefront in enhancing learning and knowledge creation and sharing. Wikis, blogs microblogs, etc. are presently helping to generate a large volume of data and making them available to a large audience. Since the Joint admission and matriculation board (JAMB) exam was handled using ICTs, students can now access their results online in less than two weeks after writing the exams and do not need to worry about their results being lost in transit after they have been posted to them by JAMB. ICTs should however not be left in Computer laboratories but should be part of classrooms to facilitate blended learning.

Business and Banking Services

Today, several business and banking procedures which offer more convenience will be impossible if there were no ICT platforms to support their execution. Banking transactions have become easier or more convenient and faster. ATM machines, electronic transfer of funds, etc. are making banking procedures faster and more convenient. Business deals and contractual agreements can be struck and executed without seeing the other person with whom you are doing business. Production processes are being automated by incorporating telecommunications systems which help to transfer information which are useful in the entire control processes. However, for Nigeria to attain a sustained successful e-business environment, the cashless society agenda of the Nigerian government should be pursued by putting in place necessary policies, infrastructures, trainings and processes that will make it work.

Health Services

ICTs have helped to enhance choices in medical care practices. Use of the Internet, social media and Web 2.0 technologies are now being used to aid provision of improved health care as family members can easily find health information and relate it to their relatives who need them. Also identification and monitoring health epidemic situations can be faster using social media and Web 2.0 technologies. Video conferencing makes operation, therapy, diagnosis and training sessions possible with the input of consultants who are several kilometres away. Transportation cost and travel risks are all eliminated. Human Capacity is developed and enhanced while facilitating efficient use of available resources. Information about recent breakthroughs in health practice and how to access them are easily available due to ICT's wide deployment and usage. It is easier to monitor public health treat in a timely and more efficient manner.

Oghogho and Ezomo (2013) gave some more specific examples of ICT usage in enhancing national development. These and many more are what Nigeria stands to enjoy with an achieved e-ready feat.

USING SOCIAL MEDIA, WEB 2.0 AND MOBILE/WIRELESS TECHNOLOGIES TO ENHANCE E-GOVERNANCE IN NIGERIA

Today, social media, Web 2.0 and several online services with user-generated content have made a staggering amount of information (and misinformation) available (Kavanaugh, et al., 2012). Proliferation and widespread diffusion and usage of mobile and wireless devices have increased the potential for citizens and government officials to be able to access the Internet anywhere and anytime (even when on the go) hence, they are able to use social media and Web 2.0 technologies to access and contribute to generating a staggering amount of information anytime, anywhere and on the go as long as there is an Internet connection. This means that government officials are presented with the potential to enhance governance if they can make sense out of the large volume of data generated using these technologies. This directly infers that social media, Web 2.0 and mobile and wireless technologies have the potential to create and sustain the desired enabling e-environment necessary for enhancing e-governance. Kavanaugh et al. (2012) in their exploratory study came up with three main findings:

- Local government uses social media without knowing its costs and benefits, or who their actual audience is, who in their organization should monitor communications, how and when they should be responding, and what effect their social media communications have on the public.
- New tools are needed to help government and citizens make sense of the overwhelming amount of data that is being generated, to model the flow of information, and to identify patterns over time.

- Digital libraries are needed to archive and curate generated content, especially for crisis and social convergence situations, but also for analyses that cover longer time frames.

These findings raise the issue that the uses of social media, mobile/wireless and Web 2.0 technologies have to be properly planned for governments to have useful dividends from them. It is not just enough to use these tools, but using them in ways that make their data useful for making real time and efficient decisions for more effective management of emergency situations, improvement on public safety, etc. is what matters. Despite the high volume of noise associated with data collected using social media, mobile/wireless and Web 2.0 technologies, critical events of interest can be identified as spikes in the social media volume (Kavanaugh et al., 2012). In this section, the Author presents some specific ways through which the Nigerian government are already using and can use social media, Web 2.0 and mobile and wireless technologies for enhancing government functions.

Identifying, Monitoring, and Responding to Emergencies in Real Time

Social media, Web 2.0 and mobile/wireless technologies have the potential to enable any government to identify, monitor and respond to several emergency issues in real time. These technologies can be used to identify, monitor and respond to critical events like earthquakes, floods, hurricanes, landslides and mudslides, tornadoes, tsunamis, volcanoes, terrorism, large scale food poisoning, thunder storm, water pollution, winter storm, extreme heat, flash mop gatherings, protests, fire outbreak or wildfires, chemical exposures, etc. These events of interest can be identified as spikes in the large volume of social media data traffic. The major advantage that these technologies have over traditional methods with respect to emergency management is their real time potential to reach a large population at much reduced cost while providing a large window of opportunity for influencing or mitigating emergency events as they occur. According to Yi et al. (2013) social media are must-have tools for Government 2.0 hence, both the U.S. and the South Korean governments look to social media for various functions, which may include, but are not limited to, communication/announcement on current events for the public, such as disease or disaster alerts and weather reporting.

Thousands of Nigerians joined and used Twitter and Facebook during and after the removal of fuel subsidy protest of 2012 (Admin, 2012, Orimisan, 2012). The massive mobilization of Nigerian citizens using social media, Web2.0, mobile/wireless technologies forced the government to reduce the earlier announced price. Social media helped to sustain the intensity of the protest all over the country hence, the government as well as citizens were well informed of events as they unfolded all over the country. Facebook is the easiest way to plan an upcoming event like a party or a rally, but Twitter is an invaluable resource if you want to follow the news in real time.

Governments have to position themselves (before the emergencies occur) to be able to use these technologies to identify, monitor and respond to emergencies in real time. Elected government officials, Public officials as well as other workers in government institutions providing services to the citizens will have to be registered and be ready users of blogs, microblogs, and other services with user-generated content. This will enable them to be able to identify spikes in the data volume when they do occur and hence will position them to respond accordingly. A good step in this direction is the recently established Public Complaints Commission (PCC) of the Federal Capital Territory (FCT) Abuja which embraced social media to receive complaints and interact with people. Any visitor to the PCC website or social media page can type in what they want to ask the commissioner, and can receive a feedback in real time.

Several natural disasters have occurred in Nigeria and the general trend has been a slow response to saving lives and alleviating the pains of Nigerians (Disaster-Report, 2012, Disaster-Report, 2013). Thousands of Nigerians have been displaced while others have been killed due to slow or no response to these emergencies. This slow response is partly due to non or late availability of information to the government response agencies saddled with the responsibility to respond to these disasters in near real time as they occur. Social media, Web2.0, mobile/wireless technologies can be used to identify these emergencies as spikes in the large data volume traffic so that the required government agencies can be directly informed using these technologies and able to respond in real time to these emergencies as they occur.

Although, we often talk about the positive potential of the use of these technologies, there is also the downside as citizens (especially youths) may be more interested in reporting an accident or a potential crime using social media and Web 2.0 technologies than reaching out to help the victims. According to Chukwuebuka (2013) youths in Nigeria on arriving at an accident scene only care about taking pictures or recording the disaster and the victims with their phones after which they gladly upload the images to YouTube, Facebook or other social media forum rather than coming to their rescue. This fact is evident in the aftermath of the June 2012 Dana Airline crash in Iju-Ishaga, Lagos, Nigeria, when thousands of young people residing in the area rushed to the scene and began using their phones to take images of the dying plane crash victims instead of rescuing the people in the plane. A similar thing occurred when almost a hundred people were burnt to death after an oil tanker caught fire in Rivers State, Nigeria.

These negative attitudes of the youths can however be channelled to yield productive results if governments can put up structures through various institutions that will be visibly present in several social media forums such that they are ready to identify, monitor and respond to these emergencies in real time as they occur.

Identifying, Monitoring and Responding to Civic-Related Situations

Social media, Web 2.0 and mobile/wireless technologies can also be used to enable governments and citizens identify, monitor and respond to civic related situations such as traffic jam, car crash, potential crime, downed power lines, etc. For example, a structure can be set up such that traffic conditions on different roads can be easily accessed by citizens and alternative routes to ease the traffic congestion can be suggested in such forums. By simply connecting to and accessing different posts on the official page of the concerned government agency in the social media forum, citizens can in real time, access information about different roads and the alternative routes. The Federal Road Safety Commission of Nigeria (FRSC) can monitor and coordinate this process along with the traffic arm of the Nigerian Police Force (NPF).

Similarly, using social media and Web 2.0 technologies, the Nigerian Police Force and other Force agencies can identify, monitor and respond to potential crimes in real time. However care must be taken to put in place a structure to filter out spam and deceptive alerts before necessary response actions are carried out. Social media and Web 2.0 technologies can be used to quickly identify militant groups and other security attacks and respond quickly to them if appropriate structures are put in place. The NPF and other armed forces can have visible presence in some social media sites and will encourage citizens to send tweets or instant messages to their official page when they notice the potential occurrence of security threats both to lives and property.

The Nigerian government established agencies or institutions responsible for environmental sanitation can also set up websites where residents can sign up to be tweeted the night before garbage and recycling collection is done in their area. These agencies can also use other social media forums for broadcasting such information on their official page and to receive feedbacks. The official pages of these sanitation agencies should be broken down into small zones such as wards or Local governments so that people living in that area will be able to easily access information about their assigned days in their area and also be able to make comments and complains back to the agencies.

The Nigerian Customs service has taken steps in this direction by setting up "The trade hub" which has various tools to help both Nigerian citizens and foreigners get the right information about their choice import or export business in Nigeria. This absolutely free service has a 24/7 customer service live chat, where people can ask an officer for any information that is not found on the portal. The trade hub provides information about all the Nigerian regulatory agencies such as National Agency for Food and Drug Administration and Control (NAFDAC), CUSTOMS, Nigerian Communications Commission (NCC), National Drug Law Enforcement Agency (NDLEA), Standard Organization of Nigeria (SON), Corporate Affairs Commission (CAC), Central Bank of Nigeria (CBN) and others along with their contact details, processes, documents, fees and processing time that an importer and exporter will need to liase with to obtain the necessary and important permits, documents and certificates required to ensure compliance (Hamzat, 2013). The Nigeria Customs also created a Facebook account for the trade-hub where it supplies daily useful information that would be helpful to business men and women doing import and export, information like the new trade and bilateral agreement between Nigeria and other countries, reduction of export dues in certain countries and so on.

Identifying, Monitoring and Responding to Citizen's Feelings on Changes to Government Policies

One way to identify, monitor and respond to the feelings of the citizens about a new policy or a modification in an old one is the use of social media and Web 2.0 technologies. Although I do not presently live in my native state (Edo state), I get much information about some new policies there from friends who make comments on such policies while I am on social media sites. I learned for the first time about the ban placed on riding motorbikes in Edo state from a friend (who was complaining bitterly of the inconveniences the ban was causing her) in a social media site.

Governments can through the use of the social media engage the citizens by creating interactive social media forums where such policies can be discussed and citizens are allowed to express their views through comments. Governments should however put up a position that they are willing to make some adjustments depending on the suggestions made by the citizens so as to encourage them to be willing to comment freely since they will have the feeling that their opinion will count. President Goodluck Ebele Jonathan operates a Facebook page where he allows Nigerians to speak out on several government policies and issues and he has responded positively to general opinion a few times by changing some earlier decisions made. For example he changed the decision to withdraw the Super Eagles of Nigeria from all International competitions for two years due to numerous appeals on his Facebook page by Nigerians (BBC News Africa, 2010). He had banned the team for two years after their poor World Cup campaign but had to change the decision due to social media influence. Indeed this clearly suggests that any sincere government can harness the feelings of her people using social media and Web 2.0 technologies if the proper structure is put in place.

FUTURE CONCERNS OF E-GOVERNANCE

Although, e-governance has numerous advantages, there are several issues that have been raised which have become national and international concerns. Cyber security issues, loss of privacy of Citizens and possibility of government to manipulate the information on their sites to suite their selfish gains are a few issues presently being considered as nations drive towards full deployment of e-governance. How do we protect sensitive information from unauthorized disclosure or intelligible interception and eavesdropping? How do we ensure the integrity of the process by preserving the accuracy and completeness of information and software and protecting data from unauthorized, unanticipated or unintentional modification? How do we ensure availability such that information and vital infrastructural services are available when required? How do we filter out the large volume of unwanted data generated by social media and Web 2.0 technologies so as to make sense of the data collected? How do we prevent citizens, government officials and employees from inappropriate and unethical uses of social media and Web 2.0 technologies? Once the trend encouraged by social media platforms where people maliciously accuse others without proof becomes established and remains unchecked, it can lead to government officials being blackmailed and accused of bribery or corruption, married men and women being accused of adultery, young unmarried singles being accused of prostitution, others will be accused of theft, murder, etc. Once it is on the social media, you are tried, convicted and sentenced even without going to court.

These and many more are questions that will have to be answered to facilitate widespread diffusion and usage of social media, Web 2.0 and wireless/mobile technologies and other ICTs for attaining a sustained e-ready environment necessary for enhancing e-governance in any country.

CONCLUSION

Efficiently implementing e-governance has the potential to create more informed citizens who are not only prepared to participate in democratic processes, but are also excited about their contribution to governance so as to achieve and sustain national developmental goals and objectives. Through a strong e-governance system, Nigeria can become a nation that thrives as it will provide a greater level of performance with convenience (both for governments and citizens), better management of time and resources, savings in cost, revenue growth, etc. The process of accessing government information by Citizens and that of Government collecting information from Citizens are greatly simplified which could translate into a decrease in corruption because public information would be available to all citizens and there would be less possibility of manipulation. Governance processes and procedures will become more transparent and all inclusive.

This chapter has considered the concept of E-readiness and the various Information and communication technologies (both established and emerging) that can facilitate the e-ready process in Nigeria as well as the necessary steps to be taken to provide all stake holders with a blue print of areas on which to focus. Necessary processes have to be put in place, infrastructures have to be acquired and trainings have to be provided to equip Nigerians with the capacity to both use and sustain the e-application platforms. After acquiring the backbone infrastructures, social media, Web 2.0 and wireless/mobile technologies were identified with relevant examples as necessary technologies that will facilitate the attainment and sustenance of the e-ready environment necessary to enhance e-governance in Nigeria. As our world becomes increasingly digital, Nigeria and other developing countries in the world need to take pragmatic steps towards bridging the digital divide while striving towards achieve an e-ready feat so as to provide the needed platform for e-governance to thrive.

REFERENCES

Adenike, A. O., & Oyesoji, A. A. (2010). The relationship among predictors of child, family, school, society and the government and academic achievement of senior secondary school students in Ibadan. *Nigeria Procedia Social and Behavioural Sciences*, 5, 842–849. doi:10.1016/j.sbspro.2010.07.196

Adeniyi, B. (2012). Galaxy Backbone ICT infrastructure rollout gets $100m boost as FG, China enter $1.1billion loan pact. *Technology Times*. Retrieved 24/06/13 from http://www.technology-timesng.com/galaxy-backbone-ict-infrastructure-rollout-gets-100m-boost-as-fg-china-enter-1-1billion-loan-pact/

Admin. (2012). *Tweet photos: Nigerians protest fuel subsidy removal Nigerians Abroad*. Retrieved 24/08/13 from http://nigeriansabroadlive.com/tweet-photos-nigerians-protest-fuel-subsidy-removal/

Akunyili, D. (2010). *ICT and E-government in Nigeria World Congress on Information Technology*. Retrieved 18/06/2013 at http://goafrit.wordpress.com/2010/06/12/ict-and-e-government-in-nigeria-prof-akunyili/

Al-Hujran, O., Al-dalahmeh, M., & Aloudat, A. (2011). The Role of National Culture on Citizen Adoption of eGovernment Services: An Empirical Study. *Electronic. Journal of E-Government*, 9(2), 93–106.

Aregbesola, I. (2012). FG partners Foreign Firm to train 5,000 Nigerians on ICT, says Minister. *Business Day*. Retrieved 26/06/13 from www.businessdayonline.com/NG/index.php/tech/78-computing/38556-fg-partners-foreign-firm-to-train-5000-nigerians-on-ict-says-minister

Baltaci-Goktalay, S., & Ozdilek, Z. (2010). Pre-service teachers' perceptions about Web 2.0 technologies. *Procedia Social and Behavioral Sciences, 2*, 4737–4741. doi:10.1016/j.sbspro.2010.03.760

BBC News Africa. (2010). *Facebook influences Nigeria football team ban U-turn*. Retrieved 26/08/13 from http://www.bbc.co.uk/news/10525699

Bonsón, E., Torres, L., Royo, S., & Flores, F. (2012). Local e-government 2.0: Social media and corporate transparency in municipalities. *Government Information Quarterly, 29*(2), 123–132. doi:10.1016/j.giq.2011.10.001

Braund, P., Frausher, K., Schwittay, A., & Petkoski. (2006). A Report on the Global E-Discussion: Information and Communications Technology for Economic Development, Exploring possibilities for Multi-Sector Technology Collaborations. *World Bank Institute Business, Competitiveness and development Team and RiOS Institute*. Retrieved 27/09/11 from http://www.riosinstitute.org/RiOSWBIediscussio.pdf

Budde Com. (2013). *Nigeria - Mobile Market - Overview, Statistics and Forecasts*. Retrieved 18/06/2013 from http://www.budde.com.au/Research/Nigeria-Mobile-Market-Overview-Statistics-and-Forecasts.html

Carmen, H., & Gabriela, G. (2011).. . *Mobile Learning Through Microblogging Procedia Social and Behavioral Sciences, 15*, 4–8. doi:10.1016/j.sbspro.2011.03.039

Chukwuebuka, U. F. (2013). The Effect of Social Media on Youth Development. *Nigeria Village Square*. Retrieved 24/08/13 from http://nigeriavillagesquare.com/articles/the-effect-of-social-media-on-the-youth-development.html

Daud, M. Y., & Zakaria, E. (2011). Web 2.0 application to cultivate creativity in ICT literacy. *Procedia - Social and Behavioral Sciences, 59*, 459 – 466.

Disaster-Report. (2012). *Natural Disasters In Nigeria 2012*. Retrieved 24/08/13 from http://www.disaster-report.com/2013/04/natural-disasters-in-nigeria-2012.html

Disaster-Report. (2013). *Natural Disasters In Nigeria 2013*. Retrieved 24/08/13 from http://www.disaster-report.com/2013/04/natural-disasters-in-nigeria-2013.html

Fedotova, O., Teixeira, L., & Alvelos, H. (2012). E-participation in Portugal: evaluation of government electronic platforms. *Procedia Technology, 5*, 152–161. doi:10.1016/j.protcy.2012.09.017

Hamzat, A. O. (2013). *Cybercrime, Social Media, Customs And The Nigeria Trade-Hub*. Retrieved 24/08/2013 from http://elombah.com/index.php/articles-mainmenu/17288-cybercrime-social-media-customs-and-the-nigeria-trade-hub

Holotescu, C., & Grosseck, G. (2011). Mobile learning through microblogging. *Procedia Social and Behavioral Sciences, 15*, 4–8. doi:10.1016/j.sbspro.2011.03.039

Howe, J. (2006). Your Web, Your Way. *Time Magazine, 168*(26), 60–63.

Internetworldstatistics.com. (2013). Internet Users in Africa 2012. *Miniwatts Marketing Group*. Retrieved 19/06/2013 from http://www.internetworldstats.com/stats1.htm

Iroko, M. (n.d.). FG Abandons $200M Rural Telephony. *Project Zimbio Inc*. Retrieved 22/07/12 from http://www.zimbio.com/Nigeria/articles/GnjWJ99uezp/FG+ABANDONS+200M+RURAL+TELEPHONY+PROJECT

Jidaw. (2013). ICT development Advisory Support and Consulting. *Nigeria Computers*. Retrieved 26/06/2013 from http://nigeriacomputers.com/tech-news/development-advisory-support-and-consulting-nigeria/#more-2046

Kavanaugh, A. L., Fox, E. A., Sheetz, S. D., Yang, S., Li, L. T., & Shoemaker, D. J. et al. (2012). Social media use by government: From the routine to the critical. *Government Information Quarterly, 29,* 480–491. doi:10.1016/j.giq.2012.06.002

NAN 1. (2013). FG to launch Fibre-optic Network connecting 27 universities. *News Agency of Nigeria.* Retrieved 24/06/2013 from http://www.nanngronline.com/section/technology/fg-to-launch-fibre-optic-network-connecting-27-universities

NAN 2. (2013). 2 mobile phone coys to begin production in Nigeria in 2013, says Minister. *News Agency of Nigeria.* Retrieved 24/06/2013 from http://www.nanngronline.com/section/technology/2-mobile-phone-coys-to-begin-production-in-nigeria-in-2013-says-minister

Nigeria Intel. (2013). Why Nigeria must embrace e-governance. *Nigeria Intel.* Retrieved 18/06/2013 from http://www.nigeriaintel.com/2013/03/28/why-nigeria-must-embrace-e-governance/

Nugultham. (2012). Using Web 2.0 for Innovation and Information Technology in Education course. *Procedia - Social and Behavioral Sciences, 46,* 4607 – 4610.

Obiozor, W. E. (2013). ICT and National Development. *Academia.Edu.* Retrieved 26/06/13 from http://academia.edu/2392046/ICT_and_NATIONAL_DEVELOPMENT

Oghogho, I., & Ezomo, P. I. (2013). ICT for National Development in Nigeria: Creating an Enabling Environment. *International Journal of Engineering and Applied Sciences, 3*(2).

Oghogho, I., Odikayor, D. C., Adebayo, A. A., & Wara, S. T. (2012). VoIP vs GSM Technology: The Way of the Future for Communication In Ekekwe & Islam (Ed.), Disruptive Technologies, Innovation and Global Redesign: Emerging Implications (pp. 280-298). Hershey, PA: IGI Global.

Onwuemele, A. (2011). Impact of Mobile Phones on Rural Livelihoods Assets in Rural Nigeria: A Case Study of Ovia North East Local Government Area. *JORIND, 9*(2), 223–236.

Orimisan, B. (2012). Fuel subsidy protests and power of social media. *Nigerian Best Forum.* Retrieved 24/08/13 from http://www.nigerianbestforum.com/generaltopics/fuel-subsidy-protests-and-power-of-social-media/

Rosa, J., Teixeira, C., & Pinto, J. S. (2013). Risk factors in e-justice information systems. *Government Information Quarterly, 30,* 241–256. doi:10.1016/j.giq.2013.02.002

Sandoval-Almazan, R., Gil-Garcia, J. R., Luna-Reyes, L.F., & Luna, D. E., & Diaz-Murillo. (2011). The use of Web 2.0 on Mexican State Websites: A Three Year Assessment. *Electronic. Journal of E-Government, 9*(2), 107–121.

Sharma, M. K., & Vaisla, K. S. (2011). E-governance applications in public healthcare for rural areas of Uttarakhand. *Elixir Comp. Sci., &. Engg., 41,* 5583–5586.

Smith, A., & Peck, B. (2010). The teacher as the 'digital perpetrator': Implementing Web 2.0 technology activity as assessment practice for higher education Innovation or Imposition? *Procedia Social and Behavioral Sciences, 2,* 4800–4804. doi:10.1016/j.sbspro.2010.03.773

Taleb, Z., & Sohrabi, A. (2012). Learning on the move: the use of mobile technology to support learning for university students. *Procedia- Social and Behavioral Sciences, 69,* 1102 –1109.

UN DESA. (2012). *United Nations E-government Survey 2012: E-Government for the People.* Department of Economic and Social Affairs (United Nations). Retrieved 18/06/2013 from www.un.org/en/development/desa/publications/connecting-governments-to-citizens.html

United Nations E-Government Survey. (2012). *Executive Summary*. Retrieved 24/06/2013 from http://unpan3.un.org/egovkb/

Wright, A., Bates, D. W., Middleton, B., Hongsermeier, T., Kashyap, V., Thomas, S. M., & Sittig, V. F. (2009). Creating and Sharing Clinical Decision Support Content with Web 2.0: Issues and Examples. *Journal of Biomedical Informatics*, *42*, 334–346. doi:10.1016/j.jbi.2008.09.003 PMID:18935982

Yi, M., Oh, S. G., & Kim, S. (2013). Comparison of social media use for the U.S., & the Korean governments. *Government Information Quarterly*, *30*, 310–317. doi:10.1016/j.giq.2013.01.004

ADDITIONAL READING

Adams, S. A. (2010). Revisiting the online health information reliability debate in the wake of Web 2.0: An inter-disciplinary literature and website review. *International Journal of Medical Informatics*, *79*(Issue 6), 391–400. doi:10.1016/j.ijmedinf.2010.01.006 PMID:20188623

Alonso, S., Perez, I. J., Cabrerizo, F. J., & Herrera-Viedma, E. (2013). A linguistic consensus model for Web 2.0 communities. *Applied Soft Computing*, *13*, 149–157. doi:10.1016/j.asoc.2012.08.009

Bhuiyan, S. H. (2011). Modernizing Bangladesh public administration through e-governance: Benefits and challenges. *Government Information Quarterly*, *28*(Issue 1), 54–65. doi:10.1016/j.giq.2010.04.006

Chen, Y. (2012). A comparative study of e-government XBRL implementations: The potential of improving information transparency and efficiency. *Government Information Quarterly*, *29*(Issue 4), 553–563. doi:10.1016/j.giq.2012.05.009

Gray, H., Thompson, C., Clerehan, R., & Sheard, J., & Margaret Hamilton. (2008). Web 2.0 authorship: Issues of referencing and citation for academic integrity. *The Internet and Higher Education*, *11*(Issue 2), 112–118. doi:10.1016/j.iheduc.2008.03.001

Juzwishin, D. W. M. (2012). Political, policy and social barriers to health system interoperability: Emerging opportunities of Web 2.0 and 3.0. *Healthcare Management Forum*, *22*(Issue 4), 6–10. doi:10.1016/S0840-4704(10)60136-6 PMID:20166516

Kuyoro, S. O., Awodele, O., Alao, O. D., & Omotunde, A.A. (2013) ICT solution to Small and Medium Scale Enterprises (SMEs) in Nigeria International Journal of Computer and Information Technology Volume 02– Issue 04.

Linders, D. (2012). From e-government to we-government: Defining a typology for citizen coproduction in the age of social media. *Government Information Quarterly*, *29*(Issue 4), 446–454. doi:10.1016/j.giq.2012.06.003

Marlin-Bennett, R. &Thornton, E.N. (2012). Governance within social media websites: Ruling new frontiersTelecommunications Policy 36 (2012) 493 –501.

Meijer, A., & Thaens, M. (2010). Alignment 2.0: Strategic use of new Internet technologies in government. *Government Information Quarterly*, *27*(Issue 2), 113–121. doi:10.1016/j.giq.2009.12.001

Misuraca, G., Broster, D., & Centeno, C. (2012). Digital Europe 2030: Designing scenarios for ICT in future governance and policy making. *Government Information Quarterly*, *29*, 121–131. doi:10.1016/j.giq.2011.08.006

Pang, A. S. (2010). Social scanning: Improving futures through Web 2.0, or, finally a use for twitter. *Futures*, *42*(Issue 10), 1222–1230. doi:10.1016/j.futures.2010.09.003

Ranerup, A. (2012). The socio-material pragmatics of e-governance mobilization. *Government Information Quarterly*, *29*(Issue 3), 413–423. doi:10.1016/j.giq.2012.02.012

Reddick, C.G. & Norris, D.F (in Press) Social media adoption at the American grass roots: Web 2.0 or 1.5? Government Information Quarterly, Available online 12 August 2013.

Sandoval-Almazan, R., & Gil-Garcia, J. R. (2012). Are government internet portals evolving towards more interaction, participation, and collaboration? Revisiting the rhetoric of e-government among municipalities. *Government Information Quarterly*, *29*, 72–81. doi:10.1016/j.giq.2011.09.004

KEY TERMS AND DEFINITIONS

Bandwidth: Determines the efficiency and speed of a user's Internet access and describes the rate at which data can be transferred to a computer or other device from a website or Internet service within a specific time. It is a term that includes a broad range of technologies which provide higher data rate access to the Internet.

E-Governance: Is the delivery of government information and services to the public using ICTs.

E-Readiness: Refers to a situation where a nation has put in place the necessary ICT infrastructures, technologies and equipment and has empowered the citizens with the economic power and relevant trainings to be able to both possess and use these technologies for accessing several e-services.

ICT: An acronym for Information and communications technology is an umbrella term that includes any communication device or application, encompassing: radio, television, mobile and fixed phones, computer and network hardware and software, satellite systems, surveillance systems, and so on, (as well as the various services and applications associated with them, such as video-conferencing, distance learning, etc.) necessary for the delivery of information in the form of audio, data, video, image, etc. from Point a sender to a receiver or receivers.

Infrastructure: In telecommunications, refers to the physical hardware and software used to interconnect computers, mobile phones, fax machines, etc. and users as well as the transmission media and equipment such as telephone lines, cable television lines, satellites, antennas, the routers, aggregators, repeaters, and other devices that control transmission paths.

Internet: Is a global system that consists of millions of people that could be private, public, academic, business, government, etc. who are linked or interconnected through electronic equipment, infrastructures, etc. based on various electronic technologies (wireless, optical or wired computer networks, mobile phones, broadband, etc.) and are able to transmit and receive information (data, voice, video, etc.) to and from each other.

Social Media: Are applications based on the Internet which are designed to facilitate social interaction and for using, developing and diffusing information through society by building on many of the same concepts and technologies of Web 2.0, with special focus on the creation and exchange of user generated content.

Telecommunications: Refer to the exchange of information by electronic and electrical means over a significant distance using devices such as telephones, telegraph, radio, microwave communication arrangements, fibre optics, satellites the Internet, etc.

Web 2.0: Is an Internet technology that supports features such as Social networks, Social bookmaking, Instant messaging, Wikis, Internet telephony (VoIP), Audio or video conferencing (NetMeeting), blogs, etc. which allow people to effectively and actively connect, participate, interact, collaborate and create and share live streaming or recorded audio, video, text, pictures, etc. information or knowledge.

Chapter 5
A Base of Knowledge, Mobile, and Web 2.0 Technologies for Connected E-Government

Muhammad Yusuf
University of Portsmouth, UK

Carl Adams
University of Portsmouth, UK

ABSTRACT

E-Government is an evolving field with continually changing practice and priorities. It is also a global phenomenon, from the richest and most technologically developed nations to the poorer and less technologically developed countries, involving a range of latest Information and Communication Technologies (ICT) and diverse methodologies. In such a dynamic field spanning all sectors of the governments and societies, it is difficult for e-government researchers and practitioners to identify the trends in the e-government activity and learn from previous cases and experiences. In this context, the aim of this chapter is to present an in-depth evaluation of e-government practice and research since 2007, to provide insight on research practicalities and emerging issues in e-government activity, and to identify the trends and technologies. The chapter also focuses on the current mobile and Web 2.0 technologies and examines the practicalities of using mobile technologies in various countries such as USA, Canada, UK, Austria, Japan, and others, as well as the practicalities of Web 2.0 technologies in some domains such as government, regulation, cross-agency cooperation, law enforcement, etc. This chapter presents a framework based on the mobile and Web 2.0 technologies in the context of e-government activity. In addition, the authors propose a framework for a government-people relationship. We hope to make a contribution for researchers, practitioners, policy makers, and people interested in e-government by providing a base of the e-government domain knowledge, practice, and framework. Additionally, the chapter illustrates how the implementation of mobile and Web 2.0 technologies support connected e-government.

INTRODUCTION

E-Government is a global phenomenon with continually changing practices and priorities. It is also a global phenomenon. E-Government is a broad area covering a variety of interdisciplinary subjects including Computer Science, Information Systems, Information Technology, Politics, Public Management, Finance, Health and Sociology. E-Government activity takes place from the richest

DOI: 10.4018/978-1-4666-6082-3.ch005

and most technological developed nations to the poorer and less technologically developed nations. The study by Bolivar et al. (2010) showed that various academic departments conducted research on e-government and noted that 22.5% of research came from Public Administration, 7.3% research originated from Marketing and Communication, 12.4% research was from Management Science, 5.8% from Library and Information Science, 15.2% research came from Public and Policy Science, 10.6% from Computer Science and Information System, 8.4% from Practitioners, 7.6% research came from Accounting, Business and Economics and 9.37% research came from other sources. Similarly, Heeks and Bailure (2007) identified that e-government researchers came from diverse departments such as: Business/Management, Public Administration, Political Science, Computer Science, Library and Information Studies, e-government, Information System, Government/Governance, Non-academic research institutions and other. They also pointed out the main literatures of e-government research which consist of e-government, Information System (including business), Public Administration, Management, Political Science, Computer Science and other (Heeks & Bailur, 2007).

E-Government is a term that appeared in the late nineties. There are various definitions of e-government. The US Congress defines it in the US 2002 e-government Act (Grönlund & Horan, 2004) as: "government supported by Information Technologies for delivering good services and information to government stakeholder effectively and efficiently." In 2004, European Union (EU) classified e-government as: "Public Administration based on Information and Communication Technologies to enhance public services and democratic processes and it supported by new skills and organisational improvement." Furthermore, the One U.S. General Accounting Office examined some of the challenging factors of e-government implementation such as: strong leader commitment, effective e-government, preserving citizen concerns, privacy and security issues, electronic records, good technical infrastructures, human capabilities for IT skills, consistent and standardized public service delivery consistently (Jaeger & Thompson, 2003).

Notwithstanding the benefits that e-government promises, it seems to present three main challenges as follows (Signore, Chesi, & Pallotti, 2005):

- Technical challenges. These include interoperability, privacy, security and multimodal interaction.
- Economic challenges. These consist of specific issues such as: costs, reusability and portability.
- Social challenges. These challenges cover social aspects such as: accessibility, usability and acceptance.

Jaeger & Thompson (2003) explain some important issues for successful e-government implementation as presented below:

- Assuring the capability exists to implement suitable technologies.
- Propagating the importance of e-government to public.
- Ensuring the public can acquire meaningful information and services.
- Creating the integration of local, regional and national e-government programmes.
- Elaborating the methods and achievement indicators to evaluate e-government performance

As a consequence, e-government implementation not only faces technical issues but also non-technical issues. This is the reason that e-government has become such a broad issue and why all manner of interdisciplinary subjects exist in order to resolve the issues and achieve the goals of e-government. Moreover, Grönlund & Horan (2004) has proposed three goals of e-government such as:

- To make government more efficient.
- To deliver better government services to citizens.
- To improve democratic processes.

E-Government practice and research also has various research philosophies and methodologies. Previous research has tried to capture this diversity. For instance, Heeks and Bailur (2007) have examined the viewpoints, philosophies, theories and methods of e-government. Their work was based on examining two journals and conference papers (Information Policy from 2002 to 2004 volume 7 to 9, and Government Information Quarterly from 2001 to 2005 volumes 18 to 22 and European Conference on e-government from 2001 to 2005). In other research, Bolivar et al. (2010) studied the methodologies used in e-government research, drawing upon 321 articles published in Journals from Information Science and Library Science also Public Administration Subjects. Another paper also discussed research approaches based on 544 papers presented to the European Conference on e-government (ECEG) from 2001 to 2009 (Bannister & Connolly, 2010).

In this chapter, we follow a similar approach to that of examining a range of published work in recognised robust research outlets to capture trends and emergent issues. For this work, we draw upon the International Conference on e-government (ICEG) from 2007 to 2010 and European Conference on e-government (ECEG) from 2007 to 2012. We focus on mobile and Web 2.0 technologies in e-government since mobile and Web 2.0 usage has significantly increased. As a result, e-government is moving to m-government and Government 2.0. M-government captures government activities using mobile technologies to achieve its goal: to improve public services, to increase transparency, efficiency and effectiveness.

Clearly, Mobile Technologies are more popular in the developing countries where the cable access to the Internet infrastructure is rather limited. Moreover, Government 2.0 suggests that government manages Web 2.0 technologies much more to support their activities to address the government goals. It has become popular since the boom of Web 2.0 or Social Network technologies such as Twitter, Facebook and MySpace but mobile and Web 2.0 technologies cannot be separated any more. Nowadays, Internet, Mobile and Web 2.0 Technologies have converged since the Internet and Social Networking became accessible both via a PC desktop and via a mobile device.

This chapter aims to make a contribution by providing a base of knowledge covering technology practicalities in the e-government domain, especially mobile and Web 2.0 technologies for practitioners, policy makers and people interested in e-government. For practitioners, policymakers and people interested in e-government, this chapter provides some insights on how e-government activity, from a global perspective, has changed and evolved and the practicalities involved in supporting mobile and Web 2.0. For e-government researchers, it provides insights on the practice of e-government research. It highlights the main approaches and areas of investigations. It also illustrates those areas where there remains the opportunity for further investigation.

The brief structure of this chapter is as follows: e-government research methodologies evaluation; e-government issues, and themes such as keywords of e-government, especially the top ten keywords. The main issues and themes that came up from Focus Group Discussion (FGD) activity will also be discussed. The following sections will then cover the use of mobile and Web 2.0 technologies, the practices and issues relating to Mobile and Web 2.0 technologies. Existing frameworks and government-people relationship framework proposed by the authors of this chapter will also be discussed. Furthermore, evaluation of mobile and Web 2.0 technologies in e-government will be presented in this chapter. Finally, the last section will provide conclusions of this study and ideas for further future work in this area.

RESEARCH METHODOLOGY

The approach taken in this research is to examine key bodies of research using a structures research method in e-government to identify the main trends and issues over time and so identify the top issues in e-government research and activity then develop a framework of e-government activity. Firstly, this chapter investigates the themes and trends based on a focussed literature review from abstract of papers from prominent conferences such as Proceedings of European Conference on E-Government (ECEG) from 2007 to 2012 and Proceedings of International Conference on E-Government (ICEG) from 2007 to 2010. Then we apply content analysis, using a systematic approach to explain quantitative data gathered from a literature review. The European Conference on E-Government (ECEG) and the International Conference on E-Government (ICEG) organized by Academic Conferences and Publishing International Limited (ACPI) are chosen because, according to Bannister and Connolly, the European Conference on e-government (ECEG) was the first Conference in Europe focussed on e-government. ECEG is also one of the most established conferences on e-government. It commenced in 2001 and has been held regularly every year until at the moment. Additionally, the International Conference on E-Government (ICEG) was determined because this conference has been, almost annually, from 2005 and the 7th Conference will be in 2013.

These are key sources of information covering e-government research and activity and consequently are likely to provide a good representation of how e-government focus has changed over time. The next stage was to collect all of the keywords from the abstracts of those papers, and count those keywords. Keywords collected because it reflects the main issues on the papers. Keywords were then grouped by word prevalence and entered into the "Wordle" software. This cloud-based software displays grouped words by images based on their frequency. The size of words indicates the amount of times that word is used in the source text. A bigger word is more referenced than smaller ones. Ignoring the word e-government, the top ten keywords were then chosen to identify the top ten trends in e-government.

The authors of this chapter then reviewed abstracts to discover the methodologies and methods used by each author. Some authors presented methodologies and methods clearly on the abstracts but some others were not so clearly stated. For example, some authors did not write explicitly about the case studies conducted but only made reference to the countries where the research was done. If so, the authors classified them as case study research. Then, the terms which were considered relevant to research philosophies, methodologies and methods were collected and how many papers were concerned with these were counted.

Moreover, Focus Group Discussions (FGD) were conducted. The aim was to understand people's perceptions and views about e-government issues and the reason why they perceived e-government in a certain way. FGDs were chosen because they represent the social constructivist paradigm and qualitative methodologies. Opinions from subjects involved in FGDs are naturally subjective, as they are based on their knowledge about Computing Technology, Public Management, Politics, Government, Education, Health, Finance and others related to e-government. For FGDs, subjects who have expertise, or are conducting research into the following areas, can make invaluable contributions to the research:

- E-Government from Computer Science & Information Systems.
- E-Government from Public Administration.
- E-Government from Marketing and Communications.

- E-Government from Management Sciences.
- E-Government from Library and Information Sciences.
- E-Government from Public and Policy Sciences.
- E-Government from Accounting, Business and Economics.
- Practitioners on government (Education/Finance/Health).
- Citizens.

Furthermore, some of the journals and conferences papers were examined to investigate the practices and issues of Mobile and Web 2.0 technologies with respect to e-government development. The existing frameworks relating to those technologies were examined. Based on this, the authors of this chapter propose a government-people relationship framework in terms of Mobile and Web 2.0 technologies. This literature review methodology was chosen because many papers have researched the practicalities and frameworks in many cases and countries. This research needed to assess how successful or problematic those frameworks were, and gather information on the practicalities of implementing them in different fields.

Finally, evaluation of mobile and Web 2.0 technologies was conducted based on parameters which found from some research in journals. This method chosen to get valid result from previous evaluation method and parameters which proposed by previous researcher.

DISCUSSION AND ANALYSIS

Research Methodologies and Evaluation

Based on the focussed literature review of the ICEG from 2007 to 2010 and the ECEG from 2007 to 2012, there were some research paradigms,

research approaches, research methodologies, conclusions and research methods used by the conferences authors. Research paradigms used at both conferences include positivist, interpretative and critical realist. A range of research approaches have been used such as qualitative, quantitative and mixed methods and are mostly based on empirical work. Overall, the trend of research approach used in the papers of ICEG is increased from 2007 to 2008, then dramatically decreased from 2008 to 2009, then slightly increased again from 2009 to 2010. Furthermore, the authors used various research approaches, including pure qualitative, mixed method and pure quantities. The case study is the most popular research methodology used by authors of both conferences. There are some other methodologies used by authors such as the empirical approach, the soft system methodology (SSM), the usability research, the comparative approach, the exploratory study, the Q methodology and hybrid methodology.

Moreover, there are many of documentary methods used on e-government. The survey dominates as the most widely used method; the questionnaire is the second widely used at ECEG 2012. The survey is a dominant method used by authors in ICEG in 2007, 2008 and 2009, but was not used at all by ICEG 2010. In other hand, Focus Group Discussion dominated in those papers submitted to ICEG 2010.

We also noticed that methods such as survey, questionnaire, interview, empirical approach and literature review or extensive literature review were used by many authors in these conferences. Both conferences also came up with various methodologies from the purist qualitative to purist quantitative (quantitative empirical) as well as qualitative and quantitative empirical in ICEG and qualitative and quantitative in ECEG as mixed methods. This combination reflects that researchers and authors in this conference used various research paradigms i.e. Constructivist – interpretive (qualitative) and positivist (quantitative empirical).

Research on e-government can be done from both angels of positivist and social constructivist paradigm or by taking the middle path. The Positivist paradigm presupposes some key factors in e-government, for instance the technology and the culture exist, and assumes that gathering of data is independent from the researcher's interest. Positivist researchers will work to build knowledge from the relation and generalisation of laws. On the other hand, social constructionist researchers set up assumptions that object's acceptance such as technology depends on individual perceptions and value about that thing. Hence it will be a subjective based on the social constructionist researchers' perceptions, values and meanings. The consequences are that knowledge will be constructed by individual interactions between each other as well as the data and the gathering of the data process cannot be independent from researcher's interest and construction.

The middle position between the positivist and the social constructionist is a compromise; a presupposition that other philosophies need to be considered. While, Heeks and Bailure (2007) analysed the papers, they did not find any concepts about research philosophy. Many researchers did not examine a base research philosophy regarding e-government research. Most methods used were unclear and had a poor epistemology as well as deductive or inductive approaches. Furthermore, just few papers clearly stated the position as pure positive; some of papers has unclear positivist and no papers existed where the social constructivist paradigm was expressed. Hence the dominant research philosophy across all papers derived from one philosophy.

Overall, the research philosophy in e-government needs to be further studied to justify e-government as a discipline. The literature review revealed changes and findings about the research philosophy of e-government since Heeks and Bailur (2007) published their paper. The dominant case study method represents social construction-ist rather than positivist. In addition, the survey was dominant alongside questionnaires, extensive literature reviews, research reports, observations, interviews, focus group deliberations, telephone interviews, etc. Another point is that only an inductive approach was identified from these conferences papers. So, there are changes in the research philosophy of e-government based on the classification of research paradigms, research approaches, research methodologies, research methods and conclusions. These findings contrast with Heeks and Bailure's (2007) results.

The results highlight a general insight that e-government has been clear and have varieties of research philosophy. The conferences authors have been pointed out that, even though it needs further investigation, that it is difficult to assess if the philosophies were managed systematically (and interrelated from research paradigms to research methods) or were instead methods used without the consideration of a base methodology, approach or paradigm of chosen method. In another paper, Bolivar et al. (2010) found e-government researchers used empirical research methods more dominant rather than non-empirical. The dominant quantitative methods are regression analysis, followed by structural equation modelling and evaluation research.

The graphic of qualitative and quantitative trends showed that qualitative methodology become decreased and quantitative methodology more increased from 2000 through to 2009. Compared with the focussed literature review of ICEG and ECEG, there are some similarities. The case study is always the dominant approach, as well as an empirical approach. In ICEG and ECEG, some authors used a mixed approach of qualitative and quantitative besides pure qualitative or pure quantitative or quantitative empirical but no trend of qualitative, trend of qualitative and trend of qualitative-quantitative based on ICEG and ECEG, so it cannot be compared to the trend at the moment.

Bannister and Connolly (2010) pointed out conceptual and case by case approaches as investigation paper are dominant but the amount of theoretical papers were very small. Comparing ICEG vs ECEG, it has the same result that case study papers are dominant.

In summary, the comparison between the conclusions of Heeks and Bailur (2007), Pedro and Bolivar (2010) and Bannister and Connolly (2010) plus the focussed literature review shows that e-government research uses various research philosophies, methodologies and methods from the extreme continuum positivist and social constructivist or pure qualitative and pure quantitative to mixed and compromise of both. E-government is emerging as a mature discipline.

E-GOVERNMENT KEYWORDS, THEMES AND ISSUES

This section examines keywords, themes and issues based on the focussed literature review. Counting the total amount of relevant keywords from papers from ICEG and ECEG conferences totalled more than 1000. These keywords were collected from abstracts of ICEG and ECEG papers

assuming that the keywords would reflect the contents, themes and issues relating to e-government.

Figure 1 shows all of the keywords of ECEG from 2007 to 2012 using Wordle. The 'e-government' keyword was ignored even though it is the most frequent in the papers. Many authors of papers mention a lot of this particular keywords because the conferences discuss e-government. Figure 1 points out that Public, Management, Government, Services, Information, ICT, System, E-Democracy, Support and Public E-Participation are dominant and larger in size rather than other words in the image. It means these words have a higher frequency in many papers in ECEG from 2007 to 2012 and indicated some issues such as Public, Management, Government, Services, Information, ICT, System, E-Democracy, Support, and Public E-Participation are the important components of e-government because many authors around the world put these term as keywords at high number at their papers on ECEG.

Figure 2 indicates all of the keywords in the papers of the ICEG from 2007 to 2010 using Wordle. There are some larger sizes of words such as Public, Information, Digital, Government, Information, Management, Service, Social, etc. It shows these words are used by authors many times

Figure 1. Keywords of ECEG from 2007 to 2012 presented by Wordle

Figure 2. Keywords of ICEG from 2007 to 2010 presented by Wordle

in their papers. It reflects the critical issues from around the world because the papers submitted to the ICEG come from around the world. E terms such as E-Democracy and Public E-Participation consists of Internet, Mobile, Social Networking Technologies and Cloud Computing. All of these technologies will support Democracy and Public Participation in the virtual world. So it may have the impact to shape Democracy and Public Participation in the future, especially in the virtual world. The shape of Democracy and Public Participation will be different between the real world and the virtual world. Activities in Democracy and Public Participation will not only divide physical activities anymore in the real world, but also embrace intelligent activities in cyberspace.

Figure 3 describes the top ten keywords appearing in the ICEG papers. 'European Union' has also been ignored as a keyword due to its high frequency. The figure shows that as at 2007, E-Democracy and E-Commerce keywords had the highest frequency. On the other hand, Governance, e-government Implementation, Public Sector and

E-Voting had the lowest frequency of keywords. In 2010, E-Participation become the highest frequency keyword otherwise E-Commerce, Governance, e-government Implementation, Public Policy and Transparency are the lowest amount of keywords. This shift in prevalence of keywords may indicate a change in the trend of e-government issues. Keywords do seem to have significant trends; the E-Participation keyword, for example, significantly increases from 2009 to 2010; the Governance keyword dramatically increases from 2007 to 2009 but then significantly decreases in 2010. The 'e-government Implementation' keyword increased from 2007 to 2008 and then dramatically decreases from 2008 to 2009. After this, its occurrence slightly increased.

Figure 4 illustrates top ten keywords appearing in the ECEG papers. As with the ICEG, the word e-government has the highest frequency but has been ignored.

There are some trends of e-government issues as shown above. E-Democracy is the most frequent keyword at the conference in 2008 but Identity

Figure 3. Top ten keywords based on ICEG from 2007 to 2010

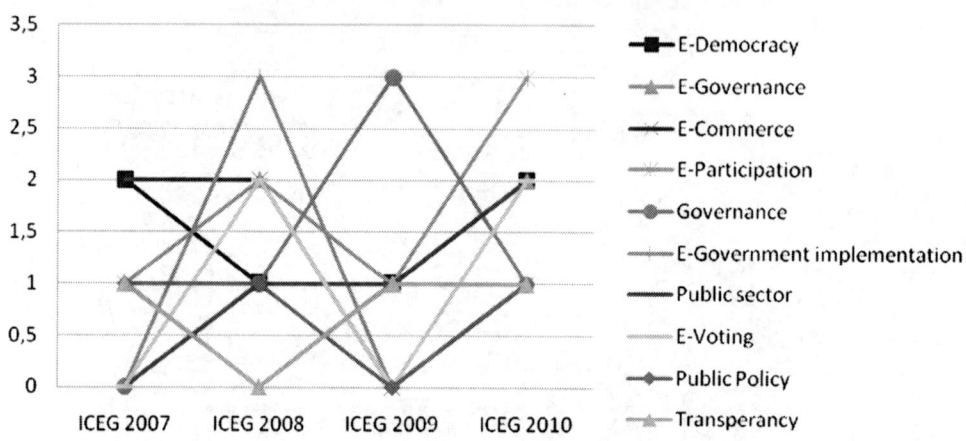

Figure 4. Top ten keywords of ECEG from 2007 to 2012

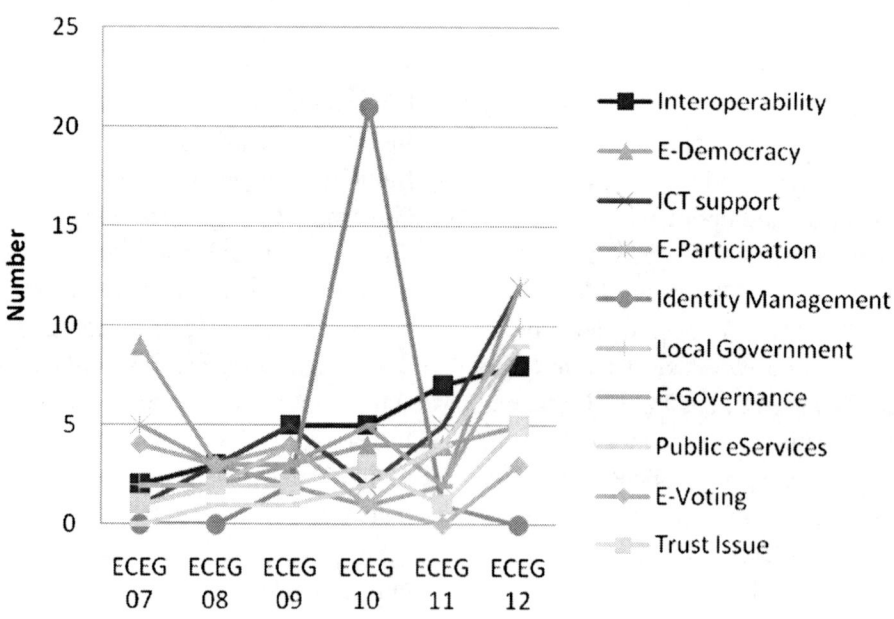

Management and Public eServices are the smallest number. Moreover by 2012, E-Participation and ICT Support are the highest frequency of keywords and Identity Management is the lowest.

There are different keywords at the highest number appearing in 2007 and 2012. On the other hand, there are similar keywords at the lowest number at 2007 and 2012. Moreover, there are some specific trends from figure above such as Identity Management is significantly increased at 2010 but decreases dramatically in 2011, E-Democracy has a dramatically decreased trend from 2007 to 2008 and E-Participation is the most significantly increased from 2011 to 2012.

Figure 5 summarizes the keywords from the ICEG and ECEG papers. There are some similari-

Figure 5. Main Issues on E-Government Research based on ICEG 2007 to 2010 and ECEG 2007 to 2012

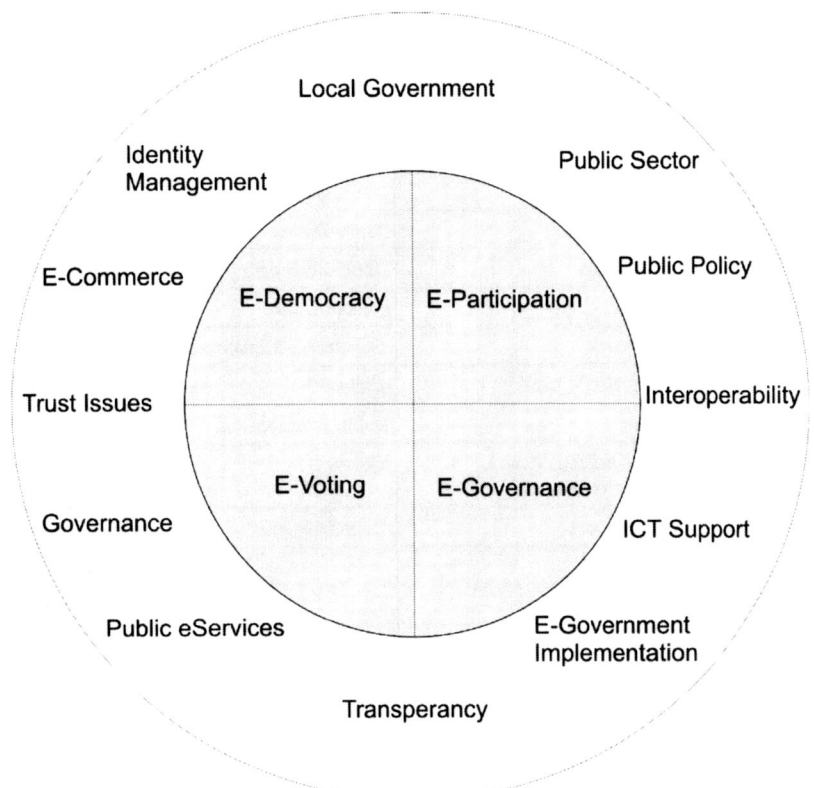

ties in keywords between the top ten keywords from ICEG and ECEG such as E-Democracy, E-Participation, E-Governance, and E-Voting. On the other hand, there are some different keywords such as E-Commerce, Governance, E-Government Implementation, Public Sector, Public Policy, Transparency, Interoperability, ICT Support, Local Government, Identity Management, Public E-Services and Trust Issues. So the same keywords are put into the centre as the main circle and different keywords into the secondary circle.

Four similar keywords represent the main pillars of society; these are Democracy, Participation, Governance and Voting. Use of E (to mean electronic) in front of those keywords points to recent and future electronic media. The Internet especially is the most important tool for supporting Democracy, Participation, Governance and Voting. The other keywords in ICEG and ECEG

show the important issues in e-government. In the secondary circle, there are important factors such as management, technology, public perception and government. From Figure 5, it appears that the fundamental issues in the area of e-government are the same from ICEG and ECEG.

The result of the Focus Group Discussions (FGD) is classification and the titles of keywords by subjects as shown in Table 1.

From the Focus Group Discussions, there are some similar titles that came up from different groups such as countries, finance, policy, research, legal, management and technology and some titles have different terms or words but same meaning. These show that most of groups have same perceptions to describe e-government activity.

Investigation results from the focussed literature review, author's construction and Focus Group Discussion (FGD) themes came up with

Table 1. List of Title based on participant of Focus Group Discussion (FGD)

Group A	Group B	Group C	Group D
Citizen	Countries	E-Government	Regional
Security	Finance	Policy	Actors
Countries	People Usability Accessibility	Democracy	E-Services
Standards and Policies	E-Government	Tools-Technology	Management
E-Government Transactions	Abbreviations	Tools-Design	Research
E-Government Activities	Ambiguous	Tools-Research	Information
Legal	Health	Tools-Practice	Technology-Usability
Technology	Bug wards	Structure-Information Needs	Technology-Security
Taxation	IT Service Related	Culture	Technology
Government Services	Future	Public Management	Legal
E-Government Portal	Research Methods and ThemesThemes	Concepts	Education
Management	Management	Communication	Economics Finance
Election	Issues		E-Governance
Characteristics of E-Government	Public		Politics
			Unclassified
			Jargon

some issues that change and do not change. Some issues that do change are specific issues on e-government such as E-Participation, Local Government, Identity Management, E-Governance and E-Democracy; some issues that do not change are management, technology, finance, politics, health, policy and governance. These unchanged issues represent the fundamental issues in e-government.

MOBILE AND WEB 2.0 TECHNOLOGIES

Technology can be used as a tool to support government activities and management in order to achieve the government's goals. For this reason, it is important to understand the function of technology as well as how technology changes over time.

This section focuses on mobile and Web 2.0 technologies for connected e-government. It examines practices and issues as well as the frameworks of mobile and Web 2.0 technologies

related to e-government. Based on the focussed literature review, there are some keywords related to mobile and Web 2.0 technologies which used by authors of ECEG papers, such as Twitter, M-Voting, M-Participation, Mobile Technologies, Social Networks, M-government and Web 2.0. In the ECEG, keywords related to mobile and Web 2.0 technologies have been prevalent since 2008. Overall, Web 2.0 has the highest frequency at six keywords and M-Participation the lowest frequency at 1.

Furthermore, Web 2.0 and M-government keywords are the most frequent keywords because these keywords always come up in almost every conference (except ECEG on 2011 for m-government and ECEG on 2010 for Web 2.0). This means that mobile and Web 2.0 technologies have been used from 2008 until now. In the ICEG, there are three keywords related to Mobile and Web 2.0 Technologies, such as "Social Network," "Web 2.0" and "Government 2.0." These keywords came up during conferences in 2007 and 2008. Overall, the most frequently keyword is "Social Network" but the less frequently keyword is "Government 2.0." The data from the ECEG and the ICEG indicates that Web 2.0, or in other words "Social Network" is the most frequent keyword in both conferences. It indicates that Web 2.0 technologies are more popular rather than mobile technologies in e-government area. This phenomenon is in line with the trend of social network technology which slightly increased in that period. It shows that Web 2.0 is not just for Facebook, but ready for the serious business of e-government. This result shows an opportunity for government, practitioners and the citizen to manage Mobile and Web 2.0 effectively regarding government activities.

In the past, m-government through mobile devices has had limited facilities i.e. the limited number of characters of an SMS. Whilst, e-mail has always been able to accommodate many more characters and multimedia content, this has come of age, as of 2009, with the advent of smart phones and tablets. The limited facilities of the old mobile devices can be solved due to the convergence of smart phones and tablets with desktop PCs.

Smartphone and tablet can be used to call and send text messages. In addition, they can be used to access Internet and related facilities such as Websites, e-mail and other multimedia content. M-government is more suited for the developing world where those areas have limited access to cable Internet, but a high penetration of mobile phones, for example, Jordan, Mexico, Indonesia, Nigeria, Philippines, etc. Many people in developing countries are more familiar with mobile phone rather than Internet. (Kumar & Sinha, 2007). Examples of mobile technologies implementation in e-government activity are illustrated below (Trimi & Sheng, 2008):

1. **My Mobile Virginia Project:** Is the first m-government project in Virginia USA which this project have various features such as weather information, legislative information, lobbyist lists, election notices, tax-related information, and tourism information. All of those features can accessed by people through mobile devices.

2. **Parking Day SMS Applications:** Are used to remind drivers in Iowa who did not park their car in the right place. The drivers will received text message from the application ANS asked them to move their cars to the proper place.

3. **A Global Positioning System (GPS):** Is used to provide a mobile traffic map to inform commuters in Seattle about traffic slowdowns, traffic lights and traffic flows. So, commuters could work out the traffic situation through that mobile traffic map.

4. **My California on the Go-System:** Helps citizens get updated information about energy warnings, traffic jams and press releases from the government office. This application can be used by citizens through mobile devices.

5. **Government of Canada Wireless Portal:** Provides information about MPs' contact information, border wait time, economic indicators, passport services and also government news releases. That portal can be accessed by Canada citizen through mobile devices.

6. **SMS Applications:** Are used by London Police Department to inform citizens about security threats and emergency alerts. So, citizen had early warning system from the police.

7. **Germany Police Used Global Positioning System (GPS) in Mobile Phones:** Monitor those suspected of involvement in crime.

8. **Mobile Devices Used by Parking Inspectors in Austria:** Check whether drivers pay for parking or not. These devices had

connection to a central parking database. So, the data gathered would delivered directly to the database server.

9. **Mobile Technologies:** Are also used in Sweden to inform vacancies. It provides parking payment system, a government inspector service, tax services and mobile healthcare providers as well.

10. **Personal Identification:** Was embedded into Subscriber Identification Modules (SIM) Card Code in Finland since every mobile phone has unique SIM Card Code. Finnish government also used Electronic ID card to make transaction through mobile phone as well as it used as a travel document.

11. **M-Government:** Has been implemented for tourist information, disaster prevention and child rearing in Japan. This government also has the Vehicle Information and Communication System (VICS) which provides some information, for instance traffic congestion, road works, car accidents, parking lots and weather information.

12. **South Korea Government Implemented M-Police System:** Has helped officers accessing information about missing cars, driving licenses, vehicles' histories and pictures of drivers through mobile devices.

13. **Hong Kong Government:** Has sent text messages to six million mobile phone users to make calm them in relation to rumours of the SARS virus health scare in 2004.

14. **Singapore Government:** Implemented text message service application to remind citizens about parking ticket, national service obligations and passport renewal deadlines

The examples above do not mention additional practices in other countries all over the world. It may also be the case that some other countries have implemented m-government but no research has been conducted in relation to its implementation. The case studies above seem to indicate that it is the USA and Europe surging

ahead with m-government implementation when compared to the countries in continents such as Asia and Africa. Results from the literature review show similar diverse application and practices in m-government activities. Further investigation about m-government implementation in various developing countries and various countries in Asia would be very interesting, particularly as it would undoubtedly result in insights into the different societal context and cultural experiences/expectations associated with e-government. Although the technologies are same, the social context and culture may give rise to a different result regarding their implementation. Kumar and Sinha (2007) illustrated some issues regarding m-government as follows:

- Mobile authentication which is important to conduct standard policy for all types of device. So, authentication should not be restricted to specific devices.

- Mobile Payments: Nowadays, mobile devices not only for call and send text message but also can used as payment devices like mobile payment implemented in Europe, US and some of Asia, so government should consider to address regulations for this.

- Location-aware applications such as Global Positioning System (GPS), Google Map, Navigation emergency 911 (e911), and other technologies will allow government and its citizens to access information based on location, impacting on the activities of both. For example, citizens in the UK can easily find the place or road by just typing the postcode or address into Google maps. A citizen who wishes to locate that address can easily use on-line navigation applications to get there.

These issues may have impacts on citizens' behaviour, government regulation and policies and social culture. Further studies are important to assess the m-government's impact on people, regulation and policies. Technological change may change society's behaviour and culture as well.

Besides Mobile technologies, governments also use Web 2.0 including social media which is very popular to support government activities. Web 2.0 or Social Network sites such as Facebook, Twitter and MySpace have been used by billions of users. Nowadays, Web 2.0 is not only for friendship, but also for business, marketing, government, politics and education. This chapter captures the usage of Web 2.0 technologies in various domains (Osimo, 2008):

1. **Web 2.0 for Government:** Osimo (2008) presented Web 2.0 used in the government activity such as: Aboliamoli.eu for facilitating Regulation and law enforcement; Alaska State agencies database for cross agency collaboration; California wildfires for Service Provision; Change.org for supporting public participation for petitioning; Cyberbullying campaign for Public Communication; and Ganfyd for Knowledge Management and Human Resources.

2. **Web 2.0 for Regulation:** Some examples show the role of Web 2.0 in the regulatory process, for instance case studies of the US Patent Office in the patenting process which filtering process for patent application can be assessed by self-appointed experts. Also, in Italy, a government-backed regulation allowing mobile operators to add a charge to each new mobile phone sold. An Italian citizen could not get clarification of what this charge was for, and collected 800,000 signatures asking his government the same question. His petition was then sent to the European Commission, who outlawed the charge, changing the regulations. Nowadays, Web 2.0 facilitated participatory process in the regulation debates.

3. **Web 2.0 for Cross-Agency Cooperation:** In most cases, cooperation between different agencies or divisions is poor and Web

2.0 can be used as an option to overcome this problem. For example, CAISI – Alaska Social Services used to coordinate various social and health service providers to give service for homeless people.

4. **Web 2.0 for Knowledge Management:** One of the example is Allen and Overy International Law Firm which have 4,500 employees and offices in 19 countries. This firm managed Web 2.0 to support knowledge. Web 2.0 also used to increase effectiveness and efficiency in managing employees which distributed in separate offices.

5. **Web 2.0 for Political Participation and Transparency:** One of the main problems for the government is low public participation. UK Prime Minister Office launched E-Petition Website to facilitate citizens who submit their petition directly to the Prime Minister and petitions can signed and seen by other people. Then, the Prime Minister's office will give its response to the petition. Many politicians also use Web 2.0 to interact with their people, especially in the campaign activity.

6. **Web 2.0 for Service Provision:** The main aim of ICT used in government is to improve government service to citizen. Web 2.0 can facilitate citizen to participate actively to overcome disaster problems such as Hurricane Katarina, the Earthquake in Njgata (Japan) and wildfires in Southern California. In UK, citizen controlled school acceptance process through Web 2.0 technologies as well.

7. **Web 2.0 for Law Enforcement:** People can upload photos of cars and bikes that are parked in disabled parking spaces and bike lanes to Caughtya.org and mybikelane.org websites. Citizens can also discuss local problems such as broken paving slabs, street lighting, etc and how local authorities solve these problems through fxmystreet.com. So, Web 2.0 encouraged citizen participation

to help government for law enforcement activity.

Results from the literature review show similar diverse areas of application and practice in Government 2.0. A key conclusion is that connected e-government implementation through Mobile and Web 2.0 technologies not only considers the technology aspect but also the non-technological aspects such as Social, Political and Cultural aspects. So, comprehensive framework that covers both aspects needed in order to improve effectiveness of e-government and avoid failed implementation.

Furthermore, this chapter will examine some of the existing frameworks regarding mobile and Web 2.0 technologies on e-government. Kiki, Lawrence & Steel (2006) proposed a response management framework of m-government. The aims of this framework are to control adoption process of new technology and manage well to reduce risk as well as guarantee effectiveness, efficiency, flexibility and transparency. The Framework consists of four main elements as follows (El Kiki & Lawrence, 2006):

- **Input:** This includes challenges and opportunities factors such as: political, organisational, administrative, developmental, technological, etc.
- **Processing:** This is related to m-government management and divided into: strategic, managerial and operational.
- **Output:** This element consists of the change and innovation aspects.
- **Outcome:** This element includes Benefits and Risks factors such as political, organisational, administrative, developmental, technological, etc.

The relationship between these points will be explained below. The government needs to manage the organisation into three levels of management - the strategic, managerial and operational levels - in order to adapt to new mobile technologies.

The response to process the level of management will mean managing change and innovation. Outcome and input have a recursive relationship which means any input will affect the outcome whilst as the outcome will influence the input of the next cycle. Outcome consist of benefits and risks and both of these should be well planned (El-Kiki, Lawrence, & Steele, 2005).

One of the Web 2.0 Technologies frameworks for e-government is A Public-Private-Citizen (PC2) Collaboration Framework. This framework contains three parties such as government (Public), Profitable Companies that support public values, and people who manage access to information and get service from Citizen Relationship Management access points (Citizen). Public-Private Collaboration means some e-government projects done by public and private partnership, for instance: In United States (US), 20 states have partnership with the National Information Consortium (NIC) to develop their e-government portal. Private-Citizen Collaboration means e-government projects are done by private company for citizens, for example: a public library user in the city of Calgary, Alberta who did not find a book on the public library Website, then they can buy from Amazon through public library Website. Citizen-Public Collaboration means that citizen can use the technology provided by government, for example: Non-emergency 311 calling service in the New York City was established on the initiative of Mayor Michael Bloomberg in March 2003 (Hui & Hayllar, 2010).

In another paper, there are two frameworks to explain efforts from the government to encourage citizens into citizen-sourcing projects. The first framework illustrates multiple dimensions to classify citizen-sourcing initiatives based on contextual components. The Nam's first framework consists of three dimensions of citizen-sourcing initiatives such as purpose (image-making), collective intelligence type (professional knowledge or innovation ideas), and strategy (contest, wiki, social networking, or social voting) and the second framework is a tool to assess performance of citizen-sourcing initiatives and this framework includes design evaluation, process evaluation and outcome evaluation (Nam, 2012).

Building upon El-Kiki, Lawrence & Steel (2004) work, Hui and Hallar (2010), and also the work by Nam (2012) and the results from the literature review, the authors of this chapter are were able to propose a government-People Relationship Framework through mobile and Web 2.0 technologies as shown in Figure 6 above. Our framework is developed based on the practicalities of m-government and Government 2.0 in some countries and areas above and contains three main parts of e-government such as government, Technologies and People.

On the left-hand side of Figure 6, the government is shown to interact with people through Mobile 2.0 technologies. The relationship from government to people is informative and directive.

Informative means the government mostly uses mobile technologies to share information with people such as weather information, tourism information, tax information, traffic jams, security threats and emergency alerts, etc. Directive means the government gives direction to people to do something. For instance, in Hong Kong, the government sent text messages to make citizen calm regarding to rumours of the SARS Health scare of 2004. The relationship of people to government is mostly passive and only to access information.

These relationships are different with the government-People relationship through Web 2.0 technologies. Relationships from government to people are informative, reactive and responsive. This means the government publishes information through Web 2.0 technologies such as Facebook, Twitter, MySpace, Wikis, Blogs, etc. and gives its reaction as well as responds to tweets or comments and statuses on Facebook. People can

Figure 6. Government-people relationship framework

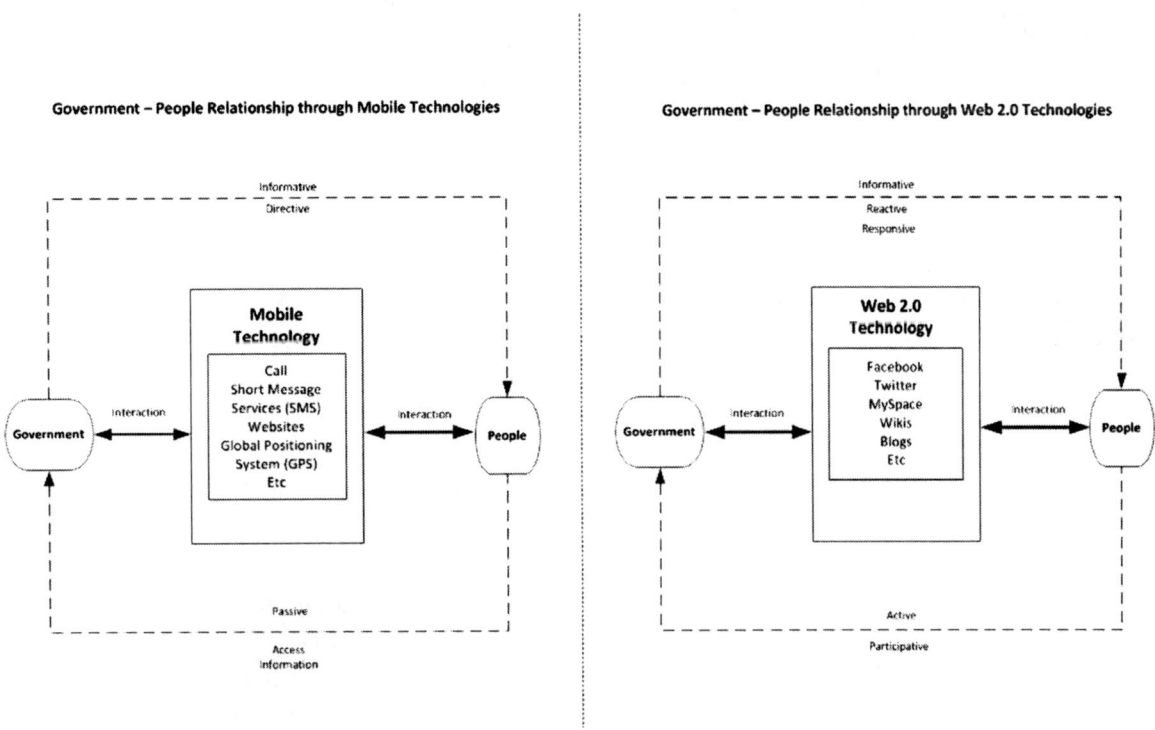

Government-People Relationship Framework
through Mobile and Web 2.0 Technologies

actively participate through Web 2.0 technologies such as making petitions, reporting, complaining, consultation, campaigning, discussing each other. So, the changing of technologies has an impact on the changing relationship behaviour between government and people.

Evaluation of Mobile and Web 2.0 Technologies in E-Government

Evaluation of information system, especially the e-government implementation, is a difficult task since evaluator should consider many aspects. The aspects consist of tangible such as cost, time and other things which can be measured as well as intangible aspects such as customer satisfaction, social and cultural impacts and other things which cannot be quantified. Alshawi and Alalwany (2009) presented some traditional approaches can be done to evaluate the e-government implementation such as Return on Investment (ROI), Cost Benefit Analysis, Payback period and present worth as well as some other approaches such as Technology Acceptance Model (TAM) and Diffusion of Innovation (DOI). Some of those approaches covered only on economic issues but some others also captured social politic issues as parameters of evaluation. Since mobile and Web 2.0 technologies are already converging, the evaluation of mobile and Web 2.0 usage in e-government can be done together. Some of the parameter to evaluate e-government are (Alshawi & Alalwany, 2009) are explained in the following paragraphs.

Technical Issues

Technical issues consist of two evaluation parameters viz: performance and accessibility. Performance parameter can be measured from effectively of service and personalized information and services. Moreover, accessibility parameter can be measured from efficiently of user interface and disability access and language translation.

Mobile and Web 2.0 technologies have good performance indicator good which users can send messages, information or complaints to government via mobile based and Web 2.0 applications. government also can give information, direction and respond via mobile and Web 2.0 applications. Nowadays, many governments provide mobile phone number, text messages number, facebook, twitter and others applications for citizen services. This mobile-based and Web 2.0 facilities are complement for landline phone facilities use personal information and services from their own mobiles. Mobile and Web 2.0 technologies are more effective because people have access to the relevant technology and can use the facilities any time anywhere. People also can interact with each other using such technologies. Mobile and Web 2.0 technologies also have many applications which are personalised informations and services. Mobile and Web 2.0 technologies are also accessible, especially for users in developing countries which have high penetration of mobile rather than Internet access via cable. There are also language translation features available, so users can set up their chosen languages on the mobile device, but not all applications have language translation. On the other hand, this technologies still have limited features and services for disability users, so this issue should be considered for technology developers. Overall, mobile and Web 2.0 technologies have been used in e-government activities effectively but it needs improve continuously especially applications for facilitating disability citizen.

Economic Issues

Economic issues are about cost saving, especially how much money saved and how many times the saving was achieved. Through mobile and Web 2.0 technologies, government can save a lot of money and time rather than only using paper based systems. In the past, governments only used paper for dissemination of information and in facilitating the citizens affairs. government spent a lot of money for paper-based process and took time. Nowadays, government can disseminate information cheaply and quickly via text message, mobile-based Website and Web 2.0. Through these medias, government also got quickly respond from people, especially via the Web 2.0 applications, such as facebook, twitter, etc.

Social Issues

Social issues consist of evaluation of openness, trust issues and user perception about ease of use and usefulness. Web 2.0 technologies such as facebook, twitter, etc which can be accessed from mobile devices as well, address the issues about openness. government can updates information or direction to citizen about some issues via facebook and twitter, then everybody can access this information and give responds as long as they are member of the facebook groups or followers in government twitter accounts. Everyone else can also access or give others comments and respond to comments as well. So, interaction in the Web 2.0 not only from government to citizen, but also citizen to citizen. Recently, many governments and people are using Web 2.0 (accessed from PC or mobile devices) for government and people affairs. This phenomenon show that Web 2.0 (includes via mobile) have addressed trust, ease of use and usefulness issues.

Evaluation of mobile and Web 2.0 technologies should be done using qualitative and quantitative methods in order to get relevant results. Some qualitative methods can be used such as in-depth interview and focus group discussion to understand what the challenges and possible solutions to overcome the problems. Moreover, quantitative methods such as statistical analysis based on questionnaire can be done to monitor mobile and Web 2.0 technologies usage in e-government activities.

CONCLUSION

This chapter has presented our research results regarding a base knowledge of e-government such as e-government research methodology and evaluation, e-government keywords, themes and issues. It has also captured the practicalities and frameworks for Mobile and Web 2.0 Technologies for e-government developments. Our analysis was based on a focussed literature review from European Conference on E-Government (ECEG) and International Conference on E-Government (ICEG). The investigation shows the diversity and breadth of e-government activities taking place around the world. One inference is that the latest technologies such as mobile and Web 2.0 are becoming an increasingly important part of e-government. The framework developed in this chapter captures the relationship behaviour between government, technology and people such as directive, active, reactive, responsive and informative.

Case studies and surveys are dominant methods used in e-government research. From the comparison between Heeks and Bailur (2007) and Pedro and Bailur (2010), and also the literature review, it can be seen that e-government research uses various research philosophies, methodologies and methods from the extreme continuum positivist and social constructivist or pure qualitative and pure quantitative to mixed and compromise of both. So, e-government is emerging as a mature discipline. Furthermore, there are e-government issues with respect to E-Participation, Local government, Identity Management and E-Governance. There are issues that do not change such as management, technology, finance, politics, health, policy, governance. These unchanged issues are the fundamental issues of e-government.

The practical use of mobile technologies in e-government in some countries suggests that the Government mostly uses them to share information and to give direction to people, but Web 2.0 technologies can encourage people to participate actively in government. So the different technologies used have different social behaviour, as showed in the Government-People Relationship Framework proposed by the authors of this chapter. Through mobile technologies, the relationship formed (government to people) are informative and directive. Directive which means government give the people a direction to do something.

These relationships are different with the government-People relationship through Web 2.0 technologies. Relationships from government to people are informative, reactive and responsive. This means the government publishes information through Web 2.0 technologies such as Facebook, Twitter, MySpace, Wikis, Blogs, etc. and gives its reaction as well as responds to tweets or comments and statuses on Facebook. People can actively participate through Web 2.0 technologies such as making petitions, reporting, complaining, consultation, campaigning, discussing each other. So, the changing of technologies has an impact on the changing relationship behaviour between government and people.

Evaluation of mobile and Web 2.0 technologies should consider three issues such as technical, economic and social issues. From evaluation of those issues show that mobile and Web 2.0 technologies are addressing those issues, but applications for disabled people need to be more developed to make mobile and Web 2.0 technologies more accessible.

FUTURE RESEARCH

Research is required in the future pertaining to assessing the impact of m-government and Web 2.0 on e-government to people, regulation and policies as well as further research about mobile and Web 2.0 on e-government in countries or areas other than Europe and US. This is because most research papers cover only European and US Case studies. Further research is also needed in the E-Participation area combining mobile and Web 2.0 technologies. E-Participation issues have increased in the e-government and e-governance areas. This also requires further investigation.

REFERENCES

Alshawi, S., & Alalwany, H. (2009). E-government evaluation: Citizen's perspective in developing countries. *Information Technology for Development*, *15*(3), 193–208. doi:10.1002/itdj.20125

Bannister, F., & Connoly, R. (2010). Researching eGovernment: A Review of ECEG in its Tenth Year. In *Proceedings of European Conference on E-Government* (ECEG) (pp. 53-62). Academic Publishing Limited.

El Kiki, T., & Lawrence, E. (2006). *Government as a mobile enterprise: real-time, ubiquitous government.* Paper presented at the Information Technology: New Generations, 2006. New York, NY.

El-Kiki, T., Lawrence, E., & Steele, R. (2005). A management framework for mobile government services. In Proceedings of CollECTeR. Sydney, Australia: CollECTeR.

Grönlund, Å., & Horan, T. A. (2004). Introducing e-gov: history, definitions, and issues. *Communications of the Association for Information Systems*, *15*, 713–729.

Heeks, R., & Bailur, S. (2007). Analyzing e-government research: Perspectives, philosophies, theories, methods, and practice. *Government Information Quarterly*, *24*(2), 243–265. doi:10.1016/j.giq.2006.06.005

Hui, G., & Hayllar, M. R. (2010). Creating Public Value in E-Government: A Public-Private-Citizen Collaboration Framework in Web 2.0. *Australian Journal of Public Administration*, *69*(s1), S120–S131. doi:10.1111/j.1467-8500.2009.00662.x

Jaeger, P. T., & Thompson, K. M. (2003). E-government around the world: lessons, challenges, and future directions. *Government Information Quarterly*, 20(4), 389–394. doi:10.1016/j.giq.2003.08.001

Kumar, M., & Sinha, O. P. (2007). *M-government–mobile technology for e-government*. Paper presented at the International conference on e-government. New Delhi, India.

Nam, T. (2012). Suggesting frameworks of citizen-sourcing via Government 2.0. *Government Information Quarterly*, 29(1), 12–20. doi:10.1016/j.giq.2011.07.005

Osimo, D. (2008). *Web 2.0 in government: Why and how*. Institute for Prospectice Technological Studies (IPTS), JRC, European Commission, EUR, 23358.

Rodríguez Bolívar, M. P., Alcaide Muñoz, L., & López Hernández, A. M. (2010). Trends of e-Government research: contextualization and research opportunities. *The International journal of digital accounting research, 10*(16), 6.

Signore, O., Chesi, F., & Pallotti, M. (2005). *E-government: challenges and opportunities*. Paper presented at the Proceedings of the CMG Italy XIX annual conference. Rome, Italy.

Trimi, S., & Sheng, H. (2008). Emerging trends in M-government. *Communications of the ACM, 51*(5), 53–58. doi:10.1145/1342327.1342338

ADDITIONAL READING

Abramowicz, W., Bassara, A., Filipowska, A., Wiśniewski, M., & Żebrowski, P. (2006). Mobility implications for m-government platform design. *Cybernetics and Systems: An International Journal, 37*(2-3), 119–135. doi:10.1080/01969720500428255

Al-Khamayseh, S., Hujran, O., Aloudat, A., & Lawrence, E. (2006). Intelligent m-government: application of personalization and location awareness techniques. *Proc. of the EURO m-GOV*.

Antovski, L., & Gusev, M. (2005). M-government framework. *Proceedings EURO mGov*, 10-12.07.

Bonsón, E., Torres, L., Royo, S., & Flores, F. (2012). Local e-Government 2.0: Social media and corporate transparency in municipalities. *Government Information Quarterly*, 29(2), 123–132. doi:10.1016/j.giq.2011.10.001

Chang, A.-M., & Kannan, P. (2008). *Leveraging Web 2.0 in government*. IBM Center for the Business of Government.

Chun, S. A., Shulman, S., Sandoval, R., & Hovy, E. (2010). Government 2.0: Making connections between citizens, data and government. *Information Polity, 15*(1), 1–9.

de Kool, D., & van Wamelen, J. (2008). *Web 2.0: a new basis for e-Government?* Paper presented at the Information and Communication Technologies: From Theory to Applications, 2008. ICTTA 2008. 3rd International Conference on.

El Kiki, T., & Lawrence, E. (2006). *Government as a mobile enterprise: real-time, ubiquitous government*. Paper presented at the Information Technology: New Generations, 2006. ITNG 2006. Third International Conference on.

European Conference on E-Government (ECEG). (2013). Retrieved from http://academic-conferences.org/eceg/eceg2013/eceg13-home.htm, Accessed Oct 2013.

Freeman, R. J., & Loo, P. (2009). *Web 2.0 and E-Government at the Municipal Level*. Paper presented at the Privacy, Security, Trust and the Management of e-Business, 2009. CONGRESS'09. World Congress on.

Gasco, M. (2012). Proceedings of the 12th European Conference on e-Government. Academic Conferences Limited

Ghyasi, A. F., & Kushchu, I. (2004). M-Government: cases of developing countries. *M-GovLab, Intl. Univ. of Japan. Electronic references Retrieved Sep, 19*, 2009.

International Conference on E-Government (ICEG). (2013). Retrieved from http://academic-conferences.org/iceg/iceg2013/iceg13-home.htm, Accessed Oct 2013.

Klun, M., Decman, M., & Jukic, T. (2011). Proceedings of the 6th International Conference on e-Government. Academic Conferences Limited

Kumar, M., Hanumanthappa, M., & Reddy, B. L. (2008). Security issues in m-government. *International Journal of Electronic Security and Digital Forensics, 1*(4), 401–412. doi:10.1504/IJESDF.2008.021457

Lavin, M. (2009). Proceedings of the 5th International Conference on e-Government. Academic Conferences Limited

Meijer, A. J., Koops, B.-J., Pieterson, W., Overman, S., & ten Tije, S. (2012). Government 2.0: Key Challenges to Its Realization. *Electronic. Journal of E-Government, 10*(1), 59–69.

Mengistu, D., Zo, H., & Rho, J. J. (2009). *M-Government: Opportunities and challenges to deliver mobile government services in developing countries*. Paper presented at the Computer Sciences and Convergence Information Technology, 2009. ICCIT'09. Fourth International Conference on.

O'Donnell, D. (2010). Proceedings of the 10th International Conference on e-Government. Academic Conferences Limited

Remenyi, D. (2007). Proceedings of the 7th European Conference on e-Government. Academic Conferences Limited.

Remenyi, D. (2007). Proceedings of the 3rd International Conference on e-Government. Academic Conferences Limited

Remenyi, D. (2008). Proceedings of the 8th European Conference on e-Government. Academic Conferences Limited

Remenyi, D. (2008). Proceedings of the 4th International Conference on e-Government. Academic Conferences Limited

Remenyi, D. (2009). Proceedings of the 9th European Conference on e-Government. Academic Conferences Limited

Rossel, P., Finger, M., & Misuraca, G. (2006). Mobile e-Government options: between technology-driven and user-centric. *The electronic. Journal of E-Government, 4*(2), 79–86.

Ruhode, E. (2010). Proceedings of the 6th International Conference on e-Government. Academic Conferences Limited

Sandy, G. A., & McMillan, S. (2005). *A Success Factors Model For M-Government'*. Paper presented at the Euro mGov.

Signore, O., Chesi, F., & Pallotti, M. (2005). *E-Government: challenges and opportunities*. Paper presented at the Proceedings of the CMG Italy XIX annual conference.

Tapscott, D., Williams, A. D., & Herman, D. (2007). Government 2.0: Transforming government and governance for the twenty-first century. *New Paradigm, 1*.

Zappen, J. P., Harrison, T. M., & Watson, D. (2008). *A new paradigm for designing e-Government: web 2.0 and experience design*. Paper presented at the Proceedings of the 2008 International conference on Digital government research.

KEY TERMS AND DEFINITIONS

ECEG: European Conference on E-Government. This is a yearly conference organised by Academic Conference Limited.

E-Government: Subject or area or domain which using technology into management of government to achieve the goal effectively and efficiently.

Focus Group Discussion (FGD): Discussion by a group of participant about some topic to share ideas between each other.

Framework: Figure consist of relationship between the real or conceptual components to describes something useful.

Government 2.0: Government supported by web 2.0 technologies to achieve the goal effectively and efficiently.

Government-People Relationship: Relationship between government to people through technology to achieve common goal between government and people.

ICEG: International Conference on E-Government. This is a conference organised by Academic Conference Limited.

Keywords: main keywords in a paper or reference which describe the content of a paper or a reference.

M-Government: Government supported by mobile technologies to achieve the goal effectively and efficiently.

Mobile Technologies: Technology that using features related to mobile devices such as: short message service (SMS), phone call, multimedia message service (MMS), mobile Internet, etc.

Research Methodologies: Step by step methods that used in the research to gain knowledge, so the knowledge can be validated.

Themes: Issues, main points or topics that describe main ideas of something or phenomenon or domain.

Web 2.0: Technology that related to Social Network Sites such as: Facebook, Twitter, MySpace, etc.

Chapter 6
Next Generation E-Government:
Reconciling the E-Participation and Data Protection Agendas

Maria Moloney
Escher Group Ltd., Ireland

Gary Coyle
Escher Group Ltd., UK

ABSTRACT

The evolving model of the Future Internet has, at its heart, the users of the Internet. Web 2.0 and Government 2.0 initiatives help citizens communicate even better with their governments. Such initiatives have the potential to empower citizens by giving them a stronger voice in both the traditional sense and in the digital society. Pressure is mounting on governments to listen to the voice of the public expressed through these technologies and incorporate their needs into public policy. On the other hand, governments still have a duty to protect their citizens' personal information against unlawful and malicious intent. This responsibility is essential to any government in an age where there is an increasing burden on citizens to interact with governments via electronic means. This chapter examines this dual agenda of modern governments to engage with its citizens, on the one hand, to encourage transparency and open discussion, and to provide digitally offered public services that require the protection of citizens' private information, on the other. In this chapter, it is argued that a citizen-centric approach to online privacy protection that works in tandem with the open government agenda will provide a unified mode of interaction between citizens, businesses, and governments in digital society.

INTRODUCTION

Social media or Web 2.0 technologies refer to a collection of technologies which help individuals to become active participants in activities like creating, editing, sharing and rating Web content as well as helping them to form social networks by interacting and linking with each other. Web 2.0 technologies include technologies such as, blogs, wikis, social networking sites, photo-sharing, video and audio sharing, podcasting and many more (Chun S. A., Shulman, Sandoval, & Hovy,

DOI: 10.4018/978-1-4666-6082-3.ch006

2010). The purpose of Government 2.0 initiatives is to use social media to facilitate communication between relevant parties of a community, thereby helping to ensure an e-community is developed more efficiently and effectively (Haughwout, 2009).

However, in the rush to embrace government 2.0, it must not be forgotten that governments must continue to provide *good governance*. Good governance, broadly speaking is about delivery of public sector services in a way that reflects as closely as possible accepted public administration values such as efficiency, security, fairness, integrity, and honesty (Bannister & Connolly, 2011). Government 2.0, while beneficial for enabling citizen participation in many ways, does not provide solutions for communicating with governments when there is a need to keep the information being submitted private and secure. Protecting a citizen's personal information is not only necessary for completing many online eGovernment services, for example submitting tax returns and invoices to the government, but it is increasingly being seen as an essential element for safeguarding the digital identity of citizens. With the ever increasing pervasiveness of ICTs in modern life, governments have a combined responsibility to: 1) engage their citizens online and 2) continue to provide good governance in the form of efficient public services. This includes services that require citizens to submit private information in order to avail of electronic public services.

This chapter explores how governments can embrace Web 2.0 technologies, while continuing to meet the challenges of protecting citizens' informational privacy when providing them with electronic public services. The next section gives an overview of how electronic government has evolved over the years to encompass government 2.0 and m-government. This is followed by a discussion of governments' obligations towards good governance and protecting their citizens informational privacy. The chapter then gives an overview of a conceptual model to guide govern-

ments when adapting their existing e-government function to incorporate a more holistic approach to e-participation of citizens. Following this, details and results of a case study, which uses this conceptual model as a guide for encouraging e-participation are provided. The chapter concludes with a section on future research.

FROM E-GOVERNMENT TO GOVERNMENT 2.0

The e-Government field emerged in the late 1990´s as a context within which to share experiences among public sector practitioners (Grönlund & Horan, 2004). However, the history of computing in government organizations can be traced back to the beginnings of computer history. A literature on "IT in government" goes back at least to the 1970s (Kraemer, Danziger, & King, 1978). This earlier literature focused on IT use within government, while the more recent e-Government literature often focuses on IT use between government departments and the larger community, such as the provision of government services to the citizens (Ho, 2002). While some earlier e-Government issues, such as office automation, may not be highly relevant to research today, many issues still are, for example decision making, service processes, and values (Grönlund & Horan, 2004).

Most definitions of e-Government go beyond the notion of simply providing services to the citizen, to include the use of IT to facilitate both organizational change and the role of government. The e-government literature acknowledges that ICTs, by their nature, can streamline how governments conduct their business and how they communicate with their citizens and clients (Ho, 2002). In this capacity, ICTs serve as useful tools for governments wishing to reform and modernize. At the same time, reforms in public administration often require a review of the administrative, policy and regulatory foundations upon which e-government is built. These reforms in e-government can often play a powerful role in changing the administration of government services and operations while simultaneously advancing economic development and the e-capabilities of citizens (Culbertson, 2005). Therefore, two strands of e-government literature (e-government for e-services and e-government for organizational change and the role of government) need to be considered together as the basis of the e-Government field (Grönlund & Horan, 2004).

Previously, the information flow between citizens and e-government was traditionally unidirectional, flowing from government websites to citizens, with limited feedback mechanisms. The advent of Web 2.0 technologies has changed this flow and has enabled citizens to participate and interact more online with both government and other citizens through such innovations as Facebook and Twitter (Chun S. A., Shulman, Sandoval, & Hovy, 2010). Facebook's Wall has, in the past, been used as a public sphere for political and social discourse. Examples of this range from the 2008 U.S. Presidential campaign where candidates used their Facebook Wall to reach and interact with voters (Robertson, Vatrapu, & Medina, 2010), to where social Web has been said to enable social action in countries such as Tunisia and Egypt where a Facebook page was said to have helped spark social unrest.

Another driver of e-government evolution has been the increase in governments opening their public sector information to the public for the benefit of its citizens. In its 'Digital Agenda for Europe' (2010), the European Commission identified the re-use of public sector information, alongside fast and ultra-fast Internet access, as key to delivering a digital single market. In late 2011, the European Commission launched an 'open data' strategy for Europe, which is expected to deliver a €40 billion boost to the EU's economy each year. The EU open data strategy will be rolled out in three directions: firstly the Commission will open its vaults of information to the public for free through a new data portal. Secondly, a level playing field for open data across the EU will be established. Finally, these new measures are backed by the €100 million which will be granted in 2011-2013 to fund research into improved data-handling technologies. These actions are aimed at positioning the EU as the global leader in the re-use of public sector information and are intended to boost the already thriving industry that turns raw data into the knowledge needed to drive Government 2.0 tools, such as smart phone apps which provide maps, real-time traffic and weather information, price comparison tools and more (European Commission, 2011).

In the UK, there is a whole series of Government 2.0 style initiatives aimed at directly increasing public participation and engagement and revitalizing the link between the citizenry and their government. The Prime Minister's Office is using social media technologies like YouTube, e-zines and e-petitions to connect with their citizens, while at a local level social media technologies involve innovative deliberative formats, which seek to enroll a range of actors in new and extended projects of government (Morisonn, 2010).

The open government initiative of the US federal government intends to use Web 2.0 technologies to achieve their aims of increased collaboration, participation and transparency in government (Haughwout, 2009). Figure 1 shows how these three principles are achieved with government 2.0 technologies to bring about a more open government.

Newman and Clarke (2009) argue that the creation of the participating citizen and the enlisting of the citizenry in e-government is both a governmental strategy and a political project. They claim that the UK government has turned this idea into a method for governing by building it into systems, expressing it in practices, and deploying it as a way of operationalising a new, individualised, consumer culture of public services.

The British government 2.0 project draws upon what Newman and Clarke (2009) describe as 'the power to constitute individuals, house-

Figure 1. Process for creating the next generation of government or open government

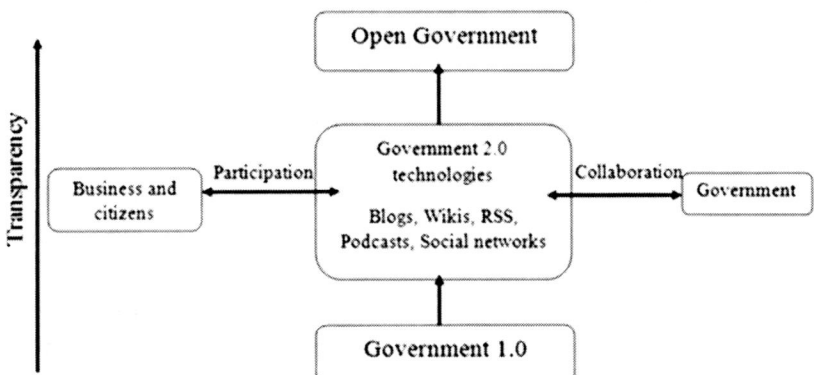

holds, communities, social entrepreneurs, NGOs, public organisations, businesses, and voluntary organisations as active partners in addressing many of the critical policy agendas that confront governments' (Newman & Clarke, 2009, p. 15). An example of this is when 'communities' are constituted and formed from the debates and practices of government. Community activists are enlisted to determine community needs that will in turn become community plans that community partnerships can develop. The idea of community acts not only as a means of mapping and managing difference, but is itself a particular resource that has special authority in terms of the attention and engagement it commands and the resources and capacities it can deploy (Morisonn, 2010).

Government 2.0 initiatives like that of the UK need extensive technology support for citizen participation. In addition, all political information available through these technologies must be easy for citizens to access to ensure a fully informed discussion and debate. This debate should, in turn, influence the collaborative decision making process between citizens and their political representatives, resulting in the inception of new political policies. This iterative process creates a type of virtuous cycle of policy and process improvement.

With regard to Government 2.0, such collaboration from citizens requires the formulation of partnerships and cooperation not only across all

levels of government but also between government and public-private partnerships, non-profit organizations, enterprise, and individuals. Government agencies should be encouraged to feed into and enlarge this virtuous cycle of improvement by disclosing information about their operations and decisions in formats that the public can readily access, understand and use. Participation of this type not only provides the government with the collective knowledge, ideas, and expertise of the population but it also enhances the government's effectiveness, improves the quality of its decisions, and promotes accountability (Chun, Shulman, Sandoval, & Hovy, 2010).

In parallel to the increase in popularity of Government 2.0 and social media technologies for government, policymakers increasingly recognise that many of the solutions to major social challenges, i.e. tackling climate change or improving public health, need to be much more local. Local solutions are frequently very effective, as they reflect the needs of specific communities and engage citizens in taking action. Added to this they are often cost-effective, since they provide a conduit for the resources of citizens, charities or social enterprises to complement those of the state. Given the growing pressure on government finances, these are important considerations (Bunt & Harris, 2010). When successfully achieved, this type of collaboration not only improves the effectiveness of service provision, it also enables the public to become proactive agents that create, organise, combine and share content. It places emphasis on the notion of *the wisdom of crowds* approach, a phenomenon explained by Surowiecki (2004), that the most reliable and accurate information is most often created within the crowd, outside the boundaries of the organisation.

Once, these virtuous cycles have been set up, service providers need to develop appropriate strategies for their continued use. Social media technologies can be used for achieving change in many aspects of an organisation, such as service provision, decision and policy making, administration, and management, but committing to this change over a longer term is often the most challenging task in which to succeed.

These governmental initiatives with social media are still in their infancy and constantly evolving and interacting via social media often introduces new challenges related to privacy, security and data management (Bertot, Jaeger, & Hansen, 2012). The current diversity of social computing environments, i.e. the need to combine Internet browsers with mobile devices, and various software products to access content. This complexity combined with the speed at which technology is changing creates a challenging environment for researchers and developers to incorporate privacy enhancements into every evolution of the technology. In such an organic and dynamic environment there is potential that private information is inadvertently made public. History shows that human fallibility has resulted in large amounts of sensitive data being inadvertently leaked to the public (Privacy Rights Clearinghouse, 2011).

Smith et al. (2011) argue that citizens' informational privacy worries are grounded in the growing "art of the possible." Citizens' privacy concerns are triggered by the seemingly boundless options for collecting, processing, distributing, and using personal information. In light of this, there is growing recognition among governments worldwide that a secure Internet that respects privacy will promote sustainable innovation (Federal Trade Commission, 2010; European Commission, 2010).

Investigations into the protection of sensitive private information within e-government systems and when providing public eservices is a growing area of research in the e-government domain. The next section examines the tension between the growing need for governments to engage with citizens by provide good, innovative and effective eservices while ensuring the private information of their citizens remains safe.

E-Participation and Privacy Protection

Public governance is a dynamic activity. Good governance encourages public trust and participation that enables services to improve; bad governance fosters low morale and adversarial relationships that lead to poor performance (The Independent Commission for Good Governance in Public Services, 2004). Generally speaking, since the mid-twentieth century, there has been a decline of public trust in government (Welch, Hinnant, & Moon, 2005; Morgeson, VanAmburg, & Mithas, 2011). This decline implies the loss of public confidence in political and administrative performance as well as dissatisfaction with public services.

Any breaches of citizens' personal information can only add to the decreasing trust in government. There have been many breaches of information privacy in recent years by not only government but also private enterprise (Pavlou, 2011). One of the more significant governmental breaches is what has come to be known as "the Snowden Affair" where a former employee of the Central Intelligence Agency (CIA) in the US, Edward Snowden, leaked details of several top-secret U.S. and British government mass surveillance programs to the press. In mid-2013, several news outlets reported that the United States, with the aid of Australia and New Zealand, have been spying on domestic and international communications on a much larger scale than previously thought. It was later revealed that France and the United Kingdom have also been spying. Based on documents provided by Edward Snowden, these media reports revealed that espionage activities conducted by US and UK intelligence agencies targeted not only foreign countries but also US citizens as well as US allies from NATO and the European Union (The Guardian Newspaper, 2013).

This surreptitious mass surveillance undermines trust in governments as they are seen by the public to be carrying out covert operations about which they do not wish the public to know. This mass surveillance operation was a major contravention to the EU data protection laws and arguably affected citizen trust in governments within Europe regarding their ability to safeguard private information. The protection of citizens' personal information has been shown to impact upon citizens level of trust in their government (Welch, Hinnant, & Moon, 2005; Cullen & Hernon, 2004).

However, if governments are to protect their citizen's informational privacy against such mass surveillance, understanding what that entails is the essential first step of the process. Research literature in the area has long acknowledged that protecting personal information is a complex and multi-dimensional issue (Information and Privacy Commissioner Canada, 2010; Etzioni, 1999). A constant challenge for privacy professionals is that the very definition of privacy is constantly being examined and redefined. In 1967, Westin (1967) declared privacy as the claim of an individual

to determine what information about himself or herself should be known to others. More recently, privacy has come to be seen as having no inherent definition, rather different social groups and disciplines have developed different meanings and interpretations of the concept (Whitley, 2009). Table 1, adapted from Smith et al. (2011), summarizes the approaches to defining privacy that can be found in various disciplines. These approaches are classified into either value-based or cognate-based definitions. The value-based definitions view privacy as a human right integral to society's value system. Historically, this was the first approach to defining privacy. Psychologists and cognitive scientists then became interested in a cognate-based conceptualization of privacy which was related to the individual's mind, perceptions and cognition rather than to an absolute value or norm. Since the state of withdrawal rests within the physical or information space, scholars argued that privacy was about control of physical space and information.

The concept of privacy as a state was introduced by Westin (1967) who defined privacy as having four distinct sub-states: anonymity, solitude, reserve, and intimacy. Later, Schoeman (1984) defined privacy as a state of limited access to a person. Laufer and Wolfe (1977) conceptualized privacy as a concept (state) tied to concrete situations with three dimensions: self-ego, environmental, and interpersonal. When privacy is viewed as a state, it is logical for researchers to consider it in terms of its role as a sought-after goal. The implication is that there must be a continuum of

Table 1. Numbers of articles defining privacy as analysed by Smith et al. (2011)

	MIS	Philosophy, Social, & Political Sciences	Law	Psychology & Marketing	Economics
Cognate based: Control	23	16	0	12	1
Cognate based: State	4	6	0	0	0
Value based: Commodity	17	10	1	2	4
Value based: Right	12	24	25	4	0

states of privacy, from absolute to minimal (Smith, Dinev, & Xu, 2011).

As previously outlined, an interesting development is the growing recognition that an individual's understanding of what constitutes the modern concept of privacy includes having *control* over personal data (Westin, 1967; Introna, 1997; Van Dyke, Midha, & Nemati, 2007). Whitley (2009) goes a step further and argues that "the understanding of...control that frequently emerges is an impoverished version based on earlier understandings of technology" (p 3). He argues that the notions of consent and the revocation of given consent form the basis for a new understanding of citizen-centric control of personal data. This perspective of citizen centric control of personal data is also prevalent in the literature of the European Union's Internet strategy for the next decade (European Commission, 2010). As a result, a number of EU funded projects are looking into the idea of giving control of personal information to the citizens that produce that information (ABC4Trust Project, 2009; PICOS, 2007; Primelife, 2008).

This move towards empowering citizens by giving them control of their personal information has implications for e-government. In order to encourage and foster e-participation, governments must find ways to engage their citizens. This can be done in a number of ways. One such possibility has already been discussed here i.e. the use of web 2.0 technologies. Another possibility, when considering information that citizens wish to keep private, is to ensure citizens are in control of their personal information when interacting with their government. Being in control will reduce their informational privacy concerns and thus provide a more favourable environment for participation (Cullen & Hernon, 2004).

Providing a framework, within which citizens not only make use of Web 2.0 technologies but can also enjoy a level of security which allows them to safely transact with their government, is an important part of a citizen-centric government 2.0 programme. From this government 2.0 framework, citizens should be able to pay taxes, send e-contracts and e-invoices and *own and manipulate* their private data records.

A CITIZEN-CENTRIC GOVERNMENT 2.0 PROGRAMME

An important element in determining the appropriate level of data openness and protection in Government 2.0 is getting the balance right between satisfying conflicting societal needs like encouraging and expanding open government while continuing to provide an efficient and secure public administration service. Figure 2 outlines these sometimes conflicting agendas for e-government.

Figure 2. Forces influencing government 2.0

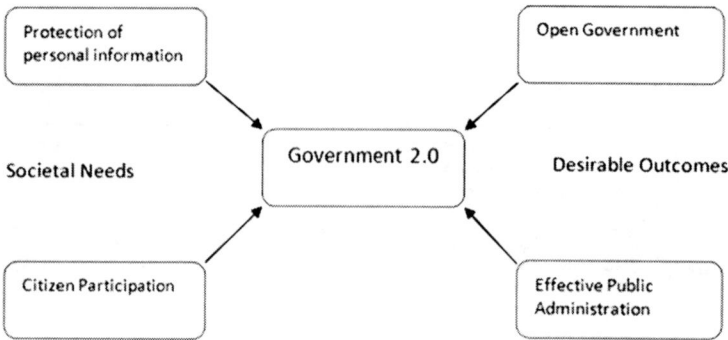

Steve Jobs of Apple Inc. was once quoted as saying "open systems don't always win," and he tried to reframe the debate regarding *open systems versus closed systems* to one of *fragmented systems versus integrated systems*, arguing that Apple iOS provides an integrated system to its customers allowing them to have a better user experience as opposed to the fragmented offering Google has with their Android system (Schinfeld, 2010).

Such an integrated system, when deployed locally by government, would provide an open space for government and citizens to interact. Additionally, due to its integrated nature, it would also have to provide a place for citizens to securely submit personal data to their government whenever necessary. It could achieve this dual purpose by authenticating all members of the integrated local system at login and by encrypting private data during transmission and when stored within the system.

Authentication is used for verifying that an agent within any system is genuine. It provides access to a digital identity or profile (Camenisch & Pfitzmann, 2007). Notions of identity and authentication are fundamental concepts in every marketplace or community. People and institutions usually need to get to know one another before conducting business. In traditional governmental/citizen interaction, people rely on physical credentials, such as a passport or a social security number as proof of identity. Proving ones identity online is still a challenge for government, businesses and citizens alike.

The Postal Service as the Continued Face of the Government

In the UK, the Government is positioning the Post Office network as "Front Office for Government" for the delivery of face-to-face government services and helping citizens interact with Government online. With 11,500 branches visited by more than 20 million customers each week and 93% of UK population living within 1 mile of a branch, the Post Office has the trust and

the reach to operate these integrated community platforms on a national scale. Furthermore, with 40% of the citizens over 55 using the Post Office weekly and 55% of branches located in rural areas, Post Offices are well placed to assist in bringing digital services to the most vulnerable members of society – the digitally and socially excluded.

By collaborating with post offices, local governments can offer a secure digital system for citizens to interact with them on multiple levels, i.e. through these integrated and collaborative platforms which both facilitate open government and public e-service provision through a secure encrypted controlled network, which uses multifactor authentication to ensure safe and secure transfer of data. This multifactor authentication is achieved by combining online user registration on the platform with physical authentication, through passport and address verification via local Post Offices (Sheedy & Moloney, 2013).

Currently, in countries across the globe, many postal operators and governments work together to use the large number of national postal outlets as delivery points for government services. There are naturally significant differences in terms of the historical, cultural and socio-economic contexts of national and local government across countries, and therefore there is no universal approach by government bodies for collaborating with the postal service in delivering government services. Postal operators for their part vary in their overall strategic vision, although generally they are looking for ways to increase retail network revenues and are therefore open in principle to the provision of more government services.

Customers worldwide want to access products and services when, where and in a way they find appropriate and convenient. The availability of postal outlets in most towns and villages across a country makes them a logical point of governmental services for citizens (Triangle Management Services Limited, 2011). In addition, many government agencies across the globe are experiencing budget constraints and resource reductions. Many of these agencies have expensive field office network structures that are often inaccessible to many citizens. Due to the extensive national network offered by the national postal service, government agencies can cut the cost of their existing field office networks by providing their government services through local post offices. The United States Postal Service Office of the Inspector General (2013), in a recent report, outlined five categories of governmental services that the postal service is currently suited to handle:

- The Postal Service could combine existing applications such as electronic postmarks with secure electronic messaging and digital-physical hybrid services to authenticate government communications and transactions.
- Through its vast retail network, the postal service could facilitate the transition of government transactions online by offering digital and in-person identification services.
- Government agencies that require front office personal contact could utilize the Postal Service's national retail network for applications, status changes, and in-person witness certifications.
- The postal service retail network could serve as an enrolment and cash redemption or cash top up channel for agencies that issue prepaid cards. The Postal Service could also provide postal money orders and its own prepaid cards on behalf of other agencies, which citizens could use for secure refunds, loan and grant proceeds, and benefit or entitlement payments.
- National efforts to expand broadband availability could be advanced by the postal service by providing convenient access points via their post offices in underserved communities, as well as aerial access to expand the broadband umbrella (US Postal Service Office of the Inspector General, 2013).

A recent study conducted for Consumer Focus Scotland (Triangle Management Services Limited, 2011) found that a key factor for success in the delivery of any government service through the postal system lies in providing the staff and licensees responsible for the front line service with the knowledge and skills to deliver accurate, timely products and services. Ongoing training, responsiveness to change and maintenance of service quality are vital to customers and the service owners, i.e. the government. Provision of governmental services has been in many countries and could continue to be for many posts a viable source of revenue, with minimal setup costs for either party, the government or the postal service.

The Postal Service and Informational Privacy Protection

Postal systems which possess the eight characteristics, outlined in table 2 below, have provided the citizens of most nations of the world with a reliable and private means of communication for over a century and a half (Church, Moloney, & Bannister, 2013). By combining these eight characteristics citizens can exercise an increased degree of choice about the privacy of their personal communications as opposed to the choice available to them currently through email services.

Three factors underpin this increased level of choice. Firstly, the legal protection of an individual's own home prevents third parties from knowing when individuals open and read their mail. It is not (easily) possible to observe an individual's actions in their own home though interested parties may be able to discover whether or not an individual's mail has been delivered.

Secondly, in traditional systems of communication, citizens can choose to remain anonymous when going about their daily lives, until they choose otherwise. An individual's method of written communication is disconnected from other areas of their lives such as shopping, meeting with friends or going to work. Knowing an

Table 2. The 8 characteristics of the postal service (Church, Moloney, & Bannister, 2013)

Universal Postal Service Characteristics	Description
Regular and reliable	Daily delivery, the location of post offices in every town
Inclusive, open and universal	The postal system is open to everybody. Anybody can send and receive mail to and from anywhere.
Delivered to a specified, private location	The customer can have mail delivered to their work or home, even if living remotely.
Delivered within the state, at agreed cost according to the wishes of the citizens	Delivered at a standard price related only to weight, speed or registration.
It is a private form of communication	The privacy of the letter was an key feature of the postal service in democratic countries
Controlled at a national level	Controlled according to the needs of each individual nation
Social infrastructure	Post offices and staff provided a centre for communication, meeting and exchange.
A universal system of accepted standards	The postal system has built up a set of standards that are recognised globally and that facilitate interoperability between all national postal systems.

individual's postal address does not link them to their other daily activities (Church, Moloney, & Bannister, 2013).

Thirdly, by being able to send and receive anonymous communication through the availability of post boxes and post office boxes, individuals can enjoy anonymous transfer of content. Suppliers do not need to know any personal information about the individual to supply products to them. The individual can anonymously request a product from a supplier and receive the goods via a numbered post office box address. Thus, at least in the area of written communication, individuals could be as private or as public as they wished. The control of this aspect of their private life remains fully with them. Figure 3 summarizes these three points.

This is in distinct contrast to contemporary and evolving digital communication. With most public digital communications systems everything is public by default. The nature of the technology facilitates tracking of individuals and makes it straightforward for organizations to access, profile and target individuals for sales, marketing or other less ethical purposes. Organizations, particularly communication and information service providers (a term which includes companies such as Google and Facebook) have access to an ever greater amount of personal information about every individual with whom they come into contact online. The ability of the legal system to keep pace with this is limited. Laws have been put in place to prevent the abuse of personal information by organizations, but enforcing these laws is often impossible due to the transnational nature of modern on-line service delivery (Edwards & Waelde, 2000). In addition to this, the current system shifts control of an individual's ability to regulate their privacy needs *away* from the individual and into the hands of service providers and suppliers, disempowering the traditional ability of citizens to control their own privacy (Church, Moloney, &

Figure 3. Three levels of privacy in the postal system (Church, Moloney, & Bannister, 2013)

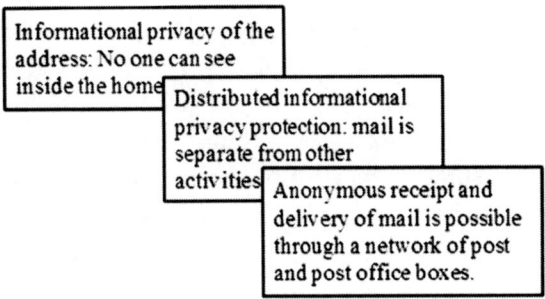

Bannister, 2013). In order for society to continue to enjoy a fully private communication system in digital form and in line with the universal postal system, attempts need to be made to preserve the three layers of informational privacy traditionally provided by posts.

The next section gives details of a proof of concept pilot project for one such integrated local platform, which was rolled out in the London borough of Tower Hamlets, in September 2012.

TOWER HAMLETS (UK) CASE STUDY

The proof of concept pilot study was conducted in collaboration with Tower Hamlets Council UK, Community Infopoint (a subsidiary of Escher Group Holdings Plc), Your Square Mile and the Big Lottery. Access to the integrated community platform was delivered through four ruggedised iPads, named "Community Infopoint kiosks." These kiosks were installed at four strategic locations in the borough of Tower Hamlets. The borough of Tower Hamlets has re-invented its network of libraries into what they are calling Idea Stores (these consist of a library, cafe, adult education classes and computer access in one location). The community platform kiosks were located in three of these Idea Stores around the borough and the fourth was located in an East London Mosque.

The proof of concept pilot ran for a total of twelve weeks from September 4th until November 23rd, 2012. Its main purpose was to identify, through local residents, local stakeholders and other interested parties, whether there was a need for such a community platform. The pilot also provided a great opportunity to learn from citizens what services and applications they would find interesting if a wider deployment of the community platform was to take place in Tower Hamlets.

A face to face visit with the Tower Hamlets Council at one of their one stop shops, on average

costs the council £9 and to service a call from a resident costs them over £4, so the council are eager to encourage citizens to engage through a web-based platform such as the integrated community platform to help reduce their costs. But in addition to the cost benefits, the platform would also help improve transparency between the council and the residents it serves. Added to this, it would help to create a community eco-system of relevant applications and interactions that would benefit all parties involved, i.e. citizens, local businesses and stakeholders.

The Integrated Community Platform

Following discussions with Tower Hamlets council, it was decided to offer an initial set of six community focussed applications on the ruggedised ipads. Two of the applications, named "I want to see" and "Have your Say," were interactive and invited users to provide feedback on various questions posed by the Tower Hamlets council or other local stakeholders.

The third application that was offered was "What's on." This application was linked to a RSS feed, which was supplied by the council to inform users of local events taking place in the community. The fourth app was entitled "Mayor's Community Champions." This app highlighted the work carried out by volunteers in Tower Hamlets during the time of the Olympics. It also provided live feeds to the Twitter and Flickr pages set up by the council to highlight volunteer activity.

The fifth app entitled "Win an iPad" was a competition app which invited users to submit ideas and suggestions for the creation of other innovative community apps that could be added to the existing set of six. The final app, entitled "Travel Alerts" was also a live RSS feed that was linked up to the Transport for London network and gave live updates of tube times in close proximity to each Community Infopoint kiosk (ruggedised ipad) location.

Figure 4. The initial set of six community focussed applications Tower Hamlets pilot study

Community Infopoint Advisors

In order to gather as much feedback and data as possible to correctly evaluate and analyse the local integrated community platform, Community Infopoint employed and trained a number of local people to spend on average 4 hours a day in each of the four locations where the ruggedised ipads were located to help showcase the community platform and raise awareness of its functionality and uses. These Community Infopoint advisors were also responsible for helping and encouraging users to fill out the short user exit survey. These surveys helped to clarify the needs of the users and the community apps that they would like to see on the platform in phase two of the platform's development cycle.

Case Study Results

The results of the case study break naturally into two sections, the feedback from local stakeholders such as local businesses and not-for-profit organisations and feedback from local residents.

Feedback from Local Stakeholders

Over the course of the 12 week pilot study, Community Infopoint met with nine local stakeholders in Tower Hamlets to raise awareness of the Community Infopoint platform and to understand how such a community platform could add value to their organisations. All stakeholders that were approached had a positive reaction to the concept of an integrated community focussed platform and all of them wanted to be considered for inclusion in phase two of the project. Due to space limitations, feedback from only three of the nine stakeholders are outlined in this section.

Toynbee Hall is a charitable foundation located in the borough of Tower Hamlets. The mission of the foundation is twofold, firstly to make East London a more attractive place to work and live and secondly to improve job prospects for local people by placing a special emphasis on financial inclusion. The foundation helps over 9,000 people each year. The CEO of the foundation felt that the integrated community platform would benefit the foundation in the following ways:

Figure 5. Six community focussed iPad applications

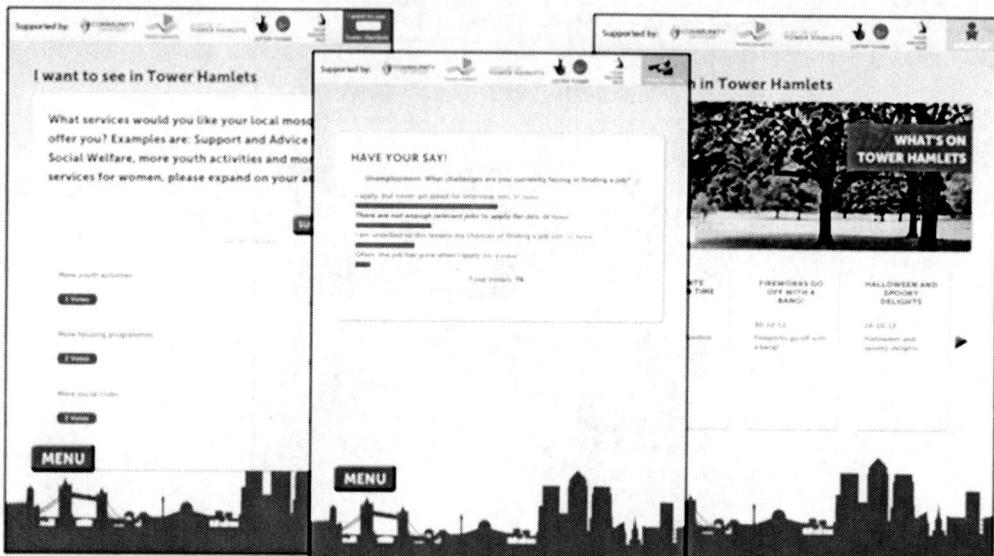

- There is currently a significant challenge facing the foundation in terms of reaching or communicating with the population of the borough due to the fact that many residents are uneducated and have a poor grasp of the English language. The community platform could be used as a channel to reach residents that need help in communicating digitally (via the help of postal workers) and those who do not have access to the internet at home. The feedback and opinions of the most digitally excluded within the community could be received in this way.
- Outworkers of the foundation could access the community platform via iPads when they are out in the community. In this way they could share the benefits of the platform with the vulnerable residents that they visit.
- The CEO was eager to locate an Infopoint kiosk within the building of Toynbee Hall.

The borough of Tower Hamlets has an unemployment rate of 13%, which is close to twice the national average of 7.6%. Within the borough,

25% of the population have no formal education and only 70% of all households have an employed person within the household. JobCentre Plus offer a specialist service for disabled people seeking employment. They provide support to disabled people who are having difficulty in finding a job because of their disability. They believe that the integrated community platform could be used as a local matching service, which would access their bank of CV's and match them to the most appropriate job opportunities for individuals. Additionally, it could be used as an information resource, informing unemployed people on how to find suitable employment. Finally, Job Centre Plus agreed to provide content and local job offers if a local jobs application was developed on the platform. They also agreed to locate Community Infopoint kiosks at local job centre reception areas.

Finally, the NHS Tower Hamlets Clinical Commissioning Group (CCG) is responsible for 36 general practices within Tower Hamlets and all of the major hospitals in the area. They consult with over 1,000 patients to help provide better services. The CCG would like to use the platform to help them attract a further 2,000 citizens to help with patient consultation and feedback. They would like to locate Community Infopoint kiosks in all 36 doctors surgeries and Accident and Emergency waiting areas. They realise that Tower Hamlets has a population in very poor health and would like to raise awareness of health and well being issues through the kiosks. They believe the Integrated Community platform fits very well with the strategic aims of PATHE engagement.

Feedback from Local Residents

Community Infopoint advisors approached all residents after they had interacted with the Community Infopoint kiosks to ask them to complete a short user survey which was available on the kiosks. The survey results provided both Community Infopoint and Tower Hamlets council with a snapshot of residents' opinions on the integrated community platform. It also provided suggestions from residents on what they would like to see in future versions of the platform to ensure the was fully inclusive as a platform. Over 1,850 customer surveys were conducted over the course of the 12 week pilot. Table 3 outlines the breakdown in age among the users of the Community Infopoint platform. The majority were between the ages of 25 and 44 years.

Almost 97% (96.8) of users of the kiosks located in the Idea Stores reported having a positive experience with the platform and 93.3% reported the same in the East London Mosque. Thirteen percent of users of the kiosk located in the East London Mosque had never used the Internet before using it on the Community Infopoint kiosk, 8.5% of users of the other kiosks had never accessed the Internet previously either.

Users were asked how they currently receive information from their local council and the majority of users receive information via the post (42.1% from users in the East London Mosque and 53.4% from the Idea Stores kiosks. The other ways of receiving information from the local council were 1) over the Web (31% and 29.8%), 2) face to face (14.3% and 8.7%) and 3) non-specified means accounted for 12.6% from the Kiosk in the Mosque and 8.1% from the kiosks in the Idea Stores.

Information regarding other community locations for the Community Infopoint kiosks was also sought from users. The Post Office was the most favoured community location with 44% of users of the kiosk in the East London Mosque identifying it as the best location and 45% of all other users of the kiosks choosing it as the most suitable location. The post was followed by locating the kiosks in local supermarkets (37.4% and 34.3% from the Mosque and Idea Stores respectively). Other favourite locations included doctors' waiting rooms and social housing offices.

The survey also enquired about whether residents would like to pay their bills via the kiosks. There was a very high positive response from users of the Mosque kiosk, 52.9% of users wanted to pay their bills through this channel. However, a lesser 33.4% of users in the Idea Stores kiosks wanted to pay bills through the platform.

With regard to phase two of the project, the survey provided valuable feedback. As expected, a local jobs app was the most desired app, with 86.3% of users from the Mosque and 72.3% from the Idea Stores wishing to see such an app in phase two. A local deal app was the second most requested with 53.6% of users from the Mosque

Table 3. Breakdown in ages among users of the Integrated Community platform

Age Brackets	Mosque Community Infopoint Kiosks	Idea Store Community Infopoint Kiosks
15-24 yrs	32.3%	40.6%
25-44 yrs	46.8%	44.7%
45-64 yrs	16.1%	11.5%

Figure 6. Information on desired services for phase 2 of the project

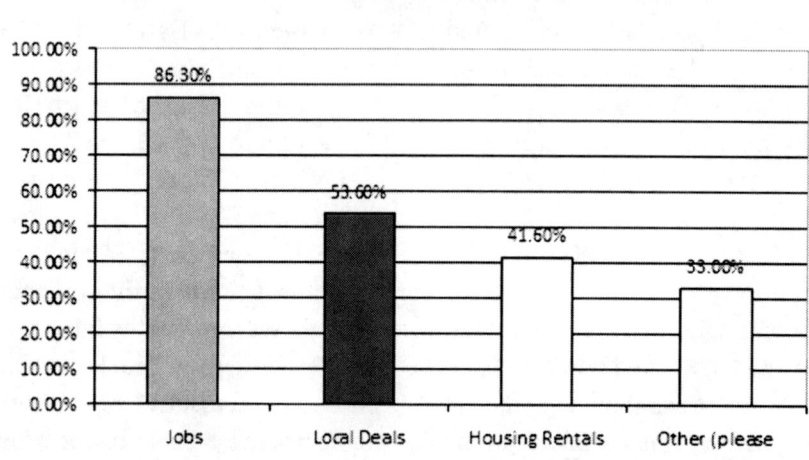

and 38.7% from the Idea Stores. This was followed by a social housing bidding app with 41.6% of users from the Mosque and 20.6% from the Idea Stores wishing to see it in phase two.

89% of users of the community Infopoint kiosk located in the East London Mosque wanted the platform to be offered in a second language and 35% of users of the Idea Store kiosks wanted the same.

November 2012 was "Money Matters Month" (MMM). This was a campaign to raise awareness about welfare reform, and the impact that it might have on the amount of benefit that Tower Hamlets residents may receive. The following answers were received via the "Have you say" app. They are presented here, in Tables 4-7, to demonstrate the low level of awareness among residents of local government initiatives.

Summary of Findings

In light of the encouraging feedback given by both local residents and stakeholders, discussions with central government and Big Lottery resulted in some recommendations being made for phase two of the Integrated Community Platform project and for Government 2.0 in general. The positive

findings from the 12 week proof of concept pilot proves a need for an Integrated community platform and the council should now seek internal and external support, both financially and for content provision, to assist phase two of the community platform roll-out.

All of the local stakeholders that gave feedback on the project were eager to develop their own content and applications for the community platform. It is now in the interest of the council to arrange a stakeholder forum to share its plans for a wider adoption of the platform and to seek funding and collaboration from local stakeholders. Such a community platform would provide many of the solutions to the recommendations laid out by the Big Society policy, Localism Act, Digital Inclusion and Digital by Default initiatives of the UK government (Mabbutt, 2010; Legislation.gov. uk, 2011; Housing Technology & Race Online, 2011). The council should seek funding from and closer collaboration with the Department for Communities and Local Government and the UK Cabinet Office.

As a result of the finding that local post offices were the most requested community location in which to host a Community Infopoint kiosk, it was suggested that the UK Post Office, which operates 21 post offices in the borough, could create a network and provide sufficient coverage across the borough to replace the One Stop Shops that being managed by Tower Hamlets Council. This would provide extra business for the post office network and significantly reduces costs for the council.

The council should make arrangements to meet with the UK Post Office to discuss the feasibility and cost implications for the Tower Hamlets post offices to offer a similar service to the one currently being provided by the One Stop Shops.

The development of a local jobs iPad app should be the first iPad app to be developed. With high unemployment in the area, a community platform providing current job opportunities would prove popular and assist residents living in Tower Hamlets to find suitable employment. The council should look to engage both Job Centre plus and other local job portals to co-develop an application specifically for the borough of Tower Hamlets.

The majority of users (52.9%) from the Community Infopoint kiosk located in the Mosque wanted the Integrated Community platform to provide bill payment services. Central Government is due to change the way social housing benefit is paid under Universal Credit.(which is a new single payment for people who are looking for work or on a low income) and is being introduced to simplify the benefits system by bringing together a range of working-age benefits into a single streamlined payment. It is intended to help claimants and their families become more independent., and claimants will soon begin receiving the benefit directly. As a result of this, it is predicted that there will an increased demand for these people to make payments in more convenient ways, namely from community locations, through use of mobile technology like smartphones and tablets and over the Internet. The council should seek to engage a number of the larger Social Housing Associations to collaborate and support the development of a payment mechanism within the Integrated community platform.

A number of Social Housing Associations within the borough were willing to host a Community Infopoint kiosk within their local offices and community centres. One of the most popular requests from users was for the development of a social housing bidding app that enables the socially and digitally excluded to access and register their interest in properties without having to continuously check for updates at one of the council's local One Stop Shops. The borough of Tower Hamlets has 62 different Social Housing organisations, the council should look to gain the full support and financial commitment from all of these organisations, in an effort to develop the most appropriate iPad app and at the same time raise awareness of the app. In this way, all 62 associations can post their vacancies in real time and provides the digitally excluded residents with the same opportunities to register their interest in available housing as residents that have Internet access at home.

Online retailing accounted for 17% of UK retail trade in 2011 and this is set to continue into the future. This is a worrying trend for both the UK high street and the small to medium sized enterprises that provide products and services to the high street. These small businesses are often collectively the biggest employers of local people in any community. 53.6% of users that completed the exit user survey on the Integrated Community platform requested a local deals app for phase two of the project. The council should consider co-developing and funding such an app to facilitate both vendors and suppliers to buy and sell locally. This would create a more vibrant and sustainable eco-system of small businesses, which would generate more local sales and ultimately more local jobs.

Tower Hamlets council are faced with further significant budget cuts over the next three years as central Government funding is set to be reduced. Upfront funding for phase two of the project is not available, therefore innovative ways for building the platform need to be created. Escher Group, the

Table 4. With which of these statements do you agree?

Answer 1: I attended a MMM session and I know what impact the reforms will have on me	14
Answer 2: I was aware of MMM sessions, but didn't go as I accessed advice elsewhere	13
Answer 3: I haven't been to a MMM session but will access advice about benefit changes	20
Answer 4: I didn't know about MMM sessions, I don't know the impact reforms will have on me	**46**
Total number	**93**

Table 5. Unemployment: What challenges are you currently facing in finding a job?

Answer 1: There are not enough relevant jobs to apply for	40
Answer 2: I am unskilled so this lessens my chances of finding a job	19
Answer 3: I apply, but never get asked for interview	**51**
Answer 4: Often, the job has gone when I apply	4
Total number	**114**

Table 6. Tower Hamlets Council is running a "Money Matters" Campaign to help residents with the changes that government is making to benefits. What do you agree with?

Answer 1: Housing benefit helps me pay my rent, and I received a letter from DWP about changes that will be made to my housing benefit.	31
Answer 2: I know how much housing benefit I may lose as a result of the government's welfare reform.	25
Answer 3: I am trying to find suitable employment, to minimize the impact of the benefit changes	**71**
Answer 4: I don't know about the changes that welfare reform will make to my benefits.	21
Total number	**148**

Table 7. Thinking of the last 12 months, which of the following statements has been true of you more often than not?

Answer 1: I budget and I run out of money by the time I get my pay/ benefits	**37**
Answer 2: I budget and I do not run out of money by the time I get my pay/ benefits	14
Answer 3: I do not budget and I run out of money by the time I get my pay/ benefits	6
Answer 4: I do not budget and I do not run out of money by the time I get my pay/ benefits	6
Total number	**63**

company behind Community Infopoint is willing to offer the Integrated Community platform as a "Software as a Service" model. In this way, a fixed monthly fee for hardware and maintenance is paid to Escher with no upfront costs. This transaction type revenue model could be based on a "per user" or "per transaction" fee structure. Even an element of both types of fee structure would sufficiently reduce costs for the council in communicating with residents as it would enable residents to communicate with the council through the digital platform rather than through existing expensive face to face or call centre channels. Obviously, further discussion is required regarding the services that the council would like to offer before a cost-benefit analysis can be carried out. The council will need to engage with appropriate government decision makers in various directorates to make a case for 1) potential cost-cutting measures that can be carried out as a result of offering such a local digital solution and 2) redirecting the money saved by this cost-cutting into further development of the community platform.

Finally, the council should actively seek funding to assist with the development and roll-out of the community platform in Tower Hamlets. The Big Lottery is soon to announce a £150million fund for "Enriching Places." This fund will provide investment for locally designed solutions in the belief that with support local communities can take charge of their future. Other funding channels could include NESTA (Bunt & Harris, 2010), UnLtd (UnLtd.org.uk, 2003) and other central government departments that invest in new and innovative projects.

BENEFITS FOR THE COMBINED GOVERNMENT 2.0 AGENDA

Providing citizens with functionality (namely in the form of ipad apps) for them to carry out specific tasks like communicating with government, job seeking and bill payments is a logical step in creating a holistic digital identity for citizens. This fully integrated approach to e-government delivers benefits to all participants of the communication and at every level. From the perspective of government, they successfully reduce: 1) their physical administration costs by streamlining their public services and 2) their carbon footprint by eliminating the paper trail; while 3) using their workforces more efficiently and 4) offering improved levels of service quality and timeliness.

From the citizen's perspective, they can safely communicate and digitally interact with not only their government but with all stakeholders, including local commercial entities and associations that linked to the system. Citizens gain increased

knowledge and expertise about digital communications, and how they receive them combined with a unified and integrated transaction and message delivery system and secure, private to facilitate the sending of financial, health and other sensitive personal data.

Central or local government would be able to regulate this local network with their own legislation as opposed to most social networks and e-commerce sites that are frequently based in the US and thus governed by US law. Additionally, to eliminate any 'big brother' concerns, a choice of storage locations should be offered to citizens regarding the location of their private storage databases within the system. Third party private companies should be allowed to provide networking and storage facilities once they meet specific requirement for ensuring the protection of citizens' information within the secure e-government network.

CONCLUSION AND FUTURE WORK

The research outlined in this chapter started with an exploration of how governments can provide a holistic approach to the protection of citizens' personal information online while meeting citizens' needs in terms of open government and user participation. Protecting the transmission of personal information when transacting with national government is important in a democracy. It serves to safeguard freedom of speech and to facilitate the digitalisation of our society. With the increase in popularity of Web 2.0 services, governments and businesses alike are looking at ways to leverage these technologies to provide more efficient and streamlined e-services for citizens. These efficient e-services are often provided, however, at the expense of the citizen who must relinquish control of their personal information and who often suffer informational privacy concerns as a result.

Reconciling these two agendas, providing citizen-centric services and safeguarding the personal information of citizens, using a holistic citizen-centric government 2.0 framework to highlight how this can be achieved was the principal goal of this research. This goal was achieved by way of a case study, which detailed the reaction of citizens to a community platform modelled on the Government 2.0 framework. The 12 week pilot study outlined in this chapter was deemed a success by all participating members of the study and phase 2 of the project is currently being planned. Phase 2 will see Infopoints being located in an increased number of strategic locations throughout the borough. There is a view to including more boroughs other than Tower Hamlets in the future, however, this will not take place during phase 2 of the project..

A lot still remains to be explored regarding the future of e-government and how governments will engage with their citizens while ensuring the citizen's personal information is constantly secure and protected. Nevertheless, this chapter serves to introduce the reader to the idea of citizens securely and privately interacting digitally with their government from within their locality, i.e. through their Infopoints in their local post office. This local interaction brings the Internet to the community, while reducing costs for government at the same time.

REFERENCES

ABC4Trust Project. (2009). *ABC4Trust*. Retrieved March 23rd, 2011, from http://abc4trust.de/

Bannister, F., & Connolly, R. (2011). The Trouble with Transparency: A critical view of openess in e-government. *Policy & Internet, 3*(1), Article 8.

Bertot, J. C., Jaeger, P. T., & Hansen, D. (2012). The impact of polices on government social media usage: Issues, challenges, and recommendations. *Government Information Quarterly, 29*(1), 30–40. doi:10.1016/j.giq.2011.04.004

Bunt, L., & Harris, M. (2010). *Mass Localism: A way to help small communities solve big social challenges*. Retrieved February 28th, 2013, from http://www.nesta.org.uk/publications/reports/assets/features/mass_localism

Camenisch, J., & Pfitzmann, B. (2007). Federated Identity Management. In M. Petkovic, & W. Jonker (Eds.), Security, Privacy and Trust in Modern Data Management. Springer.

Chun, S. A., Shulman, S., Sandoval, R., & Hovy, E. (2010). Government 2.0: Making connections between citizens, data and government. *Information Polity, 15*(1-2), 1–9.

Church, L., Moloney, M., & Bannister, F. (2013). The Sealed Letter: Safeguarding the Public System of Privacy Protection in a Digital World. In *Proceedings of the 46th Hawaii International Conference on System Sciences*. Manoa, HI: IEEE.

Culbertson, S. (2005). E-Government and Organizational Change. In M. Khosrow-Pour (Ed.), *Practicing E-Government: A Global Perspective* (pp. 83–131). Hershey, PA: Idea Group Publishing. doi:10.4018/978-1-59140-637-2.ch005

Cullen, R., & Hernon, P. (2004). *Wired For Well-Being: Citizens' Response to E-Government*. Academic Press.

Edwards, L., & Waelde, C. (2000). *Law and the Internet*. London, UK: Hart Publishing.

Etzioni, A. (1999). *The Limits of Privacy*. New York: Basic Books.

European Commission. (2010). A Digital Agenda for Europe (COM(2010) 245 final/2). Author.

European Commission. (2011). *Digital Agenda: Turning government data into gold*. Retrieved 11 3, 2012, from http://europa.eu/rapid/press-release_IP-11-1524_en.htm

Federal Trade Commission. (2010). *Protecting Consumer Privay in an Era of Rapid Change: A Proposed Framework for Businesses and Policymakers.* Washington, DC: Federal Trade Commission.

Grönlund, Å., & Horan, T. A. (2004). Introducing e-gov: History, definitions, and issues. *Communications of the Association for Information Systems, 15,* 713–729.

Guardian Newspaper. (2013, June 23). *Edward Snowden and the NSA files – Timeline.* Retrieved July 15th, 2013, from http://www.guardian.co.uk/world/2013/jun/23/edward-snowden-nsa-files-timeline

Haughwout, J. (2009). *Meeting the New Requirements for Transparency, Participation and Collaboration with Neighborhood America's Solutions.* Retrieved May 2011, from http://www.ingagenetworks.com/docs/mgmt/Wht_paper_Gov_JHaughwout_May09.pdf

Ho, A. T. (2002). Reinventing Local Government and the E-Government Initiative. *Public Administration Review, 62*(4), 434–444. doi:10.1111/0033-3352.00197

Housing Technology & Race Online. (2011). *Digital by Default 2012.* London, UK: The Intelligent Business Company and Race Online.

Independent Commission for Good Governance in Public Services. (2004). *The Good Governance Standard for Public Services.* Joseph Rowntree Foundation.

Information and Privacy Commissioner Canada. (2010). *Privacy Risk Management Building privacy protection into a Risk Management Framework to ensure that privacy risks are managed, by default.* Retrieved from www.privacybydesign.ca

Introna, L. D. (1997). Privacy and the Consumer: Why we need privacy in the information society. *Metaphilosophy, 28*(3), 259–275. doi:10.1111/1467-9973.00055

Kraemer, K., Danziger, J., & King, J. (1978). Local Government and Information Technology in the United States. *OECD Informatics Studies, 12.*

Laufer, R., & Wolfe, M. (1977). Privacy as a concept and a social issue: A multidimensional developmental theory. *The Journal of Social Studies, 33*(3), 22–42.

Legislation.gov.uk. (2011). *Localism Act.* Retrieved April 3rd, 2013, from http://www.legislation.gov.uk/ukpga/2011/20/contents/enacted

Mabbutt, A. (2010). *Big Society.* Retrieved April 3rd, 2013, from http://www.conservatives.com/Policy/Where_we_stand/Big_Society.aspx

Morgeson, F. V., VanAmburg, D., & Mithas, S. (2011). Misplaced Trust? Exploring the Structure of the E-Government-Citizen Trust Relationship. *Journal of Public Administration: Research and Theory, 21*(2), 257–283. doi:10.1093/jopart/muq006

Morisonn, J. (2010). Gov 2.0: Towards aUser Generated State? *The Modern Law Review, 4.*

Newman, J., & Clarke, J. (2009). *Publics, Politics and Power: Remaking the Public in Public Services.* London: Sage Publications.

Pavlou, P. A. (2011). State of the Information Privacy Literature: Where are we now and Where should we go? *Management Information Systems Quarterly, 35*(4), 977–988.

PICOS. (2007). *Privacy and Identity Management for Community Services.* Retrieved March 23, 2011, from http://www.picos-project.eu/

Primelife. (2008). *Primelife*. Retrieved March 23, 2011, from http://www.primelife.eu/

Privacy Rights Clearinghouse. (2011, December 16). *Data Breaches: A Year in Review*. Retrieved January 26, 2012, from http://www.privacyrights. org/data-breach-year-review-2011

Robertson, S. P., Vatrapu, R. K., & Medina, R. (2010). Off the wall political discourse: Facebook use in the 2008 U.S. presidential election. *Information Polity*, *15*(1-2), 11–31.

Schinfeld, E. (2010, October 18). *Steve Jobs: Open Systems Don't Always Win*. Retrieved March 26th, 2013, from http://techcrunch.com/2010/10/18/ steve-jobs-open-dont-win/

Sheedy, C., & Moloney, M. (2013). *Leveraging the Postal Infrastructure for the Authentication of Individuals Towards an Online Government Service Provision*. Paper presented at CRRI 21st Conference on Postal and Delivery Economics. Dublin, Ireland.

Shoeman, F. (1984). Privacy: Philosophical Dimensions. *American Philosophical Quarterly*, *21*(39), 199–213.

Smith, H. J., Dinev, T., & Xu, H. (2011). Information Privacy Research: An Interdisciplinary Review. *Management Information Systems Quarterly*, *35*(4), 989–1015.

Surowiecki, J. (2004). *The Wisdom of Crowds: Why the Many Are Smarter Than the Few and How Collective Wisdom Shapes Business, Economies, Societies and Nations*. Anchor Books.

Triangle Management Services Limited. (2011, March). *Post Offices and Local Government Services – An International Literature Review*. Retrieved February 18th, 2013, from http://www. consumerfocus.org.uk/scotland/files/2011/08/ POs-Government-Services-International-Comparisons-Final-Triangle-Report.pdf

UnLtd.org.uk. (2003). *About UnLtd*. Retrieved April 3rd, 2013, from http://unltd.org.uk/ about_unltd/

US Postal Service Office of the Inspector General. (2013). *e-Government and the Postal Service*. Retrieved February 18th, 2013, from http://www. uspsoig.gov: http://www.uspsoig.gov/foia_files/ RARC-WP-13-003.pdf

Van Dyke, T. P., Midha, V., & Nemati, H. (2007). The Effect of Consumer Privacy Empowerment on Trust and Privacy Concerns in E-Commerce. *Electronic Markets*, *17*, 68–81. doi:10.1080/10196780601136997

Welch, E. W., Hinnant, C. C., & Moon, M. J. (2005). Linking Citizen Satisfaction with E-Government and Trust in Government. *Journal of Public Administration: Research and Theory*, *15*(3), 371–391. doi:10.1093/jopart/mui021

Westin, A. (1967). *Privacy and Freedom*. New York: Atheneum.

Whitley, E. A. (2009). *Informational Privacy, Consent and the Control of Personal Data*. London: Elsevier.

ADDITIONAL READING

De Reuck, J., & Joseph, R. (1999). Universal Service in a Participatory Democracy: A Perspective from Australia. *Government Information Quarterly*, *16*(Issue 4), 345–352. doi:10.1016/ S0740-624X(00)86839-2

Fedorowicz, J., & Dias, M. (2010). A Decade of Design in Digital Government Research. *Government Information Quarterly*, *27*, 1–8. doi:10.1016/j.giq.2009.09.002

Hamilton, B. A. (2005). *Beyond e-Government: The world's most successful technology-enabled transformations. Commissioned by the United Kingdom Presidency of the European Council.* European Council.

Meneklis, V., & Douligeris, C. (2010). Bridging Theory and Practice in e-Government: A set of Guidelines for Architectural Design. *Government Information Quarterly, 27*, 70–81. doi:10.1016/j.giq.2009.08.005

Misuraca, G. C. (2009). e-Government 2015: exploring m-government scenarios, between ICT-driven experiments and citizen-centric implications. Technology Analysis & Strategic Management - Special Issue: Managing Forsight within Changing Organisational Settings, 21(3), 407-424.

Tauber, A. (2010). Requirements and Properties of Qualified Electronic Delivery Systems in eGovernment: an Austrian experience. *International Journal of E-Adoption*, 45–58. doi:10.4018/jea.2010010104

Teinowitz, I. (2011). Trust of government agencies drops, but folks still love the USPS. AOL Inc. WalletPop.com.

The Ponemon Institute. (2010, June 30th). 2010 Privacy Trust Study of the United States Government. The Ponemon Institute Research Report.

The United States Government. (2010, January). Data.gov - An Official Web Site of the United States Government. Retrieved December 10, 2012, from http://www.data.gov/

The White House. (2010). *DRAFT National Strategy for Trusted Identities in Cyberspace: Creating Options for Enhanced Online Security and Privacy. The White House.* Washington: US Government.

USPS OIG. (2013). e-Government and the Postal Service. Retrieved February 18th, 2013, from http://www.uspsoig.gov: http://www.uspsoig.gov/foia_files/RARC-WP-13-003.pdf

KEY TERMS AND DEFINITIONS

Citizen Participation: Citizen Participation, or e-participation, can be defined as the redistribution of power that enables citizens, presently excluded from political and economic processes, to be deliberately included in these processes in the future.

Digital Identity: A digital identity is a set of data that uniquely describes a person or a thing (sometimes referred to as subject or entity) and contains information about the subject's relationships to other entities.

E-Government: e-Government consists of the digital interactions between a government and citizens (G2C), government and businesses/Commerce (G2B), government and employees (G2E), and also between government and other governments /agencies (G2G).

Government 2.0: Gov 2.0 or Government 2.0 refers to government policies that aim to harness collaborative technologies to create an open-source computing platform in which government, citizens, and innovative companies can improve transparency and efficiency.

ICT: Information and Communication Technology (ICT) has been used in recent years to refer to the convergence of audio-visual and telephone networks with computer networks through a single cabling or link system.

M-Government: Mobile government, m-Government, is the extension of eGovernment to mobile platforms, as well as the strategic use of government services and applications which are only possible using cellular/mobile telephones, laptop computers, personal digital assistants (PDAs) and wireless internet infrastructure.

Open Data: Open data is the idea that certain data should be freely available to everyone to use and republish as they wish, without restrictions from copyright, patents or other mechanisms of control.

Personal information: Personal information is information relating to a living individual who is or can be identified either from the information or from the information in conjunction with other information.

Sensitive Data: Sensitive data is any type of data that needs to be kept secure. Examples of sensitive data include passwords and personally identifiable information.

Sensitive Personal Data: Sensitive personal data means personal data consisting of information that either relates to an individual's: (a) the racial or ethnic origin of the data subject, (b) his/her political opinions, (c) his/her religious beliefs or other beliefs of a similar nature, (d) whether he/she is a member of a trade union, (e) his/her physical or mental health or condition, (f) his/her sexual life, (g) the commission or alleged commission by him/her of any offence, or (h) any proceedings for any offence committed or alleged to have been committed by him/her, the disposal of such proceedings or the sentence of any court in such proceedings.

Section 3
Mobile Technologies for Smarter and Sustainable Mobile Government

Chapter 7
Mobile and Cloud Technologies for Smarter Governance

Pethuru Raj
IBM India, India

ABSTRACT

There are hordes of data-driven, context-aware, and people-centric applications and services for smarter environments such as smarter homes, governments, buildings, cities, and organizations. With the exponential growth of smart phones, there are service repositories and application stores in remote mobile clouds. Similarly, with the ceaseless advancements in the device ecosystem and in the IT field, government-specific applications will flourish and be deployed and maintained in special cloud stores, platforms, and infrastructures to be found, bound, and used by any input/output devices for a variety of everyday personal and professional purposes. Smart, sustainable, intuitive, and citizen-aware services can be dynamically created from the ground up as well as orchestrated or choreographed out of multiple atomic and discrete software services. Such composite services are directly fulfilling government activities. Thus, clouds emerge as the most common and minimum requirement for not only producing and stocking services but also for hosting application platforms. Further, clouds facilitate provisioning and renting out their configurable and customizable assets on demand. Through self-service portals, the cloud usage is to pick up fast in the days to unfold. In this chapter, the authors write about how cloud adoption is to ring in delectable transformations for worldwide governments as well as their citizens, that is, how governments can accomplish more with less, how people can experience high quality, technology-sponsored digital living, how the cloud idea becomes a centre of attraction for more ingenuity towards newer and nimbler service conceptualization, concretization, and delivery.

INTRODUCTION

Every organization aims to be smarter in its operations, offerings, and outlooks by cogently and cognitively embracing competent technologies, integrated processes, flexible and futuristic architectures, and optimized infrastructures. Governments too want to sail in that same boat to be grandiosely relevant to their constituencies and citizens by bringing in a series of innovations in their everyday administrative activities and deliveries. Governance has become a serious

DOI: 10.4018/978-1-4666-6082-3.ch007

activity and people started to expect more from their leaders. In order to cope up with the enhanced and evolving expectations and aspirations, the government officials, bureaucrats, governors and executives are increasingly cognizant of the noteworthy and novel implications of the technology choice and adoption. Incidentally there are several technologies including Web 2.0 or social Web, Web 3.0 or semantic Web, mobility, cloud technologies, analytics, integration, Bigdata, the Internet of Things (IoT) etc. that lead to different kinds of next-generation Web, cloud, mobile, real-time analytics and smarter applications. Furthermore, with the extreme and deeper connectivity technologies, not only software applications and services but also people and devices are instantaneously hooked with one another.

The world is likely to experience more shocks and stresses in the future, on scarcity and on other fronts, with increasing intensity. At the same time, policymakers, while trying to capitalize on the windows of opportunities that such crises may offer, need to try to ensure that moments of system breakdown lead to renewal rather than to outright collapse. Resilience is the quality that will determine the difference between these two outcomes.

Most governments spend a significant amount of their technology budgets towards procuring, installing and maintaining a variety of IT infrastructures, platforms, and applications. Purchasing hardware, upgrading software, and employing administrators for managing applications as well as infrastructures in an optimal state is not an easy task. There is a statistics that up to 70% of the total IT budget in any organization is being spent just for IT maintenance. Thereby developing and deploying newer and nimbler capabilities and competencies take a back seat. Also there is an exerted pressure on cutting down IT budget due to the prevailing and prolonged uncertainty in the world economy. In other words, executives expect more out of IT these days. Not only infrastructure

optimization, but also there are other factors to be given prime importance by those governments IT managers and consultants in order to smooth journey towards the ultimate vision of establishing and sustaining smarter governments (KMPG, 2010).

Data is definitely a strategic asset to be leveraged smartly in order to pursue the people agenda vigorously and rigorously. Big Data analytics is another interesting area for proactive and preemptive governance for quickly extracting actionable insights from government data heaps. Mobile phones are the latest entrants in anytime anywhere delivering a cornucopia of government services. With all these positive and progressive trends in the IT arena, cloud stands tall because cloud is the core and central technology for all other technologies to grow and glow.

Undoubtedly cloud embarkation is one proven and potential move in all kinds of enterprises including governments. The sharp infrastructure optimization being enabled by the disruptive and transformative cloud technology is keenly introduced and incorporated into every tangible domain in order to reap all its originally envisaged benefits. Governance is an important activity and the much-discoursed and deliberated cloud paradigm is bound to bring in significant advancements in the form of total transparency, accountability, responsiveness, simplicity, speed, scale, sustenance, etc. for formulating and delivering a variety of citizen-centric services. In this document, we are to see how the cloud idea is going to be profoundly paramount for the unprecedented success towards providing right and relevant services for people affordably within the stipulated and scheduled time with all alacrity and clarity. In the following paragraphs, we discuss how cloud infrastructures provides a suite of robust and resilient cloud services and solutions that are a boon and game-changer for federal and county governments across the globe in reaching out their constituencies with a bevy of generic as well as niche services

The Resilient Society

The concept of resilience is not a new one. As mentioned by Cho et al. (2011), for every service-oriented system, resilience is an important capability and competency not to be taken lightly. But it is taking on greater significance at the heart of aspirations for good governments in a period of disruption and wrenching transition. It is a brewing idea that combines two dimensions:

- "Bouncing back" from the last unexpected shock, emergency or catastrophe and,
- "Bounding forward" to anticipate, prepare for, and, as far as possible, avoid the worst excesses of the next disruption.

These two dimensions trumpet on the combination of productivity (doing things better) and innovation (doing better things). And both productivity and innovation are capabilities increasingly dependent upon the art and practice of connectedness. Putting these pieces of the puzzle together in a new narrative of "connecting for resilience" is critical for governing well amidst unprecedented risk, change, and opportunity. For the most part, the government is working harder than ever before. Pensions and other financial benefits are paid to right beneficiaries through effective identification cards, passport applications are processed online and rejuvenated investigation mechanisms to enable quicker delivery, traffic systems are monitored and managed through IT systems, schools and hospitals teach children and treat patients with the state-of-the-art infrastructures and instruments, criminals are apprehended in time through a seamless convergence amongst different and distributed departments, and justice, more or less, gets dispensed. Similarly, in dire emergencies, people and resources are mobilized often with impressive speed and scale for appropriate response and recovery. Citizens still expect that their governments have to work with such alacrity and adroitness in all the situations all the time.

The gap is growing between the nature and the scale of many of the economic, social and environmental challenges that governments are facing and trying to solve and the identified resources, institutions, and systems on which they can depend. The context for this widening gap between need and supply is that the world is in continuous transition. There are big shifts in the underlying economic, political, environmental, and social architecture of large global systems. These shifts are well documented.

- **Political Transitions:** The rapid ascents of developing countries such as China, India etc. on the world stage have precipitated a geopolitical transition to a multi-polar world in which western countries can exert relatively less influence and authority.
- **Economic Transitions:** The world's financial system is still struggling with the continuing impacts of the global financial crisis as the interconnected economies of Asia, Europe, and North America grapple with currency and current account imbalances, sensitivity to global energy prices, and the risk of financial contagion. In short, there is a subdued recovery. At the national level, most developed countries are confronting years of fiscal adjustment and austerity, while large emerging markets struggle to manage the demands of rapid growth.
- **Demographic Transitions:** Demographic trends introduce further pressures. In industrialized countries, rapidly aging populations, increased longevity, and falling birth rates are straining social safety nets. Conversely, many emerging markets, especially India and the Middle East, are struggling to create employment opportunities for a booming youth population. Globally, population growth and urbanization are driving explosive growth in unplanned and informal settlements.

- **Environmental Transitions:** Global population growth to 9 billion people by 2050 will exacerbate pressures on ecosystems, agriculture, and water resources. The world is urbanizing at a growing rate; populations in cities and regions alike are confronting record food prices, while water basins are facing shortfalls. Natural disasters will take an increasing toll as economic growth puts greater value at risk. Meanwhile, climate change, ocean acidification, and biodiversity loss are decreasing the adaptive capacity of ecosystems on which people depend.

- **Technological Transitions:** Rapidly declining IT costs, rising penetration of mobile phones and increasingly pervasive broadband Internet access have fueled an information flood that has transformed the way people interact with the world and with each other. But intensive networking, which has enabled new and more powerful forms of connection and innovation, has spawned challenges of its own. The resulting "data deluge" is often not matched by a similarly rising capacity to discern, interpret and act effectively.

In the traditional sense, resilience is a measure of how well a system (an organization or institution, an ecosystem, a city or region, or indeed a whole country) recovers from an unexpected shock or disaster. Resilience also acquires the meaning of cultivating the assets, culture, and capabilities that render systems less vulnerable to risk, more agile and adaptable, and therefore better prepared for successive waves of change and disruption. It means not only bouncing back, but also bouncing forward. Public sector institutions need to become more resilient. As rising, complex, and distributed risks disrupt the public sector's ability to deliver services that protect and enable constituents, public welfare and security will depend increasingly on the public sector's ability to sense, respond, adapt, and evolve continuously.

Public sector resilience entails ensuring continuity of operations and avoiding disruptions to critical services. It also means looking for different ways to lift both its own productivity and productivity across society and the economy as well as create effective innovation systems in service delivery that take advantage of technological and other worthwhile developments.

There are other important questions. What can governments do, and how should governments behave, to make the societies they serve more resilient? How can they invest limited resources for the greatest impact on the skills, culture, and assets of a strong economy, an inclusive and confident community, and a more sustainable natural environment? Precisely speaking, the struggle for resilience will not be won within the walls of government agencies, but in the broadly distributed communities that they serve and with which they interact regularly. In the face of powerful, uncontrollable and exogenous shocks, developing and sustaining a more resilient society becomes a principal objective for policy makers and the people. This requires two significant shifts in the public sector.

- First, decision-makers must begin to understand the aspect of resilience in a new way. Rather than a reactive posture of managing risks effectively once they materialize, resilience mandates a much more prospective, preemptive and proactive stance of forecasting risks and building the adequate and adaptive capacity to manage all kinds of unforeseen disruptions and interruptions.
- Second, decision-makers have to expand their view of how to cultivate resilience inherently. That is, instead of just focusing on increasing resilience of government institutions, it involves the appropriate empowerment of communities to become adaptive.

While improving efficiency of the limited set of levers and controls available to public institutions is a worthwhile objective, resilience as a policy objective requires new thinking and acting within government to drive innovation and productivity in the economy and in society. Central to this new task is to plunge into well-intended policymaking and to design insights-driven programs about the strength of diversified portfolios that improve overall risk-return tradeoff.

ENABLING GOOD GOVERNANCE THROUGH THE USE OF ICT

The emergence of Information and Communications Technologies (ICT) has brought in a litany of pragmatic mechanisms and means for faster and better communication, efficient storage, retrieval and processing of data and exchange and utilization of information to its users, be they individuals, institutions, businesses, organizations or governments. With growing digitalization, distribution, consumerization, industrialization and deeper connectivity via low-latency, reliable and high-bandwidth networks, the reach and richness of IT services have gone up substantially in the recent past, Business has become e-business, commerce has become e-commerce, etc. So far as governments are concerned, the coming together of computerization and Internet connectivity/ Web-enablement in association with process re-engineering promises faster and better processing of information leading to speedier and qualitatively better decision-making, policy establishment and enforcement, sensitivity to peoples' expectations and aspirations, greater accessibility, responsibility and accountability, better utilization of resources and overall good governance. In the case of citizens, it holds the promise of enhanced access to information and government agencies, efficient service delivery and transparency in dealings and interactions with their government.

With the increasing awareness among citizens about their rights and the resultant increase in expectations from the government to perform and deliver, the whole paradigm of governance has changed a lot. The government today is expected to be transparent in its dealings and deeds, accountable for its activities, effective and efficient in service delivery, and faster in its responses. This has made the use of the ICT imperative in any agenda drawn towards achieving good governance. It has also led to the realization that information and communications technology (ICT) could be used to achieve a wide range of objectives and lead to faster, inclusive, richer and more equitable development with a wider reach. While recognizing the potential of ICT in transforming and redefining processes and systems of governance, it is obvious that digital governance is the logical next step in the use of ICT in the business of governance in order to ensure wider participation and deeper involvement of citizens, institutions, civil society groups and the private sector in the decision-making process of governance.

Electronic governance is basically the application of ICT to the processes of government functioning in order to bring about 'Simple, Moral, Accountable, Responsive and Transparent' (SMART) governance. This would generally involve the use of ICT by government agencies for any or all of the following reasons:

- Exchange of information with citizens, businesses or other government departments,
- Speedier and more efficient delivery of public services,
- Improving internal efficiency,
- Reducing costs / increasing revenue,
- Re-structuring of administrative processes and,
- Improving quality of services.

The use of ICT in governance processes needs to be framed within a longer process of technology-driven public sector reforms. This process over the past decades has contributed immensely to shape up a novel vision for the public sector, where information sharing, transparency, openness and collaboration are the key concepts with tremendous organizational and policy implications. In the governance of urban areas, city managers are faced with the challenge of balancing three overriding concerns: achieving a high quality of life for all citizens, maintaining economic competitiveness, and protecting the natural environment towards its sustainability.

According to the Smarter Governments for IBM's Vision of the Smarter Planet paper (English, 2011), as local and national governments work to infuse intelligence into their transport, energy, water, telecommunications and other systems in order to stimulate world economies and benefit citizens directly, it begs the question: can the operations of government itself become smarter? Smarter government will do more than simply regulate the outputs of our economic and societal systems. It will be a smoothly functioning system itself interconnects dynamically with citizens, communities and business groups in real time to spark growth, innovation and progress in parallel. The challenges are however many from departmental silos to process delays to lack of transparency and accountability in the upper echelon of the governance. But governments around the world are really showing some real progress in these areas and aspects.

Smarter government means collaborating across departments and with communities to become more transparent and accountable, to manage resources more effectively and economically, and to give citizens the right access to information about decisions that affect their lives. In the UK, Southwest One, an innovative joint venture, is providing shared services by integrating many functions of the Somerset County Council, the Taunton Deane Borough Council, and the Avon and Somerset Police. In Albuquerque, business intelligence (BI) solution has sharply improved efficiency by 2,000% in the city's ability to generate reports and keep citizens informed. Smarter government means helping to promote economic growth by synchronizing cumbersome processes and simplifying reporting requirements, which are especially burdensome to small firms. For example, the Maryland Department of Labor, Licensing and Regulation, has enabled online renewal of professional licenses and public verification of valid license holders.

Also, the Belgian Crossroads Bank for Social Security has automated 42 services for employers, eliminating 50 social security declaration forms. As a result, 23 million declarations were made electronically in 2008, a major productivity benefit for Belgian businesses, saving them an estimated 1.7 billion Franc a year. At the most fundamental level, smarter government means making operations and services truly citizen-centric. Leading governments are integrating their service delivery, establishing offices that support multiple services and placing the most needed transactions on the Web. For example, Australia's Centrelink helps the government to provide appropriate service offerings based on citizens' life events, such as marriage, the birth of children, and the need for elder care. Kyoto, Japan, created a website that allows all people, regardless of their abilities or native language, to access city information. And then there are those times when being a citizen-centric with speed and accuracy may be a matter of life and death. During the recent wildfires in California, government agencies turned to twitter to provide real-time updates on the status of the fires directing people without power but with mobile devices to Google Maps for evacuation information. The critical imperatives for 21st century governments are pictorially illustrated in Figure 1(Cotton, 2012).

Figure 1. The critical imperatives for 21st century governments (Cotton, 2012)

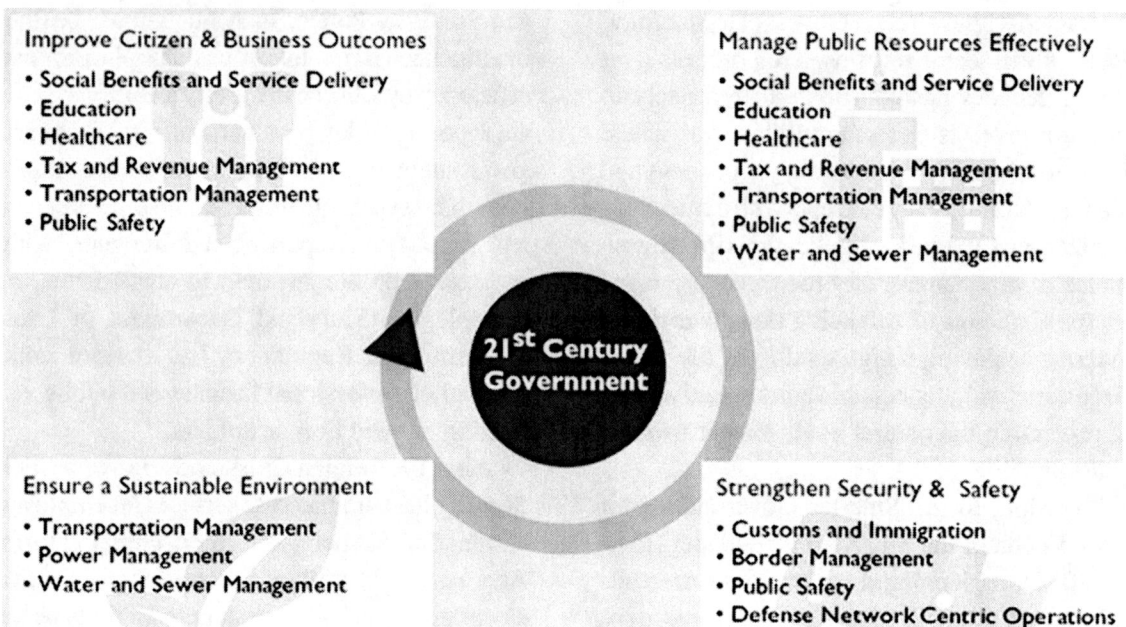

EXPLAINING MOBILE GOVERNANCE

It is all about smartly leveraging mobile technologies to extend the reach of e-governance and to empower people in their day-to-day activities and decision-making. The paper by Kailasam (2010), on Leveraging Mobile Technology to extend the reach of e-Governance, has a lot of useful information on this topic. Mobile computing has brought in a series of delectable transformations for many business domains. For the business of governance too, the aspect of mobility brings a lot of noteworthy transformations. M-governance is a part and parcel of a broader phenomenon of mobile-enablement. That is, every application and service (personal as well as professional) is being enabled with mobile access with the arrival of slim and sleek and multifaceted Smartphones. The mobile-enablement takes every sort of electronic services and makes them available to the wider community of users and commoners via mobile devices. Besides Smartphones, there are other wireless devices that are nomadic, portable, and

mobile. The device ecosystem is on the consistent climb and hence more people across the world are being empowered to access governmental services and applications. Mobile banking, payment, commerce, business, and transactions are growing faster with the enhanced accessibility, consumability and simplicity being delivered through a host of advancements in mobile computing.

Mobile services are being delivered in the wireless format and hence the capital investments to be made for communication infrastructure are relatively less compared to physical networks that once ruled the whole world for telecommunication purposes. Nowadays communication has become wireless and wireline and sometimes is a hybrid of both. Strictly speaking, m-governance is the extension of e-governance and is the excellent instrument for easing the last mile connectivity issue. There are mobile application technologies, tools, and tips in plenty these days. Every service and product organizations are being hyperactive in the mobile space. Increasingly governments are making practical strategies to leverage the mobile platforms, processes and practices to serve their citizens in a more advanced fashion. m-Governance has the innate potential to help make public information and governance services available "anytime, anywhere" to citizens and officials. Mobile services are cheaper as well as accessible in most of the rural areas. m-Governance is particularly suited for the developing world where Internet access rates are low but mobile phone usage is growing rapidly in both urban and rural areas. Globally, the number of mobile phones has exceeded the number of fixed/wired phones.

M-GOVERNANCE AND E-GOVERNANCE

M-Governance is not a replacement for e-Governance, rather it complements e-governance in many ways. E-governance is through hosting citizen services and information portal through which people from anywhere anytime any device any network could connect and get all the right and rightful services. The role of ICT in providing electronic governance in an effective as well as efficient manner is definitely immense and growing steadily. ICT is being touted as the best and cost-effective enabler of government functions. Dynamic formulation and framing of people-centric and high-quality services and innovative delivery mechanism are the ultimate goal of e-Governance. There are several government agencies distributed across and ICT plays an important role in discovering appropriate services, composing them to suit the peoples' needs and delivering them in time with all the inherent qualities are the hallmarks of digital governance. With the constant rise of peoples' expectations, national, federal, state, local or city governments are duty bound to employ proven and potential technologies to meet up the expressed and enigmatic aspirations.

In short, comprehensive and consolidated government portals and service delivery platform (SDP) infrastructures are being enabled to be accessed by mobile devices. All kinds of government services are categorized and hosted as service catalogues and repositories for people to search, pick up the right ones, and access them instantly. There are markup languages, application development frameworks, portal technologies, and localization and internationalization methods galore these days in order to facilitate mobile governance. With the bombardment of exquisite and eye-catching, trendy and handy devices, mobile governance is bound to be accepted and used in a big way in the days to unfold. ICT infrastructures and platforms are going through a variety of augmentation, acceleration and automation to deliver citizen-centric services to people on the move with all alacrity and clarity. Additionally, there are next-generation mobile solutions and services, application frameworks and architectures, service repositories, application stores, design patterns, metrics, and key guidelines by mobile product vendors and research labs.

The private sector has been greatly leveraging the use of mobile phones for delivery of value-added services for the following domains which, however, are mostly SMS based.

- Banking.
- Media.
- Airlines.
- Telecom.
- Entertainment.
- News.
- Sports.
- Astrology.
- Stock information, Movie tickets, etc.

There are also a number of good initiatives in the government sector in India using mobile computing extensively and innovatively which again are all SMS based. Some examples follow:

Food & Civil Supplies

- Tracking Lorry Movements.
- Information on availability of Ration at Food-provision Shops.
- Irrigation and Water Resources.
- Reservoir Level Monitoring.

Urban Local Bodies

- Grievance Redressal.
- Garbage Dumps and Garbage Removals.

Water Supply

- SMS Service for Water Supply Tanker.

Railways

- Ticket Booking Service.

Education

- Examination Results and Mark Lists.

Other such initiatives are in areas including Agriculture, Weather Reports, Market Prices, Seed Availability, etc.

Here are a few widely reported examples of international initiatives for tending towards full-fledged mobile governance in their regions.

M-Dubai – 4488 – Push & Pull Service

- Civil Aviation – Flight timings.
- Police – Fines.
- Notification of Expiry of Trade Licenses etc.

Singapore

- Trade Licenses.
- CPF contributions.
- Road Tax Renewal.

- Passport Renewal.
- Government notifications.
- Consumer Price Index.
- Performance of the Singapore economy.
- Court Hearing.
- Track Traffic Information.
- Live traffic images.
- Public Works monitoring etc.

Estonia

- Mobile Parking.
- Mobile Transport Ticketing.
- Mobile Payments in Shops.

South Africa

- Cell-Life Update: Using Mobiles to Fight HIV/AIDS.

Uganda

- Disease Surveillance with Mobile Phones.

Other Initiatives

- Emergency Services.
- Traffic Information.
- Payment of Government Fees.
- Children's' Absence from School.

Other examples include police department where wireless technology has always been an important part of law enforcement. In India, the Chennai City traffic police department has introduced an SMS service and a caution system for those violating traffic rules. Through this newly introduced SMS service, the public can inform the traffic control room about road accidents, vehicle breakdowns, and traffic jams. Also, the public can get detailed information before buying a particular vehicle that might have violated traffic rules by just sending an SMS with the registration and engine number of the vehicle. Health and safety inspectors can now file their reports from the field in real time using mobile or handheld terminals, eliminating paper forms and the need to re-enter the data collected when they get him back to the office. Mobile-enabled insurance systems allow drivers and vehicle owners to capture and flash his or her vehicle's pictures if it meets any accident from the incident site so that the insurance settlement happens quickly. If the mobile governance is not only about efficiency but it also allows for citizen activism. In the Philippines, citizens are able to help enforce anti-pollution laws by reporting smoke-belching public buses and other vehicles via SMS. SMS is also being utilized to get citizens involved in the fight against crime and illegal drugs.

MOBILE/WIRELESS APPLICATIONS IN THE GOVERNMENT SECTOR

Mobile governance can be pragmatically applied in many areas of public sector as summarized below. In the years ahead, there will be more applications for mobile phones, networks, applications, services, and content.

Mobile Communication (G2C2G)

Providing information to the public is not a trivial activity. It is the foundation and fundamental thing for citizens' empowerment. Without relevant information, citizens are unable to form intelligent and collective opinions and thereby are unable to act on the issues before them in time meaningfully. Mobile devices are not only communication channel but also the principal access instruments for governments to reach citizens (G2C). For example, Singaporeans could choose to receive SMS alerts for a variety of electronic services such as renewal of road tax, medical examinations for domestic workers, passport renewal notifications, season parking reminders and parliament notices and alerts. Citizens of Malta could register to re-ceive SMS notifications of court sitting/hearing deferrals, license-renewal, examination results, and direct credit payments from the department of social security.

In the United Kingdom (UK), the London police have included text messaging in their alerting service options. This service sends alerts to businesses in London about security threats including bomb alerts. The 24-hour service contacts all users in real time with a message that is sent within 30 seconds of the alert being received by the police. Despite a monthly fee for the pager/text message service and the existence of a free email service, there are more businesses that signed up for the pager/text message alerts than for the email alert system. This progressive and positive trend indicates the popularity of m-Governance services. Aside from these opt-in G2C communications via mobile phones, SMS is also being used in emergency broadcasting. At the height of the SARS incident, the Hong Kong government sent a blanket text message to 6 million mobile phones in a bid to scotch fears emanating from rumors about the intended government action to stem the disease in the bud. SMS is also a cheaper and common channel for citizens to communicate with government (C2G). The deeper connectivity amongst the various participants and constituents of mobile commerce goes a long way in visualizing newer services and delivering them proactive and unobtrusively to people in order to keep them happy. Refer to Figure 2 for emerging mobile and mobile related technologies.

Mobile Services (Transactions and Payments)

SMS and other mobile devices not only provide a channel of communication between citizens and government, but also enable government-to-citizen (G2C) transactions. The Singapore government has decided to leverage the power of SMS for its goal of increasing population. Its social development unit acts as an official dating agency

Figure 2. The emerging mobile landscape

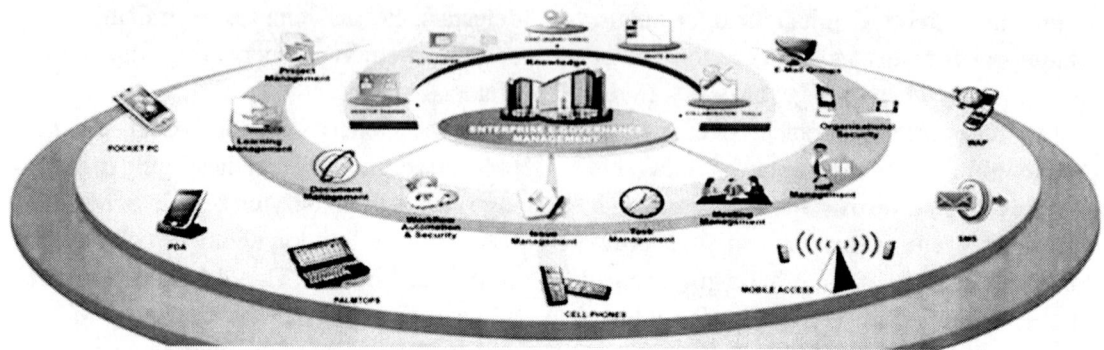

for educated single people. It gives members 40 free messages over their mobile phones to allow them to contact eligible professionals. Singapore's national library board has also introduced an SMS service that allows regular users to query the status of their accounts and books borrowed, and they could receive reminders before the due date of their book loans. They can also undertake transactions such as making book renewals or paying fines using their mobile phones. The service costs each user $5 per year. Other examples of technology-enabled initiatives in the industrialized and advanced countries include the following:

- Norway's tax collectors have introduced SMS tax returns. Taxpayers who have no changes to make to the form they receive in the post can now simply send a text message with a code word, their identity number and a pin code instead of returning the form by mail. This new service benefits the estimated 1.5 million Norwegian taxpayers who normally return this income tax form by mail.
- In Finland, SMS tickets can be used for Helsinki's public transport system. These tickets can be ordered by sending a text message and the user is billed through his or her regular mobile phone bill. The ticket itself is also delivered to the commuter by SMS. While the use of m-payment in

e-government is still limited at this point of time, it is expected that its usage will expand remarkably as mobile payment systems evolve from simple payments for digital content and services to complex integrated handset, bank and operator payment. Its use for transacting business with governments is bound to grow and glow.

- In India, train tickets can be booked using mobile phone, anywhere and anytime WITHOUT using an Internet connection. Train passengers will be able to use their mobile phones to book tickets through a simple text-messaging-based-SMS facility or a menu-based dialing service –USSD service.

Mobile Democracy

Use of SMS and mobile devices for citizens' input to political decision-making are m-Governance applications with tremendous potential to enhance democratic participation. At present, there are no significant experiments with m-democracy in developing/transitional countries. Hence evidence here is taken from experiences in the UK. Most of the UK experiments with electronic voting, including voting via mobile phones, are meant to discover and follow more convenient ways to involve citizens in national elections. Several concerns would have to be attended to before voting over mobile phones gains widespread acceptance. Questions of security and secrecy are on the top of the list. With the traditional voting method, it is sufficient to present oneself at the polling (voting) station. A m-voting system has to ensure that the message sender is a registered voter and that no one abuses the system to vote more than once or vote in place of another person. Voters in Liverpool and Sheffield in May 2002 local elections were given PIN numbers to use if they want to vote by text message.

Another issue is to make the system as user-friendly as possible. If PINs are used, chances are many would forget their PINs if they are too long. Then there is the problem of using a phone keypad to key in parties or candidate names. Finally, the voting procedure itself must allow voters at any stage to repeat the instructions and choices. In addition, the capacity of the system would need to be sufficient to deal with peak periods because congested telephone lines are as frustrating as long lines at the polling stations. However, these are 'technical' issues that may not be as difficult to overcome as voters' willingness to use mobile phones and SMS to vote grows.

Mobile Administration

M-governance also provides opportunities to improve the internal operation of public agencies. Again, there are few instances of such applications yet in developing/transitional economies. Another potential for wireless technology is that it may provide a seamless environment for government employees to stay connected from any device. Up-to-date government-to-employee (G2E) information and services can be provided at any time, whether the data they need is on the Internet, on their network, or on a portable device under their control. The other potential usages of m-administration include the following sectors:

- *Health* e.g. Monitoring Progress, Telemedicine
- *Irrigation/Water Resources* e.g. Capturing Reservoir Storages; Monitoring Releases of water through Sluices; Monitoring Minors and sub minors area under a given Canal
- *Electricity Board* e.g. Citizen Grievances; Bills Payment/Collections
- *Public Works* e.g. Monitoring
- *Urban Local Bodies* e.g. Citizen Grievances; Bill Payment/Collections; Garbage Collections

Impacts of M-Governance

An examination of a number of mobile government applications for citizens within various continents, countries, counties, and cities shows that mobile business applications may not be easily applied to governmental administration, yet there are compelling reasons for doing so. Is it unreasonable for citizens, for example, to expect technology-enabled services from their government similar to the services available to them from private sector organizations such as airlines, banks and utility companies where flight reservations, currency exchanges and bill payments are now possible without human intervention? Therefore there is an increasing need to introduce mobile technologies for governments. Mobile Government is one of the new and important developments in the realm of e-governments. The high rate of mobile phone penetration opens a new route for governments to reach their citizens fast and provide timely information to them. With the vast improvisations and innovations being incorporated in enterprise mobility, tech-savvy governments need to focus on embedding new-generation mobile technologies, platforms, processes, patterns, products, and practices in their mobile governance strategy to proactively delivering a growing array of information, transaction, knowledge and even physical services.

The promise of e-government initiative is to provide greater and quicker access to government information and service list and this tendency is fast-progressing in many developed countries as the infrastructure is not a big issue and the adoption of e-government processes and policies seem to be at a good pace. In contrast, the e-government adoption rate in developing countries is relatively lower. This may be mainly due to the lack of technical infrastructure in supporting the government's efforts as well as such factors involving cost of getting an Internet-enabled computers, tablets, and Smartphones. People are slowly yet steadily coming into the groove of electronic, mobile and ubiquitous governance. The demand and prospects for putting technology-sponsored governance is huge and so are the opportunities for the governments to reach the ever expanding network of citizens around the country. A unified information campaign could be launched by government to advertise all its SMS- based services, will raise peoples' awareness and use of these services. There is a need to enhance the electronic linkages among offices within an agency and among government agencies.

M-Governance in Emergency Situations

The Italian Ministry of Foreign Affairs, during the aftermath of the Asian Quake, sent an SMS to Italians located in the disaster-stricken area. The message was: "Answer indicating your identity, health status, and place where you are." With approximately 15000 SMSs, the ministry tried to trace Italian citizens who faced the disaster. According to the official information, several hundreds of people immediately replied that went on helping the embassy and rescue teams to list the affected people in the Tsunami area. The Italian Government obtained the list of people located in the disaster area from mobile phone companies that provided the information based on the international-roaming services. This is a real-life example of how m-governance can help in such an emergency situation. According to the National Oceanic and Atmospheric Administration's center in Hawaii, the earthquake was detected and a warning about the approaching tsunami was sent to the Pacific Tsunami Warning Center. The reason that the warnings were unable to reach the millions in the disaster region is because none of the countries had a working tsunami warning network.

The crucial part is the coordination among the technology experts, governments, and emergency response agencies not only in the tsunami hit region but all around the globe. Along with the tidal gauges and sensors in the ocean, a well-organized communication system, a well-understood the emergency preparedness and training of resort operators, fisherman and public in general have collectively helped towards the solution. Mobile applications in disaster management will be one of the most useful and critical areas of implementation. It will be useful not only for the prevention activities such as mobile alerts but also an invaluable tool for the recovery efforts of, for example, rescue team working in the fields.

It is true that the system absolutely rely on the wireless communication channel that is not widely spread in several countries. If there is any untoward incident such as earthquake, hurricane, flood, fire, etc., the communication infrastructure in that area may collapse and result in discontinuity and isolation. But for prevention and lifesaving purposes, such a system may give significant difference. The system should also be socialized to the public continuously so they can trust the information sent by the government. In villages people are more comfortable with their local language so mobile application with local language support will add value to the efforts taken for m-Governance and increase the number of citizens to avail the benefits of m-governance. By leveraging the technology and mobile government implementation, there is a hope that in the future the impact of disasters can be lessened and more lives can be saved.

M-Governance Benefits and Challenges

M-governance can bring a litany of potential benefits for the public sector, but it also faces innumerable challenges as discussed below.

Benefits. The main benefit that m-Governance brings is its boundary-breaking potential. That is, it decimates the location dependence. This truly allows people to connect at any point of time, place, device, and media. Thus mobile networks turn out to be a blessing for governments to touch base with common people, to extract peoples' feedbacks and desires and to act on them passionately. Because of its immediacy and convenience, it also reduces the barriers to public service operations encouraging citizens or service providers to make use of the technology with a purpose, plan and practicality. These core benefits can be seen reflected in a broader set of m-Governance benefits including the following:

- **Increasing the Productivity of Public Service Personnel:** M-Governance allows public servants to enter data into digital systems exactly where they are in the field. Not only does this move data-gathering closer to real-time operations, it also reduces the time public servants spend on data activities, thus releasing more of their time for value-added and service-related activities. For example, where previously reports would be noted on paper in the field and then retyped back at base, they can now be entered directly, not only removing duplication of effort but also reducing the number of data errors.

- **Increasing the Effectiveness of Public Service Personnel:** Public servants in the field currently have to make do with the data they carry around with them in their heads or in portable files. With m-Governance capability, they can take the whole of digitized governance with them into the field allowing them to make informed decisions and actions.

- **Improving the Delivery of Government Information and Services:** M-Governance can deliver data and services whenever and wherever the citizen is. This has a benefit to citizens as they can get immediate access to whatever they want no matter where they are. It also has a benefit to governments for example, in sending terror alerts or other very time-sensitive information.

- **Increasing Channels for Public Interactions:** Mobile enablement provides an additional channel for interactions for all stakeholders in governance: service consumers and providers, policy makers and peoples' representatives, non-government organizations (NGOs), and civil society representatives. This brings forth additional choice for all.

- **Lower Costs Leading to Higher Participation:** Mobile enablement goes a long way in visualizing new applications and services for people. Opinions, musings, tweets, blogs etc. besides voting through mobile devices are all going to be the trendsetters for governance and people alike. The establishment of high-bandwidth, low-latency, and dependable mobile communication infrastructures has led to a stimulating and scintillating foundation for conceptualizing and concretizing a plethora of premium and people-centric services.

Challenges. M-governance also faces a number of challenges as enumerated below.

- **Cost:** M-Governance tends to be a path-breaking channel for electronic governance thereby it will create additional costs. This will continue until mobile governance can truly substitute for other delivery channels. Such substitution will be viable for applications within governments. At least some governments have been able to adopt innovative costing strategies, for example, using fee-sharing arrangements that avoid the public sector having to provide many up-front costs.

- **Mobile Digital Divide:** Mobile phones are fast penetrating into the society. Still there is a big segment without mobile phones. This digital divide has to be closed through governmental incentives thereby the goals and objectives of mobile governance reach every individual without fail.

- **Mobile Mindsets:** Mostly mobile devices are non-serious purposes such as voice communication, texting, entertainment, etc. Mobile-enabled commerce and business activities are slowly coming up with the availability and maturity of 3G and 4G technologies. Still, it has to traverse a long way before mobile devices are being utilized for business functions overwhelmingly.

- **Trust/Security:** Security and privacy are two principal issues with mobile communication. With business transactions and payments are being accomplished through mobile phones increasingly, the security implications are grave. The security methods being followed in physical networks cannot solve all the problems prevalent in mobile networks. Thus mobile data, network, application, and infrastructure security requirements go up substantially

- **Data Overload:** Mobile devices enable people to be connected with the outside world all the time. Many types of people for a variety of purposes connect and communicate. Data services will grow manifold in different dimensions. The one unwanted result of this all the time everywhere connectivity is data overload. As mobile devices are getting integrated with other devices in the vicinity as well as with remote devices and applications in the clouds, this data overload has a chilling effect.

EMERGENCE AND EVOLUTION OF CLOUD TECHNOLOGIES

The trend-setting cloud paradigm actually represents the cool conglomeration of a number of enterprise-scale, mission-critical and state-of-the-art technologies such as Virtualization, service oriented architecture (SOA), Software as a Service (SaaS), autonomic, utility, on-demand, cluster and grid computing models, etc. Though the cloud idea is not conceptually new, this seamless synchronization has practically brought in myriad shifts on the IT domain especially on IT infrastructure optimization. The cloud-induced drastic implications on IT have started to impact hugely on the business side too. Business agility, adaptability, affordability, sustainability, efficiency, and resiliency are seeing a nice and neat reality steadily with the conscious and consistent incorporation of the disruptive and transformative cloud technology.

Cloud computing has brought in series of novelty-packed deployment, delivery, consumption and pricing models. The cloud and SOA paradigms have gelled well together in order to stunningly simplify application design, development / assembling, enhancement, management, and retirement. The noteworthy contributions of the much-dissected cloud computing are the faster realization and proliferation of dynamic, converged, adaptive, on-demand, and online IT infrastructures (server, storage and networking solutions). As IT is the best business-enabler, any differentiators being realized in IT are bound to pervade deeper towards business automation, augmentation and acceleration. The decisive distinctions are that cloud can easily and optimally guarantee most of the non-functional requirements (Quality of service (QoS) attributes) such as availability, throughput / performance, on-demand scalability and elasticity, affordability, global-scale accessibility and usability, energy efficiency etc. Computing is set to become the fifty social utility with the stability and maturity of the cloud paradigm.

Having understood the exceptional and elegant functionalities and features of cloud infrastructures, business executives and entrepreneurs have embarked on the delightful cloud journey meticulously. That is, businesses are steadily modernizing and moving their IT resources to be packaged and presented as network-accessible and cloud-hosted services that can be publicly and dynamically found, bound and used by worldwide users for a small fee. This well-planned transition could facilitate a higher and deeper reach and richness in application construction, delivery and consumability. Product vendors having found that the cloud style is a unique proposition are reflectively moving their platforms (development, execution, integration middleware, management, database systems, etc.) to clouds. Infrastructure providers are re-architecting and remedying their data centers to be next-generation cloud centers or establishing cloud centers from the scratch to optimally host a variety of ICT services and platforms of worldwide individuals, innovators, and institutions.

Cloud service providers (CSPs) are very aggressive in experimenting and embracing the cool cloud ideas and today a growing array of business and technical services are being hosted in the clouds for their global customers and consumers. In a nutshell, on-premise and local applications are becoming online, on-demand and off-premise applications. With the unprecedented advertisement and articulation of cloud concepts, the cloud movement is picking up fast and the cloud moment has just arrived. Besides the modernization of legacy applications and positing the updated and upgraded in clouds, fresh applications are being designed to be on clouds. As the cloud space is solidifying in a steady and suave manner, personal and professional services are being increasingly readied and relocated to remote cloud platforms. Further on, rich content and Internet applications, social, embedded, mobile, and other specific applications are being housed in clouds. Well-known enterprise information systems (EISs) such as enterprise resource planning (ERP), customer relationship management (CRM), human resource management, supply chain management (SCM) solutions are successfully finding their new residence in clouds. Software infrastructure solutions too are also moving to cloud environments in order to reap the unique benefits. It is not an exaggeration to write that a number of strategic and significant movements happen silently in the cloud era.

The gist of the raging cloud paradigm is to enable a clear-cut separation between IT resources and services from the underlying IT infrastructure and this in turn grandiosely empowers a variety of existing and evolving IT and business applications to be produced and provided as services to worldwide subscribers on demand. The seamless and spontaneous delivery from converged, virtualized, automated, shared, and elastic environments offers governments the ability to break down aging, monolithic, closed, inflexible IT silos while improving collaboration and meeting increasing user demands for IT-enabled, cost-effective and innovative citizen-centric services on demand. Cloud infrastructures can bring in flexibility, efficiency, and democratization around resource allocation and this results in agile IT service delivery enabling resource provisioning in minutes and the time-to-market for fresh products and services gets reduced by more than 50% . The resource utilization goes up considerably and there will be a 50% reduction in capital costs and 30% reduction in operational costs.

Cloud-Instituted Innovations

Cloud computing lays the foundation for originating a number of momentous and memorable business and technical innovations as described below.

- **Technology Cluster:** Clouds represent the seamless convergence of proven and potential technologies, tools and techniques (Consolidation, virtualization, integration, federation, composition, provisioning, etc.).

- **Heterogeneity to Homogeneity:** Clouds hide the multiplicity and heterogeneity-induced complexity of IT environments by leveraging a variety of optimized management platforms and containers, such as virtual machine monitors (VMM), power, resource, and workload management modules. A lot of internal deficiencies and discrepancies are smartly made transparent to ultimate users. Cloud is being presented to end-users as a single and simple instrument and interface for fulfilling the vast and varied compute needs.

- **Service Oriented Infrastructure (SOI):** With the faster adoption of service orientation principles, newer interaction, orchestration, and consumption models have erupted and are evolving to meet up diverse needs of users. Software as a service (SaaS) is the base, which is laying the foundation for encouraging an enormous growth of every IT resource getting expressed and exposed as a service to the general public via the Web. The growing tendency is nonetheless but IT as a service (ITaaS). As the much-anticipated service era gradually and gracefully unfolds, clouds' contribution as the elastic and epoch-making service infrastructure and

platform is really tremendous and trend-setting for the forthcoming knowledge era. Clouds will become the indisputable and insightful infrastructure for next-generation service engineering, deployment and delivery needs.

- **Business Innovations:** The indomitable cloud idea has laid a strong and stimulating foundation for emitting newer business, service, licensing and pricing models that are more tuned to changing business sentiments and customers' liking. There will be a paramount shift from the current capital expenditure (Capex) to operational expenditure (Opex). Consumption-based metering and billing will become common and casual. Ultimately cloud enterprises will see the light with the beneficial synchronization between SOA, EA and cloud infrastructures.

- **Green IT:** Due to the persistent calls from different quarters for energy efficiency and reduction of greenhouse gas emission for minimizing climatic changes, clouds are being established as the viable and valuable IT instrument for greener environments.

- **IT Optimization:** Optimization of IT development and operations is gaining traction. Clouds contribute exceedingly well to this optimization goal. In short, clouds fulfill lean, elastic, catalytic, agile, and adaptable IT. Further on, cloud enables computing to be the fifth utility. Finally IT as a Service is a foregone conclusion with the maturity of cloud standards, products and technologies.

- **Extreme Elasticity:** Capacity planning is a difficult exercise for IT as predicting exact usage and acquiring just enough IT resources to avoid excessive under or over-provisioning is really tough call in this volatile world. Other internal as well as external factors contributing for this predicament and pain are season-specific usage spikes that demand additional compute resources that otherwise remain idle. Elasticity of IT resources leads to application scalability. Clouds offer resources on demand that can settle up or down with the changing demands of businesses.

- **Tending towards the On-Demand and Shared Era:** The vision of everything on demand (computing, communication, intelligence, scalability, information, service etc.) is set to see the light when cloud reaches a level of maturity and stability.

Cloud Migration Strategy

A well-planned strategy needs to be an important ingredient while making the moves of taking IT assets to clouds. Clouds are being prescribed as the next-generation environment for hosting and delivering business and IT solutions. Enterprise IT teams are keenly formulating and finalizing appropriate plans for modernizing and moving on-premise and local applications to remote, on-line, on-demand, and off-premise environments. However clouds are not the panacea for all the IT requirements and all kinds of applications are not suitable for cloud infrastructures. Here is a list of application types that are more suitable for clouds.

- Require rapid deployment,
- Are approaching a technology refresh and/ or the end of contractual obligations to a legacy environment,
- Have variable storage needs,
- Need bursting capability,
- Use virtual servers rather than physical servers (the reliability, performance, or security of dedicated servers is considerably more expensive when procured through the cloud), or
- Are based on federal funding with "cloud first" recommendations.

The IT assets destined for cloud environments need to be analyzed in all the critical aspects before the approval for the migration. The cloud paradigm has brought in multiple advancements in the IT field so that all kinds of IT-inspired governance automation, acceleration and augmentation get a widespread recognition. This document is to illustrate the various optimizations the government sector is to receive consistently in the days to unfold with the adaption and adoption of the raging cloud idea.

BIG DATA ANALYTICS: MAKING GOVERNMENT FASTER AND SMARTER

The modern world generates a staggering quantity of data and the business of government is not an exception for this progressive trend (OECD/ International Telecommunication Union, 2011). Across the public sector, tremendous amount of data is amassed in the course of running public services ranging from sending welfare payments to the needy ones and health services for all and through to issuing passports, a number of identity cards for availing governments' citizen-centric services, employment opportunities, and driving licenses. From the rich and employed people, a variety of taxes are being taxed creating a lot of data to be preserved for analysis and for the strict compliance to governments' rules and regulations. Thus it is clear every tangible sector is producing a vast quantity of data and hence data analysis cannot be taken. All kinds of data coming from multiple sources need to be carefully collected, clustered, classified, polished, filtered, cleansed, persisted, processed, analyzed, and mined for extracting hidden patterns to create and sustain knowledge discovery and to disseminate them in time to all the authenticated and authorized stakeholders to contemplate the future course of actions with all confidence and clarity.

The term Big Data has come to refer to these very large multi-structured and multi-sourced datasets, and big data analytics refer to the process of seeking out actionable insights by combining and examining them. Regardless of the stance and stand a government takes on openness (that is, decisions on making public data free to use, reuse and redistribute), an abundance of data coupled with high-end yet affordable computing power gives the public sector a bunch of newer ways and means to organize, learn and innovate for the betterment of service assurance and delivery. The opportunity for public service transformation is becoming real and rewarding with all the noteworthy advancements in the technology landscape. For citizens, the application of data, technology and analytics can cut paperwork immensely, get questions answered more quickly, help people find and claim the benefits they are entitled to, and tune front-line services more closely to individual needs and expectations. For example, the enhanced use of data and analytics in the health arena could help ensure patients in care-homes receive the right medicines at the right time or help hospitals further personalize patient care and advice to minimize readmissions after surgery. In the welfare arena, better segmentation and personalization could help identify the support that unemployed people need and get them into long-term work (Yiu, 2012). Refer also to the report: The Big Data Opportunity, Making government faster, smarter and more personal, Policy Exchange, UK 2012.

The quality of service (QoS) attributes (reliability, availability, security, accessibility, etc.) for all kinds of governmental services and solutions being extended to citizens are to be fulfilled with ease and thereby common people are bound to receive their services with the same quality the worldwide corporations and organizations provide to their esteemed consumers and customers. At a macro level, there is sufficient scope to improve the overall efficiency of government operations and delivery systems in place, to accelerate efforts to reduce fraud and error, and to make further inroads into the tax gap (the difference between actual tax collected and theoretical liabilities). Big data technologies alone are not a silver bullet for transforming the public sector but the role and relevance of big data analytics in consonance with other associated and allied technologies is to climb up further as the society is tending towards a more knowledge-driven and market-oriented. Underlying data issues like quality, standards and bias need to be recognized and addressed. And governments must have the capability to conduct, interpret and consume the outputs of data and perform analytics to work intelligently.

Big Data Use Cases for Governments

Government is increasingly faced with growing data volumes and shrinking budgets, new use cases and legacy infrastructure. Traditional systems are not only expensive but also poorly-equipped to handle the government's twenty-first century needs for scale, cost efficiency and flexibility. Government entities from municipalities all the way up to federal intelligence agencies are building their applications on MongoDB NoSQL database to increase the speed of their application development and to provide higher quality applications (MongoDB, 2013).

Surveillance Data Aggregation

Government agencies are exploring new ways to increase national security, including innovative surveillance data collection. This data are pouring in from a variety of sources and in massive volumes. MongoDB provides an adaptable and scalable platform for aggregating various surveillance feeds and making sense of the data through real-time analyses, keeping the country safer.

Crime Data Management and Analytics

Legacy criminal record systems based on relational databases are often brittle and difficult to adapt to legal and regulatory changes. MongoDB makes it easy and cost-effective to adapt the technology as the law evolves and to incorporate information from disparate sources. With MongoDB, law enforcement organizations can deploy innovative, real-time analytics to identify offender-specific, geospatial and other crime patterns quickly and effectively.

Citizen Engagement Platform

Government entities are increasingly creating Web 2.0 engagement platforms to encourage civic participation. These platforms often include a variety of media types, metadata and social features for which relational databases are not a good fit. With MongoDB, government organizations can quickly develop, deploy and iterate on these platforms to spur civic engagement. MongoDB makes it easy to serve dynamic, personalized content, to create interactive features like contests and webcasts, and to strengthen the connection between government and the electorate.

Program Data Management

Managing program/user data and maintaining interoperability across various back office systems presents a growing challenge for municipal, state and federal agencies. MongoDB's dynamic data model makes it easy to store and access all this information in one place, increasing operational flexibility and decreasing overhead. MongoDB also enables agencies to leverage the same information in new ways, such as benefit fraud detection.

Healthcare Record Management

Healthcare records, from patient documents to procedure information, are some of the most complex and rich data in government records. A single data set can hold some records with tens of fields and others with thousands of fields. MongoDB provides a flexible platform for aggregating, storing, exposing and analyzing this data in one data store, helping reduce system sprawl and data dispersion and increasing operational efficiency (www.mongodb.org MongoDB is an open-source document database and the leading NoSQL database).

TRENDS AND DRIVERS FOR GOVERNMENT CLOUDS

There are several noteworthy trends that clearly indicate the cloud move more profitable and strategic for organizations especially for worldwide governments. The move to cloud can help governments reduce information and communications technology (ICT) costs sharply. Especially governments keep a number of data centers in order to host their services (citizen, social, legal, finance, etc.) and contents (government policies, proposals, surveys, programs, application forms, etc.). By applying cloud concepts, data center consolidation and optimization can be achieved in order to remarkably reduce capital as well as operational costs. Further on, clouds can improve agility and enable public sector organizations to do things in different ways. Government leaders recognize that ICT usually accounts for about 3% of the total government budget, so the real issue is all about using ICT as an efficiency enabler, capable of creating substantial savings across the whole budget.

Location independence is the foundation for the modern and flexible public sector workforce and a key to realize substantial cost savings. Network infrastructure should allow civil servants and knowledge workers to work seamlessly from home, from a branch office, or from a shared desk. Immediate benefits include: travel time and cost reduction, increases in efficiency, and improved work-life balance. Video conferencing through high-speed Internet and satellites allow geographically distributed government servants to interact thereby a lot of savings on travel, stay and other associated expenses can be cut down. Typically, the government owns a large number of buildings in any country. The cost of those buildings and of managing them is very significant. The government should implement a real-estate strategy to optimize the number and location of buildings, as well as to minimize operating and maintenance costs. The strategy should draw extensively on technology to ensure timely delivery of the potential cost savings from having a workforce that is no longer dependent on the facilities at any physical location. Citizens and businesses have come to rely heavily on web applications offered by the government (for example, for handling taxes and for getting customer services). Downtime is not only annoying but can also cost the government millions of dollars. For all these reasons, application security, reliability, and performance need to be continually monitored and improved. A surprisingly large proportion of the ICT budget is taken up by the maintenance tasks. Cloud technology represents a paradigm shift thereby companies focus on their core activities while outsourcing to a trusted cloud service provider (CSP).

GOVERNANCE CHALLENGES AND HOW CLOUD COUNTERS THEM

Governments today are confronting a number of serious challenges that affect their economies and their abilities to deliver core services to their citizens. They are faced with the harsh realities of swelling city populations that demand more services, aging infrastructure, declining budgets and increasing threats. As such, they are constantly looking to adopt ways and technologies that can help them address these challenges head on.

Rapid Urbanization

According to a United Nations (UN) communiqué, over half of the global population lives in urban areas and city planning officials are faced with contemplating critical decisions on how to deal with these swelling city populations. Countries such as China and India are expected to surge in urbanization, with cities in China growing from 40% to 70% of its total population by 2050. Indian cities are expected to grow from 30% in 2007 to 55% of its total population during that same period.

Addressing the multiple needs of a huge and rapidly growing population requires painstaking planning, preparation and resources. Whether it is finalizing the housing development plans, connecting more people to basic utilities such as water and electricity, or providing them a mechanism to highlight their grievances, the government and its various departments must provision resources in a way that waste is minimized and benefits to the common man as well as the government are maximized. At the same time, the resources must be allocated swiftly so that people are not underserved or the infrastructure is not over-burdened.

Given that governments, at least in well-urbanized places, use ample information and communication technologies (ICT) to conduct various functions such as procurement, loan sanctions, metering, etc., moving their current IT to the cloud would greatly benefit their attempt to meet the demands of urbanization. This is because, essentially, cloud technology standardizes and pools IT resources and automates many of the maintenance tasks done manually under the traditional or client/server model of computing. By adopting the cloud model, governments can free up a lot of time and human resources to focus on increasing their roster of services or devising even better delivery mechanisms.

Financial Pressures

With multiple regions in the world either facing financial crises or reeling under economic downturns and low GDP growth, governments are under increasing pressure to deliver their services within tighter budgets and resources. In several places, increasing taxes or other means of revenue generation seem unlikely, so they are being forced to do more with less.

The cloud paradigm offers greater advantages in terms of immediate as well as long-term cost savings for governments. Because it offers services on the "pay-as-you-go" and "pay-per-use" basis and there are no up-front costs involved in buying IT equipment. The resulting savings can indeed come in handy in these financially challenging times. Globally, city and municipal government leaders recognize the importance of cloud infrastructures, the delivery model brought in by the cloud idea and the transformation capabilities and competencies to tie budgets directly to service consumption and to lower up-front capital costs when compared to the traditional deployment and delivery.

According to a survey by the U.S.-based Public Technology Institute (PTI), 45% of local governments in the U.S. is using some form of cloud computing. PTI reported a common reason for local governments to turn to cloud was for resource savings (for example, staff time, and maintenance and support costs). A former United States Chief Information Officer had estimated that by using the cloud model, the USA. Federal data center infrastructure costs can drop by 30%, amounting to approximately $7.2 billion in total savings. In India, too, adoption of cloud computing in large projects such as the Aadhar project of the Unique

Identification Authority of India (UIDAI) and employment schemes under the National Rural Employment Guarantee Act (NREGA) can result in significant cost savings for the central government. There are market analysts and researchers forecasting a major push by various governments and their entities and elements across the globe for gradual cloud adoption in order to attain cost benefits.

Technological Obsolescence

One of the sore points of ICT advancement of several organizations, including government agencies, is the shortening upgrade cycles and the constant launch of new products and technologies. That is, there are many technologies disappearing without making any telling contributions. Further on, newer technologies are being unearthed from the ground up and even technologies are being crafted by smartly fusing existing technologies. Thus technology adoption has to be done very carefully by thorough investigations. At the same time, sticking with the same old technology for years on end, without going in for timely upgrades, runs the risk of being left behind incompetence, capability or capacity.

Thankfully, the cloud idea, which is based on the service paradigm, comes to the fore in effectively tackling this dilemma. It puts these risks in the bucket of the provider rather than the user or the user organization. Everything is cloud-based and service-enabled and since the delivery time and quality of services are regulated by well-defined service and operational level agreements, government agencies adopting the cloud can rest in peace as far as technological upgrades are concerned. All the related software and hardware upgrades are taken care of by cloud service providers. On the contrary, users of cloud-based IT and business services can avail them using the latest and cutting-edge technologies and state-of-the-art cloud data centers because CSPs have to keep their data centers and systems not only up-to-date but running as efficiently and securely as possible for a simple but compelling reason: they must stay competitive in the market. And the market for providing cloud services is growing, with more providers entering the fray. Thus the cloud paradigm takes away most of the concerns of organizations towards delivering next-generation services with all the quality of service (QoS) attributes embedded. For governments, the cloud technology is inherently capable of overcoming the persistent and perpetual concerns of governance while ensuring high-quality governance.

THE VISION FOR GOVERNMENT CLOUDS

Every initiative, technology adoption and enterprise is to begin with a well-intended and defined vision. Governments also embrace the cloud idea with a vision. In this section, the vision is being explained in detail. Computing is being consistently empowered through a litany of pioneering technologies such as Virtualization, grid, and utility, autonomic, on-demand, and service computing, ambient communication, etc. The Virtualization of IT resources such as computing, storage, network, and software enables the creation of logically partitioned and micro-manageable IT resources that efficiently share a set of physical resources in a frictionless manner. These virtual resources are then profiled, provisioned, de-provisioned, composed, managed and secured through a host of easy-to-use user as well as management interfaces. The much-desired elasticity, simplicity, sustainability, availability, consumability, and scalability of IT systems and business applications are accomplished through an adept application of proven cloud technologies. This kind of "divide and conquer" or "decompose and compose" pattern lays out a paradigm shift in IT infrastructure optimization, utilization, programming, and management.

Such transitions enable enterprise and government IT teams to carve out a number of newer and nimbler schemes and services to deliver them to their immediate constituencies and citizens in totally different ways. Cloud computing is an altogether different way to access and use ICT services in a flexible and supple fashion, buying only the services needed when they are needed. Governments, having understood its unique benefits, are envisioning cloud infrastructures exclusively for their users. However government clouds are having several functional as well as non-functional requirements that are at variance to other kinds of business clouds. Clouds are being positioned as the one-stop solution for all kinds of government-sponsored services to their constituencies such as government agencies, departments, employees, and citizens.

Several democratic governments including USA, UK, Australia, Canada, etc. are fast-creating a pragmatic cloud strategy in order to move their government IT functions to clouds. Access, usage, and procurement policies are thoughtfully changed and created in order to empower those who are in decision-making positions to give the first priority to cloud when procuring IT services. Even existing IT assets, government data centers and other government IT elements are accordingly optimized and refurbished by smartly applying cloud technologies, precepts, patterns, and practices. Data protection and privacy policies are being strengthened in view of the massive cloud adoption. Thus cloud migration has become a serious business for governments not only for achieving cost efficiency but also for producing and providing next-generation services to their people in an optimal manner.

By adopting cloud computing, the government will be able to more easily exploit and share commodity ICT products and services. This enables the move from high-cost customized ICT applications and solutions to low cost, standard, and interchangeable services. The governance culture is for a serious sweep with the embrace of the cloud theme.

BENEFITS OF GOVERNMENT CLOUDS

The cloud technology promises a number of advantages for governments and citizens. Cloud computing can help governments resolve many existing and newer challenges related to efficient, timely and massive delivery of services to their citizens. By connecting the people with their governments directly and providing a platform for sharing complaints and redressing them, the Internet infrastructure has proven that e-governance makes eminent sense in today's age of increasing transparency and growing awareness among consumers and citizens. Cloud computing can take the e-governance process to greater heights. Clouds provide a matured, centrally monitored and managed platform to be shared across a different set of government entities to share their beautiful ideas, pool resources, and collaborate with each other to take concerted actions instead of the piecemeal or secluded approach that can delay the delivery of citizen services. In short, cloud computing can be the stimulating and sustainable foundation on which governments can create a more trusted environment for digital governance while reaping the benefits of huge cost-savings and efficient and easy delivery mechanisms. Cloud computing profoundly transforms the way in which information and services are provided to and consumed by citizens.

Cloud computing can help governments of all types and sizes to continue to lower infrastructure costs, including security infrastructures; maximize limited capital and operational spending; secure the user experience; manage a multi-tenant infrastructure; align and optimize internal processes; enable usage-based per agency or department unit costing; define and rapidly deliver service-level agreements (SLAs) for applications, and help meet the demand for services and rapid service provisioning. The journey away from traditional IT infrastructures toward cloud computing and IaaS is no small undertaking for governments and their leaders at every level.

IBM CLOUD OFFERINGS FOR GOVERNMENT CLOUDS

Leading IT players such as IBM, HP, Microsoft, Oracle, and Amazon are quick in grasping the strategic value additions and the power of cloud technology in producing and sustaining next-generation IT solutions and have come out with a number of powerful cloud offerings. In this section, I would like to discuss about the various cloud solution offerings from IBM. With the mission of meeting of emerging IT needs, IBM has come out with an IBM SmartCloud Framework, which is a comprehensive cloud offering as elucidated below (Raj, 2012; IBM, 2013).

- **IBM SmartCloud Foundation:** For organizations that want to build their own private cloud, IBM SmartCloud Foundation is the family of integrated IBM cloud solutions and products for quickly building and scaling security-rich private and hybrid clouds with IaaS and PaaS capabilities.
- **IBM PureSystems:** Combine the flexibility of a general-purpose system, the elasticity of cloud, and the simplicity of an appliance. They are integrated by design and come with built-in expertise gained from decades of experience to deliver a simplified IT experience.
- **The IBM PureFlex System:** Is built on elements of the IBM Flex System. Flex System includes compute, storage, systems management, and networking components. These components are pre-configured and pre-integrated to make up three editions of the PureFlex System offerings: Express,

Standard, and Enterprise that provide "ready-to-go" cloud infrastructure.

- **IBM PureApplication System:** Is an application platform system with integrated IBM patterns of expertise that operates in a traditional or private cloud environment. This is designed and tuned specifically for transactional Web and database applications. This is workload-aware and flexible, as well as easier to deploy, customize, safeguard and manage.

- **PureData System:** As today's big data challenges increase, the demands on data centers have never been greater. PureData System, the newest member of the PureSystems family is optimized exclusively for delivering data services to today's demanding applications with simplicity, speed and lower cost.

- **IBM SmartCloud Enterprise (SCE):** An enterprise-class self-service public cloud IaaS with low costs and no licensing fees, SCE is particularly suited to economically scale enterprise IT infrastructure and to accelerate the development of new born-on-the-cloud applications.

- **IBM SmartCloud Enterprise+ (SCE+):** A fully managed and highly secure IaaS optimized for running born-on-the-enterprise production workloads (like SAP Applications) in the cloud, SCE+ offers SLAs up to 99.9% and many advantages of a private cloud such as the choice of dedicated servers and storage while providing flexible scaling and beneficial cloud economics.

- **IBM SmartCloud Orchestrator:** Is an open and integrated platform that uses workload patterns and business process automation to standardize and manage hybrid cloud environments.

- **IBM SmartCloud Application Services:** Is IBM's platform as a service (PaaS) that powers fast development and deployment of applications to the cloud with a suite of cloud-based development tools, workload patterns, middleware and databases. IBM PaaS runs on the IBM self-service public cloud (SCE).

- **IBM SmartCloud Application Workload Service (SCAWS):** Eliminates the tedium of managing middleware and infrastructure and thereby enabling IT to more easily, quickly and repeatedly deploy WebSphere-based applications to IBM's public cloud.

- **IBM SmartCloud Solutions:** Are best-in-class SaaS applications and business process-as-a-service (BpaaS) capabilities that help you accelerate innovation and focus on business goals rather than IT deployment. IBM is a global SaaS leader with over 80 applications—supported by IBM SmartCloud SaaS operation centers around the world—and delivering the enterprise-grade security, availability and elasticity you expect from IBM.

Thus, the IBM SmartCloud solutions are designed and developed to meet all kinds of cloud requirements.

GOVERNMENT CLOUDS: USE CASES

There are an increasing number of use cases for governments for overwhelmingly embracing cloud infrastructures. A variety of e-governance applications are being refactored, replenished, and renovated to be cloud-ready. Raj (2012) provides a list of major workloads for government clouds

Enterprise-Scale Business Applications

There are several customer-facing, mission-critical, packaged, and custom/home-grown business applications such as ERP, CRM, SCM, and KM applications. For these applications, reliability remains a critical parameter. To ensure reliability, organizations and enterprises traditionally provision for peak demand and disaster recovery, which leads to management complexity, cost escalation and leaves resources under-utilized and even unutilized sometimes. With the unprecedented adaption and adoption of the cloud idea, there is a silent movement among product vendors in strategically modernizing their core products into cloud-ready services so that the goals of cloud deployment and delivery can be quickly and effortlessly achieved. By carefully moving these workloads to cloud environments such as IBM SCE+, a well-managed, production-ready, highly secure and performing public cloud, businesses can steadily improve the reliability factor while lowering the capital and operational costs significantly by paying only for the resources being used.

High-Performance Computing (HPC) and Real-Time Analytics

These days, due to the maturity of several levels of connectivity and integration technologies, all kinds of physical devices at the ground level and applications deployed in enterprise as well as cloud servers are getting interlinked with one another through middleware and integration backbones. Thereby real-time interactions and notifications will see the light. With this emerging trend, the activities of data generation, capture, transmission, storage, processing, analysis, mining, dissemination, etc. are to see paradigm shifts. That is, the data volume being created by men as well as machines is to climb exponentially. Similarly as there are multiple, heterogeneous and distributed gadgets and machines participating in the integrated computing, there are different data representation, exchange and presentation formats, transmission protocol, etc.

Thus many agencies and organizations are amassing large data sets that they could process mathematically to gain critical and actionable

insights in time. However to cheaply, quickly and easily extract right and relevant knowledge out of data heaps, organizations need to look for appropriate cloud deployment models. With IBM SmartCloud offerings (private, public and hybrid cloud solutions), it is possible to spin up massive on-demand clusters of compute resources in minutes and quickly gain the information and knowledge you need to effectively meet mission goals. Analytics as a Service (AaaS) is the next-generation service type that draws a huge appreciation from executives. Cloud is the principal method for producing and providing AaaS.

Storage and Disaster Recovery

As the volume of data continues to grow unabatedly, organizations are struggling to add the system capacity needed to meet their primary storage and backup requirements. With IBM SmartCloud, governments can easily and economically access highly durable, available, and IBM experts-managed storage that can meet all sorts of data security requirements and scale with the needs of governments. Cloud-based data archival is also being prescribed as the best course of actions for many governments. Cloud-based data storage and recovery seem to be a better prospect for efficient business continuity and resiliency. Precisely speaking, cloud-sponsored government resilience services are picking up fast in order to ensure higher availability thereby all kinds of delays and breakdowns of government IT services can be avoided.

Rich Internet Applications (REAs)

There are innumerable Web applications such as B2C e-commerce and B2B e-business applications mandating unique scalability requirements since their user loads are difficult to predict in advance. That is, the capacity planning is quite cumbersome for web-based applications. However, with IBM SmartCloud, the aspect of capacity planning is fully automated by advanced algorithms, the job scheduling, load balancing, tool-based workload migration and resource provisioning at an optimal level are ensured through competent software solutions. Any unexpected spike can be smartly managed while cutting down costs, reducing risks, and guaranteeing high application performance even during heavy demands, Due to the elasticity being inherently offered by cloud services, agencies can simply remove resources when the demand subsides, ensuring high asset utilization and cost-efficient Web applications.

Platform as a Service (PaaS) and Business Process as a Service (BpaaS)

Not only infrastructures but also platforms (development, deployment, integration, management, delivery and others) are also increasingly finding their residence in cloud infrastructures. The other set of prospective entities are business processes, which are typically a dynamic pool of business and IT services. Through cloud brokerage solutions, diverse and distributed cloud-hosted applications and data sources are seamlessly and spontaneously integrated to provide government-aware processes as services. Composite applications and processes are the next-generation application building-blocks and hence the role and relevance of clouds in delivering next-generation government services are definitely immense.

Brokerage Services

With the availability of a growing array of cloud applications and services, brokerage firms got a strong foothold in the cloud space. Cloud middleware / brokers / service buses are being increasingly utilized in order to automate service discovery, message enrichment and routing, service intermediation, arbitration and aggregation, information visualization, etc. Next-generation brokerage applications will be deployed in clouds.

Government Cloud Services

The goal for the Government Application Store (Appstore) is to provide an open, visible, commoditized and cost transparent marketplace, that is the first point of call for any public sector ICT requirement create a shop window where all the relevant public sector ICT services can be found encouraging innovation, competition and new suppliers exploit pan-public sector purchasing enable the IA and security community to have access to information related to the assurance and accreditation status of the service be a key enabler for collaborative procurement, including:

- Driving up supplier performance by providing an open feedback mechanism.
- Facilitating re-use of a service to drive efficiency and cost savings.

The Government Application Store will be the market place in which public sector organizations can purchase trusted services (and in some instances trial services) from a variety of sources. Overall the Government Application Store will aim to deliver sophisticated capability, diverse services and will allow users to easily find, review, compare, purchase, commission, decommission and switch services. The Government's use of cloud computing technologies for its ICT requirements moves ICT service provision from a costly dedicated development that is often duplicated many times over, to taking the best fit the market has to offer that balances functionality, service levels and cost. This works most effectively where a mature market exists for a given service so that the business can adapt to utilize the commodity solution quickly and easily.

Government Cloud Services and Applications

Governments are contemplating several applications to be hosted in cloud environments. The primary thing is the establishment of marketplaces and cloud stores. A cloud store is a catalogue of services, many of which are going through pan-government security accreditation for public sector use. This not only makes procurement quick and simple, but also reduces the risk of ICT deployment since many of the services available offer try-before-you-buy options. The following functions are being provided by any cloud store.

- A new more powerful search.
- An ability to create an account to enable you to save and export search results to an excel file.
- Easier browsing through each of the four service Lots and their sub-categories.
- A set of filters based on the key features of each service, to help you to find the services that best meet the needs.
- Suppliers submit bids for the services they wish to supply.
- Government then evaluates and selects suppliers based on the responses provided and creates a Framework Agreement with the successful companies.
- Government then conduct mandatory assurance checks and, for the successful services, invite suppliers to put their appropriate services through an optional pan-government Accreditation Process .
- Government buyers use the cloud store to compare and select a service to meet their needs through a call-off contract.

GOVERNMENT CLOUDS: CASE STUDIES

Having understood the short-term as well as long-term benefits of the cloud paradigm, government executives, administrators, governors, and decision-makers are seriously putting policies and programs in place to smoothly realize to the envisioned service-oriented, knowledge-driven, and cloud-centric society. Several governments across the planet are visualizing the cloud era and in this section, here come a few case studies in order to emphasize the fast proliferation of the cloud concepts into the public sector domains. The cloud bandwagon is really rocking and raging.

The mandate is very simple. Those who are in charge of government IT are being asked to transform costly, closed, and inflexible legacy infrastructures, increase the productivity of the existing workforce, enable cross-agency collaboration, and improve overall organizational agility. All kinds of feasibility studies, probable risks, security concerns, economic pressures, etc. ought to be taken into consideration. In case of any natural or man-induced emerges, the cloud availability, value and viability are also understood before embarking onto the cloud journey.

The State of Texas

This state is setting a progressive example for other state governments by relying on cloud service providers to provision IT resources to dozens of state agencies. Led by the Texas Department of Information Resources (DIR), the state is creating the Texas Cloud Marketplace, a private cloud deployment model to deliver new technology while fulfilling legislative mandates. The business drivers include the following:

- Establish an IT delivery model that reduces costs, meets demand, and fulfills the state's legislative goals.

- Set up detailed operation expenditure business models to meet agency budget constraints.
- Realign current strategic initiatives and delivery models to match the alternate models established in a new roadmap and future state architecture.
- Develop a cloud service provider capability model, business architecture, and operating model for bringing state agencies into the cloud.

The Pilot Texas Cloud Offering (PTCO) project focused on infrastructure as a service, but many of the lessons learned can be generalized for government agencies adopting any cloud offering. The PTCO project was designed to allow a small group of agencies to choose a virtual private cloud-based infrastructure as a service from a marketplace of service providers made available by a cloud broker. This approach was selected as it maximizes the opportunity to produce the broadest spectrum of experiences for customers. The cloud broker helped to normalize the multiple services available, creating an "apples-to-apples" comparison in pricing and functionality as much as possible. In addition, the cloud broker provided a single, unified Web interface for end users to design, procure, provision, monitor, and govern the services. The PTCO allowed DIR and the pilot agencies to gain a greater understanding of cloud infrastructure offerings for state government and document options and issues with provider selection, pricing, access security, data security, credentialing, provisioning time frames, service levels, service remedy options, terms of use, billing models, interoperability, mobility, scalability, capacity management, provider compliance, and monitoring and licensing.

Federal Cloud Computing Initiatives (FCCI)

FCCI mission is to drive the adoption of cloud computing solutions in the federal government and to address obstacles to that adoption. The FCCI focuses on helping the government implement cloud solutions for the Federal Government that increase operational efficiencies, optimize common services and solutions across organizational boundaries and enable transparent, collaborative and participatory government. The FCCI focuses on developing best practice guidance, contract vehicles, and communication tools that help agencies more effectively adopt, implement, and manage cloud computing solutions.

The adoption of safe and secure cloud computing in the government presents an opportunity to close the IT performance gap between the public and private sectors. The FCCI helps agencies improve access to modern technology needs faster and with lower costs, allowing agencies to pay only for the resources they use in response to high and low demand, avoid the expenses of building and maintaining an IT infrastructure, and control the appropriate level of security for data and applications. The work of FCCI has furthered the government's ability to quickly and effectively acquire and implement cloud computing services, saving agencies time, effort, and crucial funding. The average IT acquisition takes months or years, and can cost hundreds of thousands or even millions of dollars to complete. With the FCCI's pre-competed acquisition solutions, Infrastructure as a Service (IaaS) and Email as a Service (EaaS) capabilities can be rapidly acquired, provisioned, and implemented. The FCCI manages several key efforts that further the government's push towards an optimized IT infrastructure. These efforts include program management and operational support for the Federal Data Center Consolidation Initiative (FDCCI) and the Federal Risk and Authorization Management Program.

CONCLUSION

The enhanced IT agility, adaptability, affordability and autonomy through cloud-enablement go a long way for worldwide state and federal governments in proactively conceptualizing, concretizing and providing a growing array of citizen-centric and creative services. Further on, by embracing the raging cloud technology, governments across the globe are able to achieve substantial cost savings in the uncertain and declining world economy. Cloud infrastructures emerge as the core and central place to put all kinds of IT assets and government applications so that anytime, anywhere, any device, any network and any media access and leverage are being fully facilitated for people. The quality of peoples' life will be sharply enhanced as services being delivered from clouds to people provide not only information and transaction facilities but also knowledge, decision-enabling and physical services. In a nutshell, by smartly applying all the matured and mesmerized concepts of the cloud technology, a number of perceptible and practical benefits can be acquired and accrued by governments to achieve the much-needed sensitive, responsive, trustworthy, and efficient delivery of their existing and emerging services to the right and rightful users. The IBM Smart Cloud Framework is to bring in a number of delectable and desirable cloud-sponsored transformations for worldwide governments in effectively reaching out their communities and citizens with a rich set of technologically richer services.

Besides the cloud paradigm, the other praiseworthy technologies such as mobility, big data analytics, social computing, and machine-to-machine (M2M) integration combine well with the cloud idea in order to deliver value-added services to people by their respective governments. Technologies are to play a very telling and compelling role in shaping up the era of computing for people. That is, businesses aspects and processes are being automated through the utilization of technologies these days and the forthcoming era is for understanding and providing peoples' needs with technology-sponsored precision and perfection to the core.

REFERENCES

Cho, A., Willis, S., & Stewart, M. (2011). *The Resilient Society Innovation, Productivity, and the Art and Practice of Connectedness. Cisco Internet Business Solutions Group*. IBSG.

Cotton, B. (2012). *Century Governance Transforming IT Infrastructures to Meet Critical Imperatives. Frost & Sullivan*.

IBM. (2013). *IBM Cloud Services and Solutions*. Retrieved from www-03.ibm.com/systems/cloud/

Kailasam, R. (2010). m-Governance - Leveraging Mobile Technology to extend the reach of e-Governance, 2010. Academic Press. English, M. (2011). Government –shaping or catching the new wave? IBM Global Business Services.

KMPG International. (2010). *Dynamic Technologies for Smarter Government. Unlocking Knowledge in the Web 2.0 Age*.

Mongo, D. B. (2013). *MongoDB is an open-source document database and the leading NoSQL database*. Retrieved from www.mongodb.org

OECD/International Telecommunication Union. (2011). *M-Government – Mobile Technologies for Responsive Governments and Connected Societies*. Author.

Raj, P. (2012). *Cloud Enterprise Architecture*. CRC Press. Retrieved from http://www.peterindia.net/peterbook.html

UK Government. (2010). *Government ICT Strategy: Smarter, cheaper, greener*. UK Government.

Yiu, C. (2012). *The Big Data Opportunity, Making government faster smarter and more personal. Policy Exchange Report*.

ADDITIONAL READING

Center for Digital Government. (2006). *Governing Faster*. Smarter, The Promise of Useful Data Close at Hand, Strategy Paper.

Deka, G. C. (2011), ICT its Role in e-Governance & Rural Development in International Conference on Advances in Computing and Communications (ACC 2011) April 22-24, 2011, Retrieved August 2013, Available at: http://www.springerlink.com/content/w860888557757h14/.

Deka, G. C. (2012), Prospects of Cloud Computing in Education and e-Governance-An Analytical Study ISBN: 978-3-659-21026-6, LAP LAMBERT Academic Publishing, AV Akademikerverlag GmbH & Co. KG, Heinrich-Böcking-Str. 6-8, 66121 Saarbrucken, Germany.

Ferro, E., Caroleo, B., Leo, M., et al. (2013), The Role of ICT in Smart Cities Governance, International Conference for E-Democracy and Open Government 2013. \

KEY TERMS AND DEFINITIONS

E-Governance: It is the use of ICT such as Internet, Local Area Networks, mobiles etc. by Government to improve the effectiveness, efficiency, service delivery to promote democracy.

G-Cloud: G-Cloud is Internet based communications which enables public bodies to host, select and use computing services from secure, resilient and cost-effective IT service environment. The G-Cloud program's core benefits are cost reduction, improved services, faster delivery and progressive sustainability agenda. The citizen-centric data, services and applications will be stored, maintained and delivered from G-Cloud and a comprehensive government portal will be a part and parcel of this cloud-based service delivery. The latest trend on Government cloud seems to be following the agency-by-agency scenario.

ICT: Information and Communications Technology or Information and Communication Technology(ICT), is often used as synonym for Information Technology(IT) that stresses the role of unified communications and the integration of telecommunications(telephone lines and wireless signals), computers as well as necessary software, middleware, storage, and audio-visual systems, which enable users to access, store, transmit, and manipulate information. The term ICT had been used by academic researchers since the 1980s, but it became popular after it was used in a report to the UK government by Dennis Stevenson in 1997 and in the revised National Curriculum for England, Wales and Northern Ireland in 2000. The term ICT is now also used to refer to the convergence of audio-visual and telephone networks with computer networks through a single cabling or link system (UK Government, 2006). There are large economic incentives to merge the audio-visual, building management and telephone network with the computer network system using a single unified system of cabling, signal distribution and management.

Mobile Government: M-Government, is the extension of e-Government to mobile platforms, as well as the strategic use of government services and applications which are only possible using cellular/mobile telephones, laptop computers, personal digital assistants (PDAs) and wireless Internet infrastructure. m-Government can help make public information and government services available "anytime, anywhere" and that the ubiquity of these devices mandates their use in government functions such as use for sending short message service, or SMS, in the event of an emergency.

Mobile Service Delivery Gateway: A platform called Mobile Service Delivery Gateway (MSDG) will be set up on which all ministries and departments will be able to start their services, according to the document posted on the department's website for public comments. The distribution and retail infrastructure of mobile operators is likely to utilize for collecting and processing payments for mobile services, which can be leveraged for receiving payments for public services and delivery of basic financial services under the financial inclusion initiatives. Also in the plan is integration with Aadhaar so as to process the request for service made through mobile faster and easier. The policy envisions that the service will be delivered via SMS support by every handset using technologies such as 3G to deliver richer experience to those who can afford the technology.

Open Government: Open Government is the governing doctrine which holds that citizens have the right to access the documents and proceedings of the government to allow for effective public over sight. In its broadest construction it opposes reason of state and other considerations, which have tended to legitimize extensive state secrecy. The origins of open government arguments can be dated to the time of the European Enlightenment: to debates about the proper construction of a then nascent democratic society. Among recent developments is the theory of open source governance are application of the free software movement to democratic principles, enabling interested citizens to get more directly involved in the legislative process. Recently "Gov 2.0" (Government 2.0) has grown from being a name coined by William Eggers to now becoming the umbrella term for serious change in government, and not just here in the United States, but around the world. However, the openness and transparency that Gov 2.0 efforts around the world advocate for; driven by mobility and the cloud-have allowed people to be heard. To hear each other, those people whether in government or civic life, in business or entertainment.

Social Computing: Social computing is an area of computer science that is concerned with the intersection of Social behavior and computational systems. It has become an important concept for use in business. In the weaker sense of the term, social computing has to do with supporting any sort of social behavior in or through computational systems. It is based on creating or recreating social conventions and social contexts through the use of software and technology. Thus, blogs, email, instant messaging, social network services, wikis, social bookmarking and other instances of what is often called social software illustrate ideas from social computing, but also other kinds of software applications where people interact socially. In the stronger sense of the term, social computing has to do with supporting "computations" that are carried out by groups of people, an idea that has been popularized in James Surowiecki's book, The Wisdom of Crowds. Examples of social computing in this sense include collaborative filtering, online auctions, prediction markets, reputation systems, computational social choice, tagging, and verification games. The Social Information Processing page focuses on this sense of social computing. Social computing has become more widely known because of its relationship to a number of recent trends. These include the growing popularity of social software and Web 2.0, increased academic interest in social network analysis, the rise of open source as a viable method of production, and a growing conviction that all of this can have a profound impact on daily life. A February 13, 2006 paper by market research company Forrester Research suggested that: "Easy connections brought about by cheap devices, modular content, and shared computing resources are having a profound impact on our global economy and social structure. Individuals increasingly take cues from one another rather than from institutional sources like corporations, media outlets, religions, and political bodies. To thrive in an era of Social Computing, companies must abandon top-down management and communication tactics, weave communities into their products and services, use employees and partners as marketers, and become part of a living fabric of brand loyalists."

Chapter 8
Scalability and Sustainability of M–Government Projects Implementation in Developing Countries

Olalekan Samuel Ogunleye
Meraka Institute, South Africa

Jean-Paul Van Belle
University of Cape Town, South Africa

ABSTRACT

Mobile technology has played a crucial role in facilitating democratic change in many of the developing countries. Many countries have attempted to implement Mobile Government (m-government), which is a form of electronic government, using mobile and other latest technologies such as social media as the most fundamental infrastructure for implementing such changes. However, m-government projects' scalability and sustainability are amongst the key issues relating to the use of Information and Communication Technologies (ICT). This chapter attempts to discuss the scalability and sustainability of m-government projects in the context of developing countries. The aim is to provide a broader understanding of the inherent issues surrounding scalability and sustainability of m-government projects: in general terms and also in relation to mobile phone-based projects for governments' service delivery. In order to understand these issues, definitions of these two concepts are provided and various e-government maturity models are discussed. This is then followed by an overview of the challenges of scaling up and sustaining the m-government projects in developing countries, and lastly, an elaboration of how sustainability and scalability can be achieved is also presented.

INTRODUCTION

The high penetration rate of mobile telephony in the developing countries has raised some hope in terms of government service deliveries. Most of the developing countries who missed telephony revolution, due to lack of infrastructure and required investments, have now participated in the mobile revolution directly.

Governments in the majority of developing countries have poor reputations with respect to service delivery which in many cases involves

DOI: 10.4018/978-1-4666-6082-3.ch008

repetitive and manual operations at government offices (Bassara *et al.*, 2005). Low throughput coupled with traditional communication channels are expensive and require intensive human processing. Also, the lack of a single point of contact with the government has been identified as the one of the key challenges facing service provision of traditional governments processes (Mansoor & Rohan, 2010).

In order to overcome the limitations of traditional government and to improve the quality of service delivery, many of these governments, including the South African government, have started moving towards new ways of implementing government services, for example service processes, service flows, approaches to service delivery and service delivery philosophies (Valentina, 2004). The South African government proposes a transition to e-government in order to enhance access to and delivery of government services to the citizens, businesses, employees and other government departments twenty-four hours a day and seven days a week through a single government portal using modern Information and Communication Technologies (ICTs) (Blessing *et al.*, 2007).

Challenges, such as limitation to fixed line Internet access by many citizens, faced by the government in implementing E-government to ensure delivery of services, have led some governments to shift their attention to m-government as the ultimate target of E-government (Sharma and Gupta 2004). This is due to the development in mobile technology and the mobility of people with respect to the use of mobile devices and technologies that surround it (Sharma & Gupta, 2004). In addition to this, Song (2005) advocates going beyond E-government and recognizes the potential of E-government for the transformation of government services.

This chapter attempts to discuss the scalability and sustainability of m-government projects in the developing countries. The aim of this chapter is to provide a broad understanding of the issues around scalability and sustainability of m-government projects in general and mobile phones based projects for service deliveries in particular. In order to understand these issues, definitions of these two concepts will be provided; these definitions are followed by an overview of the challenges of scaling up and sustaining m-government projects in developing countries and lastly an elaboration of how sustainability and scalability can be achieved will be provided.

E-GOVERNMENT: A DEVELOPING COUNTRIES PERSPECTIVE

Governments in developing countries have considered e-government as the means of delivering government services to citizens. Generally, e-government is defined as the provision of government services through the use of Information and Communication Technology (ICT) Infrastructure (Sprecher, 2000). It is about using tools and systems made possible by ICT infrastructure to provide better public services to citizens and businesses. It is widely used in most developed countries (e.g. most England, France, Germany, Canada and USA) and has also been implemented in some developing countries (e.g. South Africa, Kenya, and Mauritius) (Coordinating, 2001). Majority of the citizens in developed countries have access to most of the service provided by the governments via e-government. This is due to the fact that access to Internet connectivity is not a barrier and most citizens can afford it. This has made e-government implementation and usage a success in developed countries. Therefore, service delivery by governments of those countries is enhanced which consequently improves standard of living of the people in those countries.

However, in developing countries it is a challenge to provide ICT infrastructure in remote or rural areas where majority of the citizens need access to government services. Hence, implementation of e-government, which is dependent on

ICT infrastructure, is likely to benefit only those who have access to Internet who are minority of the population in most developing countries. Even the countries that have successfully deployed e-government face subsequent problems such as accessibility of those services by citizens. The limitation of access however is different from country to country and this may include availability of information relating to online governmental services, satisfaction with the quality of services provided via the internet, citizens' proficiency in the use of personal computers, and the availability of personal computers and Internet connectivity. However, low availability of personal computers and fixed Internet accessibility are seen as the basic limitations in deploying e-government services and implementation in developing countries.

A decent design and implementation plan is required to make e-government an accomplishment. Due to this, the European Union (EU) has formulated policies in the field of e-government to support concrete initiatives to improve public services (Coordinating, 2001). The EU publishes an e-government action plan which is used as a guide to ensure that developed e-government systems benefit everyone. This, however, is only useful and applicable to ordinary citizens of developed countries as they have required and appropriate infrastructure. For example, it has been noted that one of the things e-government action plan seeks to ensure is that no citizen is left behind, by promoting Inclusive e-government. However, in developing countries, Inclusive e-government is not achievable because of lack of infrastructure to poor or rural communities (Botha *et al.*, 2010).

E-Government has to be more than just providing information to citizens via websites, as this is the current service provided by the governments of most developing countries. An engaging or participative kind of implementation is required in order for the citizens to be able to afford the opportunity to benefit from the system. Hence,

in most developing countries, this has been well thought-out in implementation as their e-government systems are interactive. As a result, people who are connected to Internet can do effective and beneficial processes. An example is the tax e-filling systems implemented by the South African Revenue Service (SARS) in South Africa.

Several researchers have explored the field of e-government in improving the delivery of public service. Most researchers indicate the fact that e-government has failed to meet expectations due to many factors, the most important of which is lack of ICT infrastructure in rural and remote areas. This is a challenge in developing countries. This is also one the factors that have hindered e-government to reach its full potential. The architectural scope of e-government, shown in figure 1, was proposed in (Kumar *et al.*, 2008).

The architecture above shows different realms through which services are delivered by the government to the citizens. The realms are namely Government-to-Government (G2G), Government-to-Employees (G2G), Government-to-Business (G2B) and Government-to-Citizens (G2C). In each of these domains, e-government has benefits to provide up to some degree. For example, G2B is effectively enabled on e-government because most businesses have access to Internet in developed and developing countries. However, in developing countries, most small business enterprises that are owned by people in urban and rural areas do not have access to Internet and they do not benefit from e-government. In different manner in (G2C), majority of the citizens in remote and rural areas do not have access to the internet hence, they do not benefit from the implementation of e-government. This presents an opportunity of using a different platform to satisfactory and effectively delivers public services to citizens. This new platform which compliments e-government is presented in the next section.

Figure 1. E-Government Maturity Model

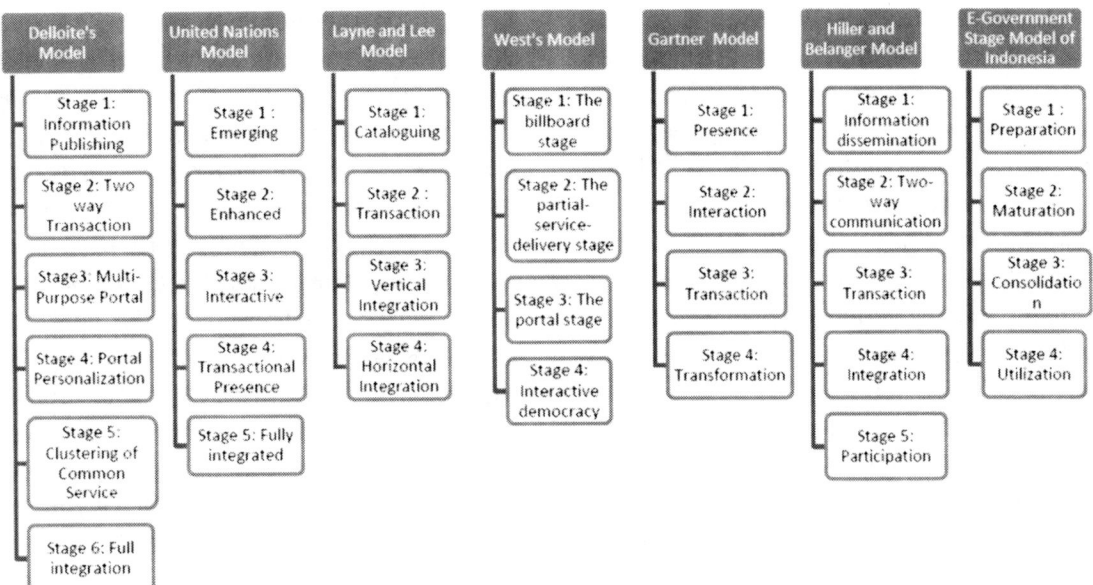

E-GOVERNMENT TO M-GOVERNMENT: A DEVELOPING COUNTRIES PERSPECTIVE

There have been rapid growths in E-government services and technologies. The application of innovative Information technologies, especially web-based applications, to improve the basic and primary activity of governments is the purpose of spreading the related activities to the e-government. The development of mobile technologies and harmonious technologies in the modern era of information technology have created a new direction in e-government which called is a term known as m-government (Kushchu & Kuscu, 2003). Although, e-government is a passage to m-government, m-government is in its early stage of implementation and it has implemented completely in nowhere. However, different factors such as technical infrastructure, information infrastructure, mobile telephone penetration rate, social conditions, security situations, and political decisions are some of the consideration for transition from e-government to m-government (Welch & Wong, 2001).

M-government can be seen as an enabler for e-government in order to streamline the service delivery to citizens through different tools and platforms. Furthermore, m-government is the use of mobile and wireless communication technology in government for service and information delivery to citizens and organizations; e-government service improvement is the goal of the m-government (Nava & Dávila, 2005).

E-government and m-government are related subjects, but m-government is considered a better choice for general information and services presentation to the citizens because in m-government, accessing the information and services in any time and any place is possible through wireless tools connected to the Internet (Ghyasi & Kushchu, 2004.).

There is an important question: whether m-government is a substitute or replacement for e-government? However, m-government implementation is more important for a nation's development, especially in developing countries where mobile devices are seen as the computers of the masses. It is not supersede for e-government. Rather, it is a complementary for e-government activity. The value of m-government is supporting of movement of citizens, businesses, and governments. As an example, m-government support the citizens to request for government service which otherwise they would have gone to the government offices to queue for (Kushchu & Kuscu, 2003).

The cooperation of m-government and e-government is important, especially for developing countries that have not been able to make heavy investment on e-government implementation. Nowadays, m-government is inevitable. The Impact of wireless tools and wireless network has enabled the developing countries to activate the employees of governments more, through the preparation of real time and up to date information. In addition, m-government increases the communication between the citizens and the governments. New found services as location-based services - services that are related to the location of users, are inspiration for m-government which increase the value added of services delivered to the citizens by the government (Sadeh & Norman, 2002).

E-GOVERNMENT MATURITY MODELS

Maturity models are designed purposefully to evaluate the maturity of a service, an object, a performance, etc., based on one or more set of conditions, including competency, capability, and level of sophistication. Maturity models are established to assist an organization as a basis for assessing as well as comparative measure for the organizations improvement (De Bruin et al., 2005).

Maturity models also define the development of explicit object over time in an organization, so that the organization can recognize which activities in each area possess desire to accomplish possible outcomes. Maturity models are descriptive and normative, but not prescriptive in nature. It defines each maturity level without recommending how to get there (Tapia, 2009).

Furthermore, several researchers (Siau & Long, 2005; Andersen & Henriksen, 2006; Kim, 2010), have discussed various benefits of using a maturity model in government service delivery: (1) Maturity models play significant roles as a roadmap in guiding the governments in long-term plans. (2) Maturity models describe the theoretical strategies about important requirements in each maturity stage that allows the citizens, businesses and employees to understand the government activities relating to service deliveries. (3) Maturity models can be used as communication tools to validate government potential capabilities. Citizens, businesses and employees will recognize in which maturity levels the government is currently at, and government, on the other hand will increase their competences to improve the services to the citizens.

A description of various maturity models in the area of e-government and how the reviews of this are directly linked to m-Government will be presented in this section. This is because m-government and e-government are not two separate entities, but m-government is a complimentary to e-government. The difference is the use of mobile devices and mobile technologies so that the citizens can access government services anytime, anywhere via electronic means. We have also tried to search the literature about mobile maturity model, but until now, there is no specific scientific research of mobile maturity model available in the field, related to this research work.

In this section, a summary of different e-government maturity models is presented in figure 1 below. More detailed explanation of each of the existing model can be found below - some of these models were developed by individual researchers, while some other were developed by government agencies and various research institutions and consulting companies.

Delloite's Six-Stage Model

In recent times, most of the governments focus on provision of services in comprehensive way. The Delloite group argues that e-government is an evolutionary transformation that affects the way in which governments manage and deliver service to the citizens. Delloite argue further that this also affects every aspect of how an organization delivers service to the citizens, in the form of using technology, business process and human resources, etc. The central point of Delloite argument is that customers (i.e. Citizens) should be seen as the central that makes a citizen-government relationship to be more inclusive and direct. Due to this, Delloite group (DelloiteResearch, 2000) proposed six stages of e-government maturity model as follows:

- **Stage 1. Information Publishing:** It presents a one-way communication; each government agencies establishes its own website to provide their self-information. By publishing the information on the website, the citizens can reach the government more easily and will therefore reduce the number of phone calls from the citizens who need information about government services.
- **Stage 2. Two Way Transaction:** Here, citizens can submit their personal information and transact information with individual government departments with secure websites. The citizens are also able to have electronic interaction with the government services. Furthermore, the security is a concern in this phase; each department should be able to keep all the information private and free from piracy through some security features.
- **Stage 3. Multi-Purpose Portal:** In this phase, a portal will allow the citizens to use a single point of entry to send and receive information across multiple departments. It is a concept to meet broader user needs both within and outside government services.

- **Stage 4. Personalization of Portal:** Here, citizens will be able to customize portals. Sophisticated web application implementation will be required in order to allow users to customize their portals with their desired features.
- **Stage 5. Clustering of Common Service:** In this Phase, citizens will see the services delivered to them as a unified package through the portal. Real transformation of government structures takes shape. All services clustered along common lines by government.
- **Stage 6. Full Integration and Enterprise Transformation:** At this stage, government services have been transformed into integrated technology. The government provide sophisticated, full service centre, personalized to each user's needs and preferences.

Hiller and Belanger Five-Stage Model

Hiller and Belanger (2006) define a five-stage e-government maturity model as follows:

- **Stage 1. Information Dissemination:** In this stage (Hiller & Belanger, 2006) argue that the government provides basic information to the citizens. They can also make provision for some government publications that have the information that the citizens may be interested or that will be useful to them. However, it is the responsibility of the government to ensure that the availability and accuracy of the information being provided.
- **Stage 2. Two-Way Communication:** This stage allows two-way communication between government official and citizens to be developed. Email system and data-transfer technologies are provided. Citizens can fill in information, make a request and have a feedback by email.
- **Stage 3. Transaction:** Here, interaction between user and government becomes

more interactive. This includes online transactions. Citizens conduct financial transactions completely online, such as pay fine, taxes or renewing of licenses.

- **Stage 4. Integration:** This stage provides fully integrated government services, both vertically-integration between different level of government (intergovernmental integration), and horizontally-integration between another department or non-governmental agencies (intra-governmental integration).
- **Stage 5. Participation:** At this stage, web-based public service is transformed into web-based political activities. Citizens can be involved in political participation such as online voting, online opinion surveys, online public forums etc. This stage is also concerned with high privacy and it requires high technology to support it.

Furthermore, (Hiller & Belanger, 2006) address the privacy issues in e-government implementation. The principles of privacy are used to represent the best practices in self-regulation

UN Five-Stage Model

The United Nation (UN) described five stages of e-government maturity model. Each of the stages is a gradual process for quantifying progress in order to achieve success (UnitedNations, 2001). These five stages are discussed below:

- **Stages 1. Emerging:** A limited Web presence is established. Basic and static information of government provided through a few independent official sites.
- **Stages 2. Enhanced:** The Web presence established begins to expand the content into dynamic website. Information is regularly updated as the number of official websites increase. Hyperlinks to other departments, government publications and newsletter are then made available.

- **Stages 3. Interactive:** At this stage, interaction between government and citizens is realizable. Users (i.e. the citizens) can access broader range of government institutions and services. They can download forms, contact the official and make appointment(s). The content is regularly updated.
- **Stages 4. Transactional Presence:** This represents the stage where government transforms itself by engaging in two-way interactions online for 24/7. Complete and secure transactions are provided. Secure sites, digital signatures and user passwords are also present. User can pay online for the financial transaction services received from the government.
- **Stages 5. Seamless /Connected / Fully Integrated:** At this stage, all services across departmental boundaries are fully integrated, all in an integrated package. The services are clustered together along common needs; it provides services across the different lines and level of department with the highest level of integration.

West's Four-Stage Model

West (2004) investigates whether the interactive features of Internet is useful to improve service delivery, democratic responsiveness and public participation. He highlights four stages of e-government transformation as follows:

- **Stage 1. The Billboard Stage:** This is the first stage where government sets up a basic websites containing some static mechanisms to display information as same as billboards. The government reports, publications are accessible by the citizens but they cannot interact with it, so there is no two-way communication at this stage.
- **Stage 2. The Partial-Service-Delivery Stage:** During this stage, citizens can access and manipulate information. They can

also search informational databases if there is any. Some online services are made available to the citizens so that they have access to some of the services they need if not all.

- **Stage 3. The Portal Stage:** This stage provides fully executable and integrated service delivery. All different levels of government are fully integrated. This will enable the government to increase the citizens' ability to find government's information and services. Security and public privacy is of great concern at this stage. Translation options in multi-language are also made available in the case of a multilingual country.

- **Stage 4. Interactive Democracy with Public Participation and a Range of Accountability Measures:** At this stage, government moves from service-delivery model to system wide political transformation. The websites offer customize personalization and push for technology, such as emails and electronic subscriptions, provide feedback, make comments and enhance democratic responsiveness. These characteristics help citizens to have interactive and engage two-way communications between citizens and the government.

Gartner Four-Stage Model

Gartner Research (2000) also discuss e-government maturity stage model and show the progression of the model in a connected environment. The model proposed four stages maturity model as follows:

- **Stage 1. Presence:** This represents the initial stage at which point the government establishes an online presence to provide elementary information about the government.

- **Stage 2. Interaction:** At this stage, some features are extended, such as downloadable forms, basic search capabilities, link to the other agencies or relevant sites and email address for interactions.

- **Stage 3. Transaction:** At this point, the online transaction is enhanced with some security features such as online payment, tax filling, receiving licenses or renewal of licenses. This stage focuses on self-service application so that citizens can access it online.

- **Stage 4. Transformation:** This represents the stage where the government delivers fully integrated services by providing a single point of contact to the citizens in order to provide full communication between the official, citizens or other non-governmental organizations.

Indonesian Four-Stage Model

In order to meet the needs of e-government at the national level and to improve transparency and accountability of good governance, the Indonesian Ministry of Communication and Information Technology published a master plan to guide the development of e-government, both at the central and local government level (Rose, 2004). The report suggests four stages of e-government development (MCIT, 2003) as follows:

- **Stage 1. Preparation:** In the preparation stage, the following steps are adhered to:
  - Each government agency (central or local government) should establish the website to provide basic information.
  - Training must be provided to government officials regarding to the e-government,

- ○ Provide public access such as Multipurpose Community Center (MCC), Internet kiosk, etc.
- ○ Socialize electronic information to the citizens to create public awareness about e-government.
- ○ Develop e-leadership for supporting the development of e-government
- ○ Prepare the supporting regulations.
- **Stage 2. Maturation:** At this stage, the website is developed into more interactive session. Search engine, and email are provided. Citizen and government will be able to engage in two-way communication between each other, there were also a hyperlink with other government agencies to enhanced interactive session with other government agencies.
- **Stage 3. Consolidation:** Citizens will be able to engage in financial transaction. The services should be trustworthy and confidential, with reliable security provided. At this stage, integration of application and data with other government agencies (interoperability) are put in place.
- **Stage 4. Utilization:** This level ensures full integration and utilization of the application between government to government (G2G), Government to Business (G2B) and Government to Citizens (G2C). At this stage, the government provides the best service to the citizens.

Layne and Lee Five-Stage Model

Besides the model proposed by institution and government agency, some individual researchers also proposed some stages of the e-government maturity model based on their research work. Layne and Lee (2001) propose a maturity model of e-government in terms of complexity and different level of integration. These stages are listed below:

- **Stage 1. Cataloguing:** At this stage, preliminary efforts are made to setup an online presence for the government. The websites delivers static and basic information. The functionality is limited to online presentation of the government information.
- **Stage 2. Transaction:** Here, the capability of the government is extended which allows citizens to transact. This stage can be called transaction-based e-government, by putting live database links to online interface so that the citizens are able to do some simple online transactions such as pay fines, renew their licenses and filling some forms.
- **Stage 3. Vertical integration:** At this level, the integration are divided into two, vertical and horizontal. Vertical integration connects different levels of governments with different services within similar functionality. This stage initiates the transformation of government services rather than automation of the existing processes. Central database or a connected Web of database is used, it is expected that different levels of government, will be connected and communicate each other so that the results of transactions can be interchanged from one system to another.
- **Stage 4. Horizontal integration:** This stage integrates different levels and across different functions of government. Varying functions of separate systems and different functional areas communicate with each other and share information among them to provide citizens unified services. The integration across different functions enables one department to automatically checks against data in other functional departments.
- **Stage 5. Integration Stage:** This stage integrates different levels and across different functions of government. Varying functions of separate systems and differ-

ent functional areas communicate with each other and share information among them to provide citizens unified services. The integration across different functions enables one department to automatically checks against data in other functional departments.

The widely used e-government models described above are based on different point of view of different researchers (Layne &and Lee, 2001; West 2004; Hiller & Belanger, 2006). Also consultant companies such as (DelloiteResearch, 2000) and (GartnerResearch, 2000) have conducted some significant amount of research on E-government. The e-government maturity model that was developed by the (UnitedNations, 2001), based on the research in various countries, especially in the developing countries is also discussed. This is also followed by the Indonesian e-government stage model which was developed and implemented based on its own research (MCIT, 2003).

Layne and Lee (2001), (Hiller & Belanger, 2006) and (West, 2004) present a similar model. They however combine the stages of model with major types of electronic relationships between government and different level of constituents. In the first relationship, they showed that government directly delivers services to the citizens and government to the individual in the political or democratic process which is basically known as government to citizen (G2C) relationship. The second relationship that was shown is the government to business (G2B) in which major portion of online transaction involve the business stakeholders such as payment of tax online, booking of license online etc. the third relationship is to another government agencies or government employees. Through this, collaboration will be formed to provide services to one another, as explained in the stage model as vertical and horizontal integration. This is called government to Government (G2G) relationship.

On the other hand, some researchers are now extending the concepts and theories of e- government to include mobile services with the aim of providing effective and efficient services to both the citizens, the business and the government agencies and employees as well as convenient access to those government services through mobile and wireless technologies (Sandy & McMillan, 2005; Georgiadis & Stiakakis, 2010). This is an argument we have started witnessing. However, there are limitations in the number of research in m-government maturity model. Some of the models available are described and analyzed.

M-GOVERNMENT MATURITY MODELS

As mentioned previously, there is limited literature on m-government maturity models. Most of the available researches (Fasanghari & Samimi, 2009;Mengistu et al., 2009;Alijerban & Saghafi, 2010) develop the model based on the review on e-government developments. Also, m-government and e-government are not two separate entities. Therefore, we could only find very little about m-government maturity model in the literature. These models were developed from the previous e-government model earlier explained with the utilization of mobile devices and technology as the focus of the models.

Alijerban and Sahafi Maturity Model

Alijerban and Saghafi (2010) identify different models of e-government maturity model, and propose six stages of m-government Maturity model which are described below:

- **Stage 1. Presence and Disseminating Information:** All sites could be observed by mobile phone. Some basic services could be present, such as a weather, news, and access information
- **Stage 2. Interaction:** At this stage, user can download information via mobile phone, give comments in the website, re-

Figure 2. M-government Maturity Model

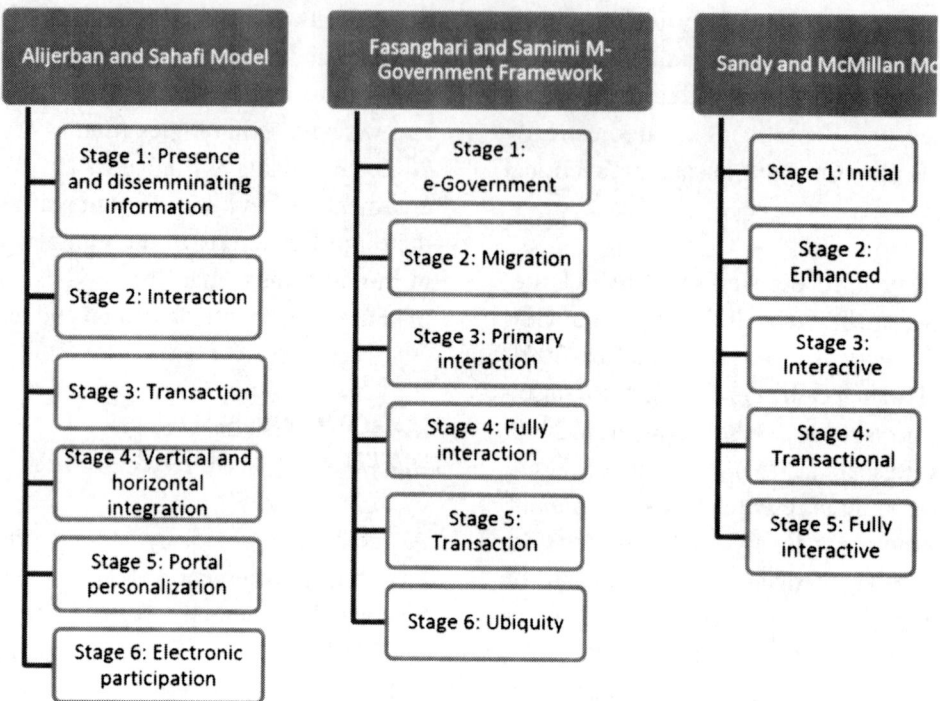

ceive feedback and provide location-based services.

- **Stage 3. Transaction:** Financial transactions and positioning services are presented at this stage. Security is a big concern for the government to overcome. Personal detail information and privacy of the user should be kept safely.
- **Stage 4. Vertical and Horizontal Integration:** An integrated mechanism of communication between different departments and different governments are presented at this stage. The vertical and horizontal integration is not applicable through mobile technology, if the country has not reach the integration stage in their e-government implementation.
- **Stage 5. Portal & Personalization:** The portal allows citizens to use single window to send and receive information, processing financial transaction, and personalize their user-interface. The services are given

based on customer needs and placed in different categories

- **Stage 6. Electronic Participation:** At this stage, transparency, accountability should be implemented. If a government has not reached this stage, democracy cannot do its real role, because the government is not accountable. For example, E-voting can be done technically in stage 4, but not completely realize without transparency schema.

Fasanghari and Samimi M-Government Framework

To develop m-government framework, Fasanghari and Samimi (2009) also identify various e-government maturity model as a foundation. They proposed a m-government model which consist of six stages or five phases.

- **Stage 1. 0th Phase:** This is an initial phase. Citizens can access government services with e-government infrastructures and landlines phones
- **Stage 2. 1st Phase:** Access to the government information is made available via mobile phone at this stage. Therefore the migration of e-government to m-government is needed.
- **Stage 3. 2nd Phase:** Primary interaction to the website via mobile phone, search capability is present at this stage.
- **Stage 4. 3rd Phase:** Here, citizens can fully interact/communicate with the government services through their mobile phones. At this stage, the public service delivery is more convenient than previous stage.
- **Stage 5. 4th Phase:** At this stage, citizens can perform transaction on the services that the government offers. Online interaction between the citizens and government official can now be put into effectiveness.
- **Stage 6. 5th Phase:** This is the stage where government services are in ad-hoc situation. The government services are delivered based on real-time situation (such as information about earthquake, terrorist attack) without any request from the citizens.

Sandy and McMillan Five Levels of Functionality in E-Government

Sandy and McMillan (2005) explain that service delivery in the field of mobile and Web presence correlates positively with the desired level of m-government sophistication. They identify five stages m-government model with the critical success factor for each level.

- **Stage 1. Initial:** This stage provides initial wireless access and non-interactive responses such as response to complaints or questions from the citizens.
- **Stage 2. Enhanced:** This is the level that presents updated information for the

citizens, such as weather forecast, policy changes or traffic conditions.
- **Stage 3. Interactive:** This stage provides interactions between citizens and government service providers. Searching features is available that will allow citizens to search for the specific database based on their needs and interests. Citizens can fill in the forms, download forms, and submit them from mobile devices or wireless connections.
- **Stage 4. Transactional or Mature Interface:** This provides unique or single-entity interactions for mobile and wireless device users. A single mobile government agency is used for the m-government application. At this stage, a simple and non-critical payment interaction is provided.
- **Stage 5. Fully Interactive:** At this stage, provision is made for high security feature for mobile wireless transaction made through the citizen for government's services such as payment, ordering and billing. This stage offers a 24/7 services and can be accessed anywhere from a mobile wireless device with secure identification and authorization.

(Sandy & McMillan 2005) further identify the critical success factors for each stage. Cost, business re-engineering, education, acceptance, security, and access are six factors they endorse for the success of government services using m-government.

SCALABILITY AND SUSTAINABILITY: AN M-GOVERNMENT PERSPECTIVE

Scalability and sustainability can have different meanings for different people in different contexts. Therefore, before going into the details of the issues of scalability and sustainability of m-government projects in the developing countries context, these terms are defined in the context of the study presented in this chapter.

Scalability

Scaling up a project involves duplicating the output of the pilot project over a wider area in order to extend and/or expand the outcomes of the pilot to more beneficiaries (InfoDev, 2003). This is however, unique from a mere replication of the pilot project. InfoDev argued that a distinction should be made between scaling up a pilot project and reproducing a project because reproducing a project might be concerned with testing research questions for further evidence in a different setting, but not the whole implementation of such project to reach a greater number of potential users. The term scalability in this context relates to m-government project scalability and should not be confused with scalability as an attribute of an Information Systems which refers to the competency of an application to handle work load (Kimaro & Nhampossa, 2005). Furthermore, Walsham and Sahay (2006) argued that scaling is not only about numbers; it also involves socio-technical networks and question of what and how to perform the scaling. Also, there are two aspects of scalability in the context of this study. These are (a) internal scalability which relates to the organizational capability and dynamics and (b) external capability which relates to the demand for services.

Sustainability

Sustainability of projects is a major challenge in many developing countries. This is due to the fact that a large number of projects implemented at huge costs often tend to experience difficulties with sustainability. And due to this, all major donors, such as the World Bank, and the bilateral aid agencies have been expressing concerns on this matter (Adil, 2000). The term sustainability is used in various contexts, different disciplines

and it has varying meaning for different factions (Kimaro & Nhampossa, 2005). Ali and Bailur (2007) explain that although the term sustainability is very common, finding its definition is problematic. Sustainability is concerned with the possibility that the benefits from an intervention will be maintained at an appropriate level for a reasonably long period of time after the withdrawal of support from contributor or sponsor (InfoDev, 2005). It can also been seen as the ability of a project to maintain its operations, services and benefits during its projected life time. From these definitions, one key aspect is that of the withdrawal of sponsor. In general, project sustainability can be defined as the percentage of project initiated goods and services that are still being delivered and maintained after many years of termination of implementation of the project; the continuation of local action stimulated by the project and generation of successor services and initiatives as a result of project built initiatives (Adil, 2000).

In the context of developing countries, reliance on sponsors or donors support is almost total. This lead to the following question: How can sustainability be envisaged in a context where there is a reliance on sponsors or donors? InfoDev (2003) proposed that sustainability definition will be considered in the context of mobile phone-based applications but without a major emphasis on the withdrawal of sponsors. However, Adil (2000) explained that what is important to note is that if a government for reasons better known to itself, decides to provide support to a particular project and maintain its sustainability without regard to its economic capability (which is and should be case in the context of m-government), then that is a choice that the government has made and that the issue of sustainability of such an activity should be seen purely from the perspectives of a decision taken by such a government in order to deliver services to its citizens. Sustainability is therefore defined as the likelihood that the benefits from mobile phone-based applications (in this instance for m-government) will be maintained at an appropriate level for a long period of time. This implies that m-government sustainability concerns itself with:

- Level of continuation of delivery of m-government services.
- Changes stimulated / caused by the m-government project implementation.
- New initiatives caused by the m-government project.

CHALLENGES FOR M-GOVERNMENT SCALABILITY AND SUSTAINABILITY

Although, scalability and sustainability in developing countries can be achieved, the realities of developing countries implementing and adopting m-government become very complex and difficult process. Crawford and Lester (2004) argue that developing countries continue to face dramatic shortages in resources, trained personnel, retention mechanisms, and the culture and systems necessary to address overwhelming needs. Based on this, it is evident that the effective use of ICTs (which mobile technologies are part of) in developing countries is likely to be hindered by factors such as: poor infrastructure, lack of appropriate skills and other resources, lack of government commitment, socio-economic and political instabilities, illiteracy, poverty, etc. These have implications on sustaining and scaling ICT projects for development in general and m-government projects for improved services deliveries.

In the case of m-government sustainability, for example, Bassara *et al.* (2005) identified the following factors as being big challenges to achieving sustainability:

- Inadequate infrastructure.
- Inadequate human resources capacity.
- Inappropriate policies and strategies to manage the sustainability problem.

Apart from the above mentioned contextual challenges, there are possibly additional factors and technical challenges that can influence sustainability and scalability. InfoDev (2005), for example, the identified the following as possible factors affecting sustainability: (1) Social support, (2) Technological soundness, (3) Government commitment, (4) Commitment of other stakeholders, (5) Economic viability, (6) Financial viability, (7) Institutional support, (8) Environmental impact and protection, (9) Resilience to exogenous factors such as price variability and market access, natural disasters and unstable security in the project area, (10) Replication of the project approach as an indicator of sustainability.

Furthermore, additional challenges could include the lack of a business model that can ensure financial sustainability of m-government projects in general and the lack of a general framework of reference that can be used to address the complexities of scaling up and sustaining m-government projects for service deliveries in a developing country's environment. How can these many challenges be addressed? This is the topic for the next section.

Addressing the Challenges of Scalability and Sustainability

Analysis of the definitions of scalability and sustainability of m-government projects reveals that there are key aspects that arise and that need to be well understood and dealt with at project level. From the definition of sustainability the following elements are highlighted (Kimaro & Nhampossa, 2005):

- Identifying and managing the risks that threaten long terms viability.
- Maintaining the project after the withdrawal of sponsors.
- Maintaining the benefits for a long time (Ensuring project viability).

- The following elements were derived from InfoDev (2003) proposed scalability definition:
- Replicating the output of the pilot project over a wider area.
- Extending the outcomes of the pilot to more beneficiaries.

It is evident that there are possible risks associated with each of the above mentioned elements. In this regard there is a need to have certain measures at the project level that will ensure that these risks are dealt with effectively. These measures are likely to have positive or negative impact on the process of sustaining and scaling up m-government projects. The primary concern would therefore be to identify the risks associated with (1) Maintaining the benefits for a long time, (2) Maintaining the project running with or without sponsors, (3) Replicating the output of the pilot project in other areas, (4) Extending the outcome of the pilot project to more beneficiaries; while the secondary concern would be to establish measures that will help in dealing with the identified risks.

The ability to extend the outcome of the pilot project to more beneficiaries will imply that the pilot project was successful. In this regard, Ferguson and Ballantyne (2002) note that effective pilot implementation is a pre-requisite for project sustainability and scalability. However, possibilities of failed pilot implementation are not to be excluded. Ferguson and Ballantyne (2002) also explain that the failure of pilot project should not necessarily lead to closure but rather it should be a learning process and an opportunity to redefine processes; and that in some cases "downscaling can be more appropriate in terms of sustainability." It is therefore critical to ensure pilot project success and, in case of pilot project failure to consider downscaling as a project sustainability measure. This is because ICT-enabled implementation has some specific characteristics and effects on the people (i.e. the citizens) and the organizations that engaged in it (i.e. the government).

Sustainability should not be looked at from a financial perspective only because it encompasses other dimensions. To this end, Ferguson and Ballantyne (2002) argue that sustainability also involves a look at the social, environmental and economic implications of actions that are taken by organizations which implies finding solutions to problems at a local level on a scale that citizens can understand.

There are several other aspects of sustainability that can be identified and that should be addressed appropriately. Pade *et al.* (2006) identifies the following categories of sustainability that can affect ICT projects in different ways in the context of developing countries (these however apply to m-government project as well): Social and Cultural Sustainability, Institutional Sustainability, Economic/Financial Sustainability, Political Sustainability, Technological Sustainability, and Harmonious Development. Therefore, it is imperative that m-government projects sustainability be viewed beyond the financial sustainability context and that these different aspects of sustainability are taken into account and addressed.

Sustaining and scaling up of m-government projects has been described in the previous section as being a big challenge specifically in developing countries. Mobile phone-based applications for service deliveries are not an exception to this reality. When envisaging sustainability and scalability of mobile phone-based application projects for service deliveries it would therefore be required that steps to deal with the above described challenges are taken and that actions and procedures to address the identified risks associated with sustaining and scaling up m-government projects are practically planned for.

In the context of this study, and with regard to the use of mobile phone-based applications to enhance the quality of service deliveries, it is imperative that the difficulties in the dynamics of mobile phone-based projects be well understood so as to be able to gain knowledge about the consequences of the various sustainability and scalability influential factors. This will eventually contribute to the development of more sustainable and scalable mobile phone-based projects for m-government. In this study the assumption is that the prospects of developing more sustainable and scalable mobile phone-based projects can materialize as follow:

- Evaluate the contextual factors that are likely to impact on attaining sustainability and scalability objectives.
- Assess the interdependencies between these factors and other possible influential elements being it at project level, technology level, organization level or stakeholder level.
- Evaluate existing mobile phone-based projects for m-government, their potential use and acceptance in a developing country context, and assess the implications for sustaining and scaling up these projects.
- Use appropriate theories and models to develop a deep understanding of the challenges of and the possibilities of sustaining and scaling up mobile phone-based projects for m-government.
- Develop and implement practical sustainability and scalability guidelines and good practices requirements to be adhered to at the different levels (project level, stakeholder level, technology development level and technology implementation level).

M-GOVERNMENT PROJECTS' SCALABILITY AND SUSTAINABILITY IN DEVELOPING COUNTRIES

Information and Communication Technologies for Development (ICT4D) refers to the use of Information and Communication Technologies (ICTs) in the fields of socioeconomic development. The philosophy behind this is that more and better information and communication utilization will further the development of a society. However, the use of mobile cellular device as part of ICT4D initiatives has proven to be a success as the rapid circulation of mobile telephony has made it possible for poor people to have easy access to useful and interactive information (Alfred S.S. *et al.,* 2010). For instance, in India, the total number of mobile phone subscriptions reached 851.70 million in June 2011, among which 289.57 million came from rural areas, with a higher percentage of increase than that in urban area (India, 2011).

Therefore, the idea of m-government Scalability and sustainability in developing countries are key issues in the field of Information Technology and Communication (ICT) including Information Systems (IS). This means that these two perspectives are to emphasize the need to build requisite infrastructure so that m-government projects can reach viability. Sustainability and scalability of m-government depend on the ability of the government to provide value added services that will satisfies the quest of the citizens for efficient and effective service delivery, which relates to the fact that effective and efficient m-government program depend upon national strategy, infrastructure development and support, appropriate technology platform, low access cost and increased awareness most especially for the citizens in underserved areas.

Furthermore, for m-government project implementation to make sense they must be inclusive and have a national spread. However, to scale-up m-application initiatives have proven hard (Blessing *et al.,* 2007). When planning m-government project implementation, it is important to take note of the architectural framework in order to understand and capture the existing knowledge about processes and work-flows that formulates m-government scalability and sustainability. What is working, what is not working, how could things be done better? Then the next line of action is to map the solutions into an abstraction that then can be translated into implementation (Johan, 2008). Finally, government ministries and agencies related to the planning of m-government project implementation should be highly involved in the planning and implementation phases. This will ultimately lead to a change in attitude towards provision of services and transform their models of providing public information to citizens. This means that scalability and sustainability that form one of the real successes in terms of m-government project implementation can only come from universal access to government services through mobile cellular devices and technologies while the project implementation is continuous (Blessing & Vesper, 2006).

ENVISAGING SCALABILITY AND SUSTAINABILITY

Sustaining and scaling up m-government projects in general has been described in the previous section and it being a big challenge in developing countries specifically. When envisaging sustainability and scalability of mobile phone-based projects for government service delivery, it would therefore be required that steps to deal with the above described challenges are taken and that actions to address the identified risks associated with sustaining and scaling up m-government projects are proactively planned for.

In the context of this chapter, and with regard to the use of mobile phone-based applications for improved government service delivery, it is critical that the complexities of the dynamics of m-government project implementations be well understood so as to be able to understand the implications of the varying sustainability and scalability influential factors. This will eventually contribute to the development of more sustainable and scalable m-government projects for improved service delivery in the future. In this chapter, the assumption is that the prospects of developing more sustainable and scalable m-government implementation can materialize as follow:

Evaluating the contextual factors that are likely to impact on attaining sustainability and scalability objectives suggests the following:

- Assess the interdependencies between these factors and other possible influential elements being it at project level, technology level, organization level or stakeholder level.
- Evaluate existing m-government projects that have been implemented, their potential use and acceptance level among the citizens who are the users and asses the implications for sustaining and scaling up these projects.
- Use appropriate theories and models to develop a deep understanding of the challenges of and the possibilities of sustaining and scaling up m-government projects for efficient, effective and improved service delivery.
- Develop and implement practical sustainability and scalability guidelines and good practices requirements to be adhered to at the different levels (project level, stakeholder level, technology development level and technology implementation level).

FRAMEWORK FOR ADDRESSING THE CHALLENGES OF SCALABILITY AND SUSTAINABILITY

Reviewing the literature on m-government scalability and sustainability, it can be determined that no integrated framework for ensuring scalability and sustainability of m-government is available. Therefore, establishing a framework to address this subject would facilitate academic research as well as knowledge transfer that will address practical issues and also enable more focused approach. The proposed framework in the table below outlines the factors that are essential to address the subject of sustainability of m-government implementation.

We propose four perspectives on m-government sustainability: Policy, Technology, Financial Management, and Citizen. These four perspectives are based on the necessary sustainability measures in terms of m-government implementation. As m-government project implementation progresses through the various stages of maturity, these requirements must be addressed adequately and simultaneously. If there is no proper scalability or sustainability measure put in place, the probability of a failed m-government project implementation is high. The framework is proposed based on our adoption of (Alijerban & Saghafi, 2010) model through the course of our research study. Therefore, as m-government progresses through the six maturity stages these framework must be addressed adequately and simultaneously.

The aim of the framework is to explore the necessary requirements for m-government as a success factor in government service delivery effort through the use of mobile devices and technology. The development of this framework is based on (Alijerban & Saghafi, 2010). However, the focus of the proposed frameworks comes largely from a technology perspective. The horizontal axis of the framework outlines the consecutive stages of m-government implementation while the vertical axis applies the three different technology require-

ment perspectives. This results in five successive steps per technology requirement which build upon their respective predecessor indicated by their alignment within the framework.

Horizontal Dimension: Maturity Stages

The six maturity stages build upon each other, for example to get to a transactional capability (Stage 3), two-way Mobile-enabled communication (Stage 2) between government and citizen must be possible. The first four maturity stages describe a government's evolution in providing electronic services. The first stage, presence, recognizes government's initial ability to provide one-way information electronically to citizens. The latter stages progress through Two-way communication, e.g. establishing communication via Mobile devices and online forms through mobile devices, through to citizen centric, integrated electronic service delivery, i.e. Transaction, then across multiple government's agencies and departments, Integration. The fifth stage, Personalization, was added to reflect government platforms allowing users to participate politically, e.g. by applying for government's services at citizen's convenient times. The last stage, electronic participation

Table 1. Framework for m-government project implementation scalability and sustainability developed from (Alijerban & Saghafi, 2010)

Stages of M-Government							
		1 Presence	2 Interaction	3 TransactionTransaction	4 Integration	5 Personalization	6 Electronic Participation
	Policy	Policy on presence	Policy of Interaction	Policy on TransactionTransaction	Policy on information sharing	Policy of personalization	Policy on informational self-determination
Scalability and Sustainability Requirements	Technology	Appropriate mobile technology	Secure communication	Integrity of transaction and storage	Data access right management	on time and anytime accessibility	Citizen controlled management
	Financial Management	Funds provision	Accountability	Minimal transaction charge	Consultation	Justification for transaction	Citizen controlled management
	Citizen	Citizen Awareness	Trust	Choice of TransactionTransaction	Consultation	Management right	Control

reflects government platforms allowing users to participate politically, e.g. by voting or posting comments.

Vertical Dimension: Scalability and Sustainability Framework

The Proposed framework adopts a multi-perspective approach - Policy, Technology, Capital and Citizen for establishing scalability and sustainability of m-government project implementation. To address the implementation of this framework, four things need to be considered alongside each other for enhancing the scalability and sustainability of m-government: Policy Requirements, Technology Requirements, Financial Requirement and individual empowerment otherwise known as Citizens requirements. These four perspectives are selected for the proposed framework.

Policy Requirements

The policy perspective, in the *Policy* row, defines the development of scope and clarity of protection through the maturity stages. This begins with the existence of basic policies on m-government project implementation. These may restrict what types and amount of data is collected about citizens as well as stipulate the need to inform citizens about this. However, policy at this stage does not yet cover the need to address how the collected data is used. What type of transaction will be taking place via citizen's mobile devices in order to communicate with the government? Also policies such as cost involved in applying for any of government services through mobile means needs to be put in place. This comes into play in the second stage, with policy adding greater protection of citizens' privacy by specifying how data is used when two-way communication occurs. Government accountability comes into play in the third stage as protection of data becomes part of the government's mandate. This is broadened in

the fourth stage where policy needs to define how citizen information is shared between government organizations, possibly reaching down to the level of the roles and responsibilities of the information handler. In the Participation stage policies need to stipulate the complete informational self-determination of the citizen by addressing how a citizen can obtain full control over the data the government has collected about her/him and how the government uses this data.

Technology Requirements

The Technology perspective is the manifestation of the application of technology. The proposed framework expands this notion to the concept of regulating technology's ability to help collect information indiscriminately and without the citizen's knowledge, so that it occurs in an appropriate manner where both the government and the citizen will have mutual benefit and understanding. In the second stage, where two-way communication is relevant, technology such as encryption and secured storage needed to be to protect interactions between citizen and e-government system.

Financial Requirements

This perspective requires that in order to implement a sustainable m-government project, the government has to look at the infrastructure and financing requirements that are required to meet the needs of the implementation. It is imperative that government and various stakeholders involved in m-government project implementation understand the total cost of ownership before the decision to invest in such a project is made.

Furthermore, there is the need for sufficient human resources to manage the implementation which will require some amount of financial resources to pay them. As with most of the developing countries, there is huge dependent on donor funding to implement project such as this. The question however is if donor funding disappears what happens to the infrastructure and human resources?

Citizens Requirements

The Citizen perspective is generalized to convey important citizen characteristics which support the advancement of technology usage in m-government project implementation (Belanger & Hiller, 2006). These characteristics must be established through the policy, and practice of government including the use of technology. Citizens' awareness of privacy issues, and the government's policy towards it, structures the basis for the citizen's stake in m-government project Implementation. Once trust by citizens in the m-government project has been established, the choice of how to implementation m-government system needs to be addressed. The next action is therefore to consult citizens, educate them and empower them to decide on which data the government collects about them and how it is used at a high level of granularity while the last stage gives full control over this matter to the citizen.

CONCLUSION

The use of mobile technologies in the public sphere has increased and will continue to increase participation in the democratic processes. Making use of mobile devices and other latest technologies allow people to connect, interact and collaborate with governments and vice-versa in many spheres within the governments' functionality. By understanding how m-government is adopted within a governmental environment. some understanding can be gained as to its scalability and sustainability as a collaboration platform that will enhance services delivery by governments and services consumption by the governed. The prospective success of m-government projects implementation at each stage of intricacy can, however, be easily shown through the frameworks to support the scalability and sustainability as being dependent on meeting a combination of policy, technical, financial and citizens requirements. This chapter has hopefully suggested a way forward.

REFERENCES

Adil, K. M. (2000). *Planning for and Monitoring of Project Sustainability: A Guideline on Concepts, Issues and Tools.* Retrieved May 15, 2013, from http://www.mande.co.uk/docs/khan.htm.

Alfred, S. S., & Elizabeth, K. (2010). Contribution of Mobile Phones to Rural Livelihoods and Poverty Reduction in Morogoro Region Tanzania. *Electronic Journal on Information Systems in Developing Countries, 42*(3), 1–15.

Ali, M., & Bailur, S. (2007). *The challenge of sustainability in ICT4D - Is bricolage the answer?* Paper presented at the 9th international conference on social implications of computers in developing countries. Sao Paulo, Brazil.

Alijerban, M., & Saghafi, F. (2010). M-government Maturity Model with Technological Approach. In *Proceedings of the 4th Conference on New Trends in Information Science and Service Science.* Gyeongju, South Korea: IEEE.

Andersen, K. V., & Henriksen, H. Z. (2006). E-government maturity models: Extension of the Layne and Lee model. *Government Information Quarterly, 23*(2), 236–248. doi:10.1016/j.giq.2005.11.008

Bassara, A., Wisniewski, M., & Zebrowski, P. (2005). *USE-ME.GOV – A Requirements-driven Approach for M-GOV Services Provisioning.* Business Information Systems.

Belanger, F., & Hiller, J. (2006). A framework for e-government: privacy implications. *Business Process Management Journal, 12*(1), 48–60. doi:10.1108/14637150610643751

Blessing, M. M., & Vesper, O. (2006). *Bring M-Government to South African Citizens: Policy Framework, Delivery Challenges and Opportunities*. Paper presented at the Second European Conference on Mobile Government. Brighton, UK.

Blessing, M. M., Vesper, O., & Wallace, T. (2007). Enabling M-government in South Africa. In Mobile Government: An Emerging Direction in E-Government. IGI Global.

Botha, A., & Makitla, I. et al. (2010). *Mobi4D Platform*. IEEE.

Coordinating, N. E. C. (2001). *M-Government: The convergence of wireless technologies and e-Government*. Retrieved May 21, 2013, from www.ec3.org/Downloads/2001/m-Government_ED.pdf

Crawford, T., & Lester, W. (2004). *Information Management challenges and opportunities for community based organisations serving people living with HIV/AIDS*. Academic Press.

De Bruin, T., Freeze, R., & Uday, K. (2005). Understanding the Main Phases of Developing a Maturity Assessment Model. In *Proceedings of Australasian Conference on Information Systems*. Sydney: Academic Press.

DelloiteResearch. (2000). *At the Dawn of e-Government - eGovernment Resource Centre*. Retrieved 15th September, 2012, from http://www.egov.vic.gov.au/pdfs/e-government.pdf

Fasanghari, M., & Samimi, H. (2009). A Novel Framework for M-Government Implementation. In *Proceedings of International Conference on Future Computer and Communications*. Los Alamitos, CA: IEEE.

Ferguson, J., & Ballantyne, P. (2002). *Sustaining ICT-Enabled Development: Practice Makes Perfect?* Academic Press.

GartnerResearch. (2000). *Gartner's Four Phases of E-Government Model*. Retrieved 15 Spetember, 2012, from http://www.gartner.com/id=317292

Georgiadis, C. K., & Stiakakis, E. (2010). Extending an e-Government Service Measurement Framework to m-Governement Services. In *Proceedings of Mobile Business and 2010 Ninth Global Mobility Roundtable (ICMB-GMR), 2010 Ninth International Conference on*. Athens, Greece: IEEE.

Ghyasi, A. F., & Kushchu, I. (2004). *m-Government: Cases of Developing Countries*. Paper presented at the European Conference on e-Government. London, UK.

Hiller, J. S., & Belanger, F. (2006). A framework for e-government: privacy implications. *Business Process Management Journal, 12*(1), 48–60. doi:10.1108/14637150610643751

India, T. R. A. O. (2011). *Highlights of Telecom Subscription Data*. Retrieved September 16, 2013, from www.trai.gov.in/WriteReadData/trai/upload/PressReleases/835/Press%20Release%20June11.pdf

InfoDev. (2003). *Improving Health, connecting people: The role of ICTs in the health sector in developing countries*. Author.

InfoDev. (2005). *Harnessing ICTs to fight poverty and promote development: A research strategy and work plan 2005-2007*. InfoDev.

Johan, H. (2008). *Mobile phones for good governance – challenges and way forward*. Retrieved September 19, 2013, from http://www.w3.org/2008/10/MW4D_WS/papers/hellstrom_gov.pdf

Kim, D. Y. (2010). E-government maturity model using the capability maturity model integration. *Journal of Systems and Information Technology, 12*(3), 230–244. doi:10.1108/13287261011070858

Kimaro, H. C. & Nhampossa, J.L. (2005). Analysing the problem of unsustainable health information systems in less-developed countries: case studies from Tanzania and Mozambique. *Information technology for development 11*(3), 273-298.

Kumar, A., Sukanta, M., & Sahu, K. (2008). *Challenge of Wireless and Mobile Tecnologies in Government.* Academic Press.

Kushchu, I. & Kuscu. (2003). *From E-Government to M-government: Facing the inevitable.* Dublin, Ireland: MGovLab.

Layne, K., & Lee, J. (2001). Developing fully functional E-government: A four stage model. *Government Information Quarterly, 18*(2), 122–136. doi:10.1016/S0740-624X(01)00066-1

Mansoor, A., & Rohan, D. (2010). *An M-government Solution Proposal for Dubai Government.* Paper presented at the 9th WSEAS International Conference on Telecommunications and Informatics. Sicily, Italy.

MCIT. (2003). *e-Government Lembaga.* MCIT.

Mengistu, D., Hangjung, Z., & Jae, J. R. (2009). M-government: Opportunities and Challenges to Deliver Mobile Government Services in Developing Countries. In *Proceedings of Computer Sciences and Convergence Information Technology, 2009. ICCIT '09. Fourth International Conference on.* Seoul: IEEE.

Nava, A. S., & Dávila, I. L. (2005). M-Government for Digital Cities: Value Added Public Services. In *Proceedings of the First European Mobile Government Conference Mobile Government Consortium International.* Academic Press.

Pade, C., Mallinson, B., & Sewry, D. (2006). *An Exploration of the Critical Success Factors for the Sustainability of Rural ICT Projects – The Dwesa Case Study.* SAICSIT.

Sadeh & Norman. (2002). *M-Commerce: Technologies, Services and Business Models.* John Wiley and Sons.

Sandy, G. A., & McMillan, S. (2005). A Success Factors Model For M-Government. In *Proceedings of Euro MGOV 2005.* Brighton, UK: MGOV.

Sharma, S. K., & Gupta, J. (2004). Web Services Architecture for M-government: Issues and Challenges. *International Journal of Electronic Government, 1*(4), 462–474. doi:10.1504/EG.2004.005921

Siau, K., & Long, Y. (2005). Synthesizing e-government stage models - a meta-synthesis based on meta-ethnography approach. *Industrial Management & Data Systems, 105*(3), 443–458. doi:10.1108/02635570510592352

Sprecher, M. H. (2000). Racing to e-government: using the Internet for citizen service delivery. *Government Finance Review, 16*(5), 21–22.

Tapia, R. S. (2009). *Assesing business-IT alignment in networked organizations.* (PhD Thesis). University of Twente, Eschende, The Netherlands.

UnitedNations. (2001). *Benchmarking e-Government: A Global perspective: Assesing the UN member states.* New York: United Nations Department of Economic and Social Afairs.

Valentina, N. (2004). E-Government for Developing Countries: Opportunity and Challenges. *Information Systems in Developing Countries, 18*(1), 1–24.

Walsham, G., & Sahay, S. (2006). Research on Information Systems in developing countries: Current Landscape and Future Prospects. *Information Technology for Development, 12*(1), 7–24. doi:10.1002/itdj.20020

Welch, E., & Wong, W. (2001). Global information technology pressure and government accountability: The mediating effect of domestic context on Website openness. *Journal of Public Administration: Research and Theory, 11*(4), 509–538. doi:10.1093/oxfordjournals.jpart.a003513

West, D. M. (2004). E-Government and the Transformation of Service Delivery and Citizen Attitudes. *Public Administration Review, 64*(1), 48–60. doi:10.1111/j.1540-6210.2004.00343.x

KEY TERMS AND DEFINITIONS

E-Government: The provision of government services through the use of Information and Communication Technologies (ICT).

Maturity Level: This is a model that is intentionally designed to evaluate the maturity of a service, an object, a performance, etc., based on one or more set of conditions, including competency, capability, and level of sophistication.

M-Government: The use of mobile and wireless communication technology in government for service and information delivery to citizens and organizations.

Scalability: This refers to the ability to duplicate the implementation a pilot project over a wider area.

Sustainability: This refers to the possibility that the benefits from an intervention will be maintained at an appropriate level for a reasonably long period of time after the withdrawal of support from contributor or sponsor.

Chapter 9
The Case of the Mexican Mobile Government: Measurement and Examples

Rodrigo Sandoval-Almazan
Universidad Autónoma del Estado de México, Mexico

Yaneileth Rojas Romero
Universidad Autónoma del Estado de México, Mexico

ABSTRACT

The mobile government has become a reality in a large majority of countries around the world. However, the use of mobile apps (small software programs for use on mobile devices) to link government Websites and information is a recent trend that is becoming of interest to citizens and public officials. The uses, advantages, and disadvantages have recently become a study field for several scholars around the globe. The mobile government is not new for e-government scholars; however, the explosion of apps and the increase of smart phones have created a new trend in the mobile government field. In order to understand these phenomena in the Mexican society, the authors have gathered data from different sources: government departments, business enterprises, and citizen organizations. Based on this information, they analyze the impact of apps across the country and suggest a classification method that can be used for a better understanding of this new field. In this chapter, the authors discuss five small case studies, which they consider good examples to follow by different government organizations. To accomplish this objective, they divide this chapter into seven main sections. After the introduction, the authors provide a literature review, describe the method of study and classify the apps, discuss the findings with the model application, present the case studies for government apps, discuss ideas for future research on government apps, and then in the final section, they present final remarks and conclusions of the investigation.

INTRODUCTION

The electronic government (or e-government) has become widely accepted for most governments around the world. The use of technology for public administration processes, tax processing systems and political participation such as discussions, feedback and openness has become a constant transformation at different government levels. In this context, software applications, better known as *apps*, have become a disruptive technology for governments. The commercial use of this technol-

DOI: 10.4018/978-1-4666-6082-3.ch009

ogy in smart phones has increased exponentially and the numbers in use are astonishing high; even though the governments' adaptation has been slow.

In July 2013, Apple Corporation celebrated five years of the *iTunes App Store* with 850,000 apps available in the store and accounted for more than 50 billion of apps, being used over 800 apps per second (Apple, 2013). Android, with 700,000 apps available in the Google Play Store, counted more than 1.5 billion apps' monthly downloads (Developer, 2013). The digital ecosystems, promoted by Microsoft, Apple, Amazon and Google, are increasing their content into islands that threat the Internet freedom and the governance of the net (Berners-Lee, 2010; Iansiti & Levien, 2004).

These new challenges menace neutral networks but provide great opportunities to create competition among companies and improve communication with companies-consumers and citizens-government. In this area of opportunity, new paths of research have been made along the way. One of them is the use of mobile devices that are linked with government apps to share, to exchange and to collect information.

The use of mobile devices has increased in the last years and governments around the world take advantage of this kind of features and communication possibilities (de Kool & van Wamelen, 2008; Sandoval-Almazan & Gil-Garcia, 2012). At the same time, the information wave has emphasized the need of a more open and transparent government (Hans J. Scholl & Luna-Reyes, 2011). For example, the US government has made available about 75,718 data sets of raw data and 68,147 geospatial datasets to the public through the data.gov website. Many other states are following the US initiative, making raw data available to the public (Amaravadi, 2005). The incorporation of new technologies, smart phones and computers has allowed information to be distributed much faster at a very low cost and with a great range to all segments of the population. Small applications running on web pages or in mobile devices

(mobile apps) have become a wide-adopted way of interaction. It is expected that mobile computing will replace desktop computing by 2015 (Milam & Avery, 2012). Again, just in U.S. portal, www.data.gov, about 1,500 web and mobile applications (apps for short) have been developed, more than 200 by citizens themselves. We believe that these trends together have the potential to transform the relationships between government and the public.

Major changes in technology, citizens and politics are changing behaviors. They are also changing the process of doing politics and the relationship between politicians, public servants and citizens (Geiger & von Lucke, 2012).

Despite this emerging context, we still know very little about the level of adoption of apps in government websites and their uses. The purpose of this paper is to provide a methodology for the categorization of government apps, looking for an initial categorization of practices among top-rated governments in terms of the use of information technologies, as well as to provide some examples of current app uses in the Mexican government.

In the Mexican case, the publication of government information through electronic media is mandatory by law since 2002 because of the "Transparency and Access to Public Government Information Law," which focuses on providing a secure access to everyone to federal, local and municipal government information. They handle the information and they need to support this information in order to accomplish this objective. Our research is taking advantage of this official policy and we tried to collect our data – apps – from the government and companies related to government functions.

We divided this chapter into seven main sections. This introduction constitutes the first section. The following three sections present a literature review, describe the method we followed to classify the mobile apps and discuss our findings. The following section introduces several case studies of different mobile apps in the Mexican government.

Then, we address ideas for future research in the field of mobile government and associated apps. Finally, we make some remarks and conclusion of this topic in the last main section.

LITERATURE REVIEW

The purpose of this literature review is to introduce previous knowledge and research about the evolution of mobile apps and mobile government through the last years. We divided this section into three parts: apps or software app; technology for government; and mobile government. Finally, we present and support the research model with some theoretical ideas in the last section.

The term "small software application," better known as "app," was coined in 1985 (Holwerda, 2011). The principles and basics of this kind of small software comes from the need of reducing battery use and the optimization of computational processes (B P Lientz, Swanson, & Tompkins, 1978; Bennet P Lientz & Swanson, 1981). Later, this idea of computational processing was used for the interface and users tasks (Beath & Walker, 1998; Kersten, Kersten, & Rakowski, 2001). Despite the shy start of apps' software, a new trend was launched with the rise of the Web 2.0 that included the interaction and user relationship with websites (O'Reilly, 2005).

This evolution into a new version of interactive websites i.e. Web 2.0 enabled the use of more technology such as the apps. Most of them evolved into mobile apps because of their size which allowed them to be display using mobile phones or the new tablets (Lugano, 2008; Yamakami, 2007). Mobile internet has changed consumer habits and the behavior of the final user allowing instant connectivity and immediate update of information (Johnson, 2010). Later on, the development of mobile learning applications started (Pocatilu, 2010). The next stage was the development of apps for commerce and gaming, creating a new market environment for all kind of purposes (Anthes,

2011), and leading the change for more content sharing, content creating and content delivery for governments (Murugesan, Rossi, Wilbanks, & Djavanshir, 2011).

The use of technology in governments has a long history (Garson, 2003). However, the use of apps is more recent. Yamakami (2007) discusses this change of perspective and the possible evolution of apps in the new government environment. Other studies from different scholars have taken several directions (Herrick, 2009; Woods, 2009). One of the most important contribution comes from de Kool & van Wamelen (2008) who explore the context of this evolution of the government to mobile devices and their relationship with the society and the government impact, through their analysis of the Web 2.0 tools in the Netherlands. This study highlights the new face and the change of the government using this kind of interactive tools, later on focused on mobile devices.

In the mobile government research path, wireless technologies and the Web 2.0 started to develop the basic form of the mobile government. The preliminary work of Townsend (2002) developing government opportunities for this technology unfolded this ignored path of research. On the other hand, research from Kushchu (2003), (2007) has become a starting point for analyzing this topic. Kushchu's book (2007) introduces the e-government research. Different authors and perspectives of the book - mobile city, usability, strategy, design and business models - determine some trends of the present challenges. A complementary perspective of this field of research is the Mobility Response Model developed by Kushchu and Borucki, and validated with five case studies (four in the UK and one from Hong Kong). They show a more efficient government process using this mobile technologies (Borucki, Arat, & Kushchu, 2005).

One step forward in the m-government research is the work of Trimi & Sheng, (2008) analyzing the consequences of this new trend on leading governments around the world. The contribution of Traunmüller (2011) adding the social media and innovation into the framework of m-government increases the potential and locates the discussion and use of this field in a different dimension. The idea of a more user friendly government, personalized services and knowledge management available from a cell phone is placed and analyzed in this stage (Sandoval-Almazán, Gil-Garcia, Luna-Reyes, Luna-Reyes, & Murillo, 2011). Some consultants and scholars have started to look at the m-government as an important place to make business and develop new ideas including apps, strategies and software (Eggers & Jaffe, 2013; Eggers, 2013; OECD & ITU, 2011).

A first attempt to develop a theoretical framework for the mobile government comes from Antovski & Gusev (2005). It incorporates five principles: interoperability, security, openness, flexibility and scalability. The following year, a more refined model was presented by Antovski & Gusev (2006). More research has been made to introduce different models and frameworks (Fidel, Scholl, Liu, & Unsworth, 2007; Maumbe & Owei, 2006; Sheng & Trimi, 2006). Another framework for the implementation of m-government was developed by Emmanouilidou (2010) incorporating the W3C mobile best practices. The implementation of the incorporation of specific issues on the mobile government increased the use and perspective, such as security (Athanasios Karantjias, Spyridon Papastergiou, 2009), enterprises and rural efforts (Ntaliani, Costopoulou, Manouselis, & Karetsos, 2009) and software development (Kesavarapu & Choi, 2012).

In the case of Seattle, the implementation revealed that the implementation of an m-government represented several challenges in the organizational front (Fidel et al., 2007; H.J. Scholl, Fidel, Mai, & Unsworth, 2006). An important issue for the m-government is security and the case of Agroportal established some risks and challenges on this topic (Chatzinotas, Ntaliani, Karetsos, & Costopoulou, 2006). A different perspective is the Greek case that links government

and business cooperation using m-government (Kapogiannis, Touzos, & Kreps, 2006). Finally the work of Walravens, using the m-government implementation in cities such as New York, developed a complementary perspective of this tool on governing cities and increased efficiency (N. Walravens & Ballon, 2011; Nils Walravens, 2012).

From the perspective of the apps world, these new mobile technologies have invaded all platforms: such as the IOS, Android and Microsoft Windows (Johnson, 2010). Many similar devices have revolutionized the way information can be distributed (Hosmer, Jeffcoat, Davis, & McGibbon, 2011). Smartphone users are increasingly shifting the use of apps as "gateways" to Internet services rather than traditional web browsers (Xu et al., 2011) creating an impact on citizens, business and governments (Murugesan, Rossi, Wilbanks, & D Javanshir, 2011). Mobile technology offers many advantages for governments over the traditional methods of information dissemination. The offer of apps grew with the Apple-iTunes Market, the Android market, the Windows Phone Marketplace and the Blackberry App World (Anthes, 2011).

Many governments are encouraging their agencies to write their own applications in areas of impact focused on topics such as feedback from citizens for administration, citizen participation (Traunmüller, 2011), G2G inter-agency collaboration (Beer, Kunis, & Runger, 2006), public services (Estevez & Janowski, 2007) and privacy (King, Lampinen, & Smolen, 2011).

The use of mobile apps has promoted research in different areas such as user behavior (Xu et al., 2011), feedback and usability (Fu et al., 2013; Tang, Hsiu, Huang, & Chen, 2013), user characteristics (Wu, Ozok, Gurses, & Wei, 2009) and citizen engagement (Raths, 2011). Also the promotion of competition of quality apps (Carlson & Eyler-Werve, 2012) and different platforms implementation for government apps (Wei, Gao, Jia, & Yang, 2010). Mobile applications have the potential to improve service delivery, as well as efficiency and efficacy in government operations.

Furthermore, mobile applications create the capability of interaction, participation and transparency (Wamelen & Kool, 2008) providing the ability to access critical operational information regardless of location (Hosmer et al., 2011).

Most government apps need to identify goals: transparency by making open data accessible with the possibility of report flaws in data sets, allowing feedback from citizens regarding administration, collaboration among agencies, delivery of government services in a more efficient way and businesses innovation (Carlson & Eyler-Werve, 2012). User-friendly design (user view also an adequate restructuring and a consistent outline), knowledge enhancement (facilitate orientation in public life based on life events), knowledge collection (collecting parts of the knowledge which are implicit) and information about the data source (Traunmüller, 2011).

Apps could expand the potential for e-government to allow citizens and businesses to access content, create content, send communications to government agencies, adding in some cases knowledge enhancement and allowing citizens to create ecosystems interested in solving important public problems (Sandoval-Almazan, Gil-Garcia, Luna-Reyes, Luna, & Rojas-Romero, 2012). The government repositories such as Data.gov - in US, UK, India, Spain - and IT Dashboard or Challenge. gov are examples of open government data and the use of data for government. However, they need a way to be advertised and spread. A solution is to develop an app and share it with citizens (Goggin, 2011).

Apps are related to government repositories, open government ideas and democratic principles (O'Hara, 2012). The group Apps for democracy is an example of this idea, when developers and programmers help governments to achieve their purposes using application technology and delivering information through cell phones and Internet (Labs, 2012). This effort is combined with citizen contests in order to develop apps for certain purposes such as public services, public transportation or government organizational problems (Carlson & Eyler-Werve, 2012; Milam & Avery, 2012). Finally the combination of social media and apps have become the perfect combination to promote, exchange and collaborate between citizens and governments (Abramowicz et al., 2005; Lerman, 2007; Traunmüller, 2011).

Alongside this literature review, we can find some links among the application technology - apps - with the e-government field of research and the mobile government dynamic field. Most of these connections are specifically related to develop tools and frameworks to communicate, share and promote collaboration among government information and citizens. We found that this connection has three common aspects, 1. Apps research and development refer to government efficiency and accountability; 2. Apps for mobile government refer to specific issues such as economy, development and citizens problems and 3. Recent apps development is focused on open government and transparency. We are going to consider these links in our research model in the following section.

RESEARCH DESIGN AND METHOD

The purpose of this research is to understand the use and evolution of e-government apps in the Mexican case. Several scholars are starting to understand the impact of apps in society and specifically in the government field. Even though, this kind of research needs a different focus to design, collect and analyze data (Thomas & Streib, 2005). In order to conduct this exploratory research we follow the approach of Internet-Mediated Research (IMR) which consists of "the gathering of novel, original data to be subjected to analysis in order to provide new evidence in relation to a particular research question" (Hewson, 2008 p. 58). One of the problems of doing internet research is to define the sample, especially when the object of study is as new as the apps development (Jones, 1999). A combination of systematic research along with the data collection and rigorous analysis can develop a more accurate design of research for the internet field (Estalella & Ardevol, 2011).

We divided our research into three main stages. The first stage was to develop a model to classify and understand the Mexican apps; the second stage was to collect data and the final stage was to analyze and discuss data. For the purpose of this chapter, we chose the most relevant cases, which we present as examples.

Research Model

The first stage was to develop a model in order to classify and determinate if the found apps contributed to the mobile government. We combined two models, the civic apps model (Carlson & Eyler-Werve, 2012) and the model of m-government implementation (Antovski & Gusev, 2006). The first model has three main goals: accountability, government efficiency and economic development and consists of seven components. The second model consists of five principles: interoperability, security, openness, flexibility and scalability. We combined the principles and components to create

a model of apps and mobile government that allowed us to assess the Mexican case, adding one additional component and some aspects that we could evaluate. A combined model allowed us to create a more integrated perspective of the mobile government. We choose the objectives of the civic apps models as inspirational and long vision goals. We achieved these goals with the principle applications – security, openness, flexibility – in order to create a model that measured and promoted a mobile government. The combination of principles to promote goals is a logical path for expanding and understanding such phenomena. The result appears in Table 1, showing the components of the model and the variables.

Data Collection

The problem of apps data collection is that most of them are disperse in the Mexican government or in the market. There is no single place to find out about apps and to determine if they match the criteria to be considered as government apps. In order to find out about this application software in a systematic approach, we follow three main steps:

Table 1. Model of mobile government apps

Components	Variables
Accountability	1. Raise awareness of available open government data sources 2. Focus energy on building apps on open data 3. Security
Government Efficiency	1. Inter-collaboration 2. Apps that benefit people and businesses 3. Crowd source data publishing priorities 4. Knowledge enhancement and collection 5. Interoperability
Economic Development	1. Drive innovation 2. Build a community of practice around 3. User-friendly design
Transparency/Openness	1. Reports a data flaws & feedback from citizens 2. Open data accessible 3. Flexibility 4. Interoperability

- We decided to look for current applications in the 31 portals of the Mexican state government and in the federal entity websites of the Mexican Republic.
- We visited and analyzed the federal government portal www.god.mx that offers a search service specialized on Mexican government and the personalized access to procedures and services.
- Finally, we performed a search on the iTunes App Store and on the Google Play Store.

We collected data from February to July 2013 and created a database with the cases. Finally, we found 320 apps. We analyzed the apps but not all met the necessary criteria to be considered in our sample. In order to determine whether we could consider them as government apps, we decided to evaluate them one by one. We found 74 apps, which were developed by the government, independent agencies and secretariats. They are published on their own sites and available for downloading in the apps marketplaces such as iTunes App Store and Google Play Store. We also found that 49 apps were available in the iTunes App Store and 25 apps in the Google Play Store.

Data Analysis

In order to classify and categorize the 74 apps available in the Google Play and iTunes App Store, we evaluated one by one according to our model. We created an array of components and aspects to be evaluated, giving one point for every complied feature and a zero if it did not comply with the feature to be assessed. We added the scores of each of the targets by component, first individually and then together. The results are shown in Figure 1. Our results were that the 74 evaluated apps met an average of 6.2 of the 14 evaluated variables. Only 12 apps gathered all the evaluated points representing the 16.21% of the total number of apps. We describe our main findings in the following section.

Figure 1. Main findings - Mexico apps 2013

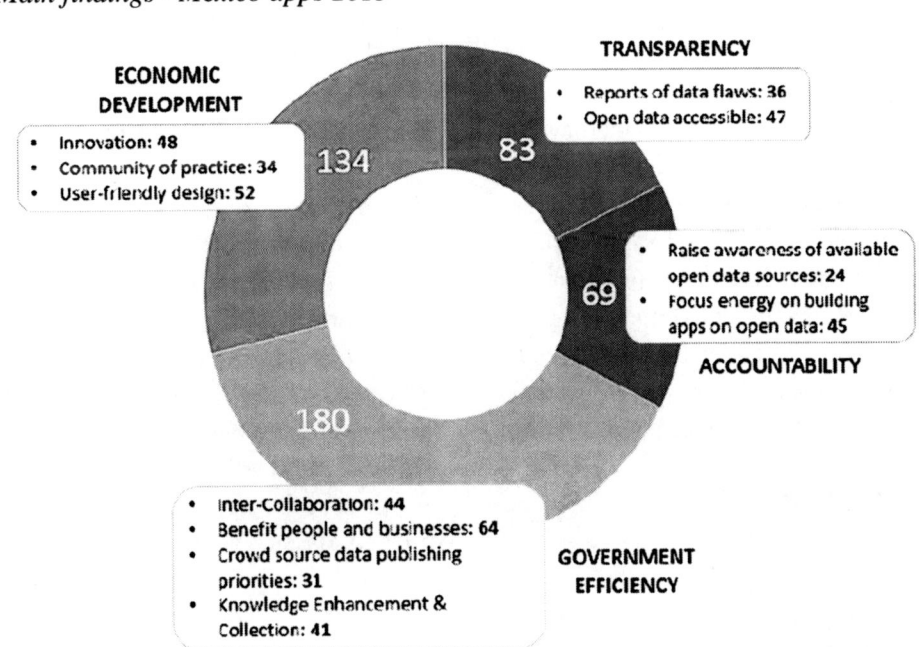

MAIN FINDINGS: MEXICAN GOVERNMENT MOBILE APPS

This section presents an overview of the use of apps and some examples of apps available for citizens in Mexico. We collected the 74 apps for this survey from February to July 2013. We present our findings using the previous model description: the first section describes the accountability component and the apps related to this component; the second section presents the government efficient component; the third section addresses the economic development apps and the final section describes the transparency and openness apps. We present a summary of the 74 Mexican apps we collected in the sample in Table 2. A summary of the findings can be seen in Figure 1 at the end of the section.

Table 2 briefly describes the 74 collected apps for this research. Several comments arise from this initial exploration: Municipalities have not developed any apps and the federal or local government (state governments) uses most of them and they interact with citizen using this kind of resources to reduce time and costs for government services.

Accountability

This component is the lowest one from the entire model. It only reached 69 points, divided by 24 apps that rose awareness on open data sources and just 45 apps are focusing their energy to build open data in the government. There are no apps related to interoperability nor flexibility or reusable information for the government. Some examples of this component are the following apps: Quién es quién en los precios (Who is who on prices) and 072 Movil (Emergency mobile number 072).

Government Efficiency

The aspect of government efficiency is one of the most covered aspects in the assessment of the apps reaching 180 points. Specifically, it is the factor that allows people and businesses to get the benefits, for example, in the tourism, hotels and restaurants industries. The inter collaboration principle reaches 44 apps that have developed some features related to this effort. An interesting point, the crowdsourcing and data publishing could have been the most relevant one but it only reached 31 apps data that are close related to this idea. Finally, the apps related to knowledge are 41. Some examples of Mexican apps for government efficiency are Bus Sonora; Reporte MH (Miguel Hidalgo, Mexico City delegation report for public services), Reporte Vial Puebla (traffic report of the state of Puebla).

Economic Development

In addition, the 134 points covering aspects such as the user-friendly design and the drive innovation are involved in the economic development. We found 48 apps related to innovation and 52 apps with a user-friendly design. Finally, we only found 34 apps of the category "building community of practice." Examples for this component are Conaculta Mexico (App for the Mexican Cultural Office); Taxiaviso (Taxis warning for Mexico City); Reporte Vial Puebla (Puebla's traffic report).

Transparency and Openness

We only found 47 apps related to open data and 36 apps able to report data flaws or problems on building apps. We did not find any app related to government security so far. The best examples for this component are Gobierno Movil (M-government local app), Senador Isais (Apps from a member of the senate) and Chihuaha en tu móvil (Chihuahua State App).

Some of this apps share common features. In order to classify them, we considered the most predominant feature or objective of the apps. Table 3 shows the apps and their classification in the component. We found that 12 apps ac-

Table 2. Sample of 74 Mexican apps

App Name	Agency / Organization	Federal	State	Short Description
Info DF Movil	Mexico City Government		x	Information about city services
ICAP	Government of Coahuila & ITELTEQ		x	Electoral college system of citizen participation and monitoring, Coahuila trainers, Coahuila state election
GuíaPemex	Petróleos Mexicanos	x		It will find information about 10 thousand Pemex gas stations in the country in which gasoline and diesel is sold, as well as 644 restaurants, 268 hotels and 44 spas selected by experts
Agente Vial Movil	Government of Jalisco		x	Relevant information about offenses listed in Highways and Transportation, Act of Jalisco
Jalisco Móvil	Government of Jalisco		x	Virtual window of procedures and related information, the Government of Jalisco
Ayuntamiento de Merida	Government of Merida Yucatan		x	Mission, Vision and Values of the City of Merida, Yucatán
Congreso Estado de San Luis P.	Government of San Luis Potosi & yellowlink		x	Information about the congress and legislature of the government of San Luis Potosi
Profeco en 30	Procuraduría Federal del Consumidor	x		A tool designed by the Federal Consumer Protection Agency and available for consumers in order to report non-compliance from the service providers with the provisions of the Federal Consumer Protection
Jalisco Móvil_	Government of Jalisco		x	Virtual window procedures and related information, the Government of Jalisco
Diputados	Federal Government of Mexico	x		This application is intended to educate the public by providing access to information through Liaison Units Obligees by the Federal Law of Transparency and Access to Public Government
Huixquilucan	Government State of Mexico & IBloomMx		x	City Information Huixquilucan, Mexico State.
Sefiplan David	Government of Veracruz & omnius		x	This is an example of the basic course in Android Sefiplan Veracruz
JaliscoFactBook	Government of Jalisco		x	Information on development indicators, and public investment projects Jalisco state and its municipalities. Besides data on population and housing census 2010 of all federal entities Mexico.
SIAM - Estatus tramites	Ministry of Economy	x		Mining System Administration - Consultation procedures
SIAM - Act y recargos	Ministry of Economy	x		Mining System administration of the Ministry of Economy - actualiaciones calculator and surcharges.
COPAES Móvil	Autonomous University of Chihuahua		x	You can check in a practical and simple individual accredited academic programs in Institutions of Higher Education in Mexico face modalities, semi-presenial and virtual.
CARNET Móvil	Autonomous University of Chihuahua		x	You can make your attendance records valid event for the cultural card "Universidad Autonoma de Chihuahua" very simple and dynamic.
Cuajimalpa de Morelos	Government of Morelos & IBloomMx		x	Official bulletins, notices and news of this delegation.

continued on following page

Table 2. Continued

App Name	Agency / Organization	Federal	State	Short Description
Descubre Puebla	Government of Puebla & Estrategia 360		x	This is an application where the Government of the Municipality of Puebla, invites you to visit the main tourist attractions of the town, this guide covers everything a traveler needs to know: museums, galleries, cultural venues, restaurants, hotels, shopping options, nightlife and more.
Revista Enterate	Government of Mexico & CFE	x		The ENTTÉRATE Magazine is an internal publication of CFE, the Peninsular Division, which disclosed the most important events related to this dependence, without any political, social or financial gain.
Durango Turistico	Government of Durango & FIPADE		x	Secretary of Tourism of the State of Durango offers the application that will touch your instincts and take him to a bleak journey through the beautiful and morepicturesque places in our state, a beautiful place where time embellishes things, where modern merges with the story, the real heart of Northern Mexico.
Adeudo Vehicuar	Government of Veracruz & omnius		x	App vehicular query Debit Sefiplan Veracruz
Partido Verde Ecologista	iBloomMx		x	Check out latest news from the Green Party. Watch videos with official content of the Green Party. Be informed about who the Green Party. Exclusive downloads, maps and locations of ecological sites in the country.
SNIIM Móvil	Ministry of Economy	x		Mobile version of the National Information System and Market Integration. It is an application of the SE to help us get the price and presentation of agricultural, livestock, fishery, etc products. by different price points.
Fernando Mayans	Daniel Islas Rodriguez	x		Information about Senator of the Republic. John Graham Casasús Clinical Hospital. General Medical Specialty in gastrosurgery. Politician. I have been Local and Federal Deputy for the state deTabasco.
Aire DF	Government of Federal Distric & Jorge Cornejo Martínez		x	Application Air DF is designed to present information in air quality in Mexico City. Uses data Air Monitoring System of the state of air quality and the intensity of ultraviolet radiation.
Consulta de Licencia Federal	Ministry of Communications and Transport	x		Implementation of Consultation of Federal Licenses. With it you can check the validity of any federal license issued in Mexico checking the QR (QR Code) code printed on the back (for printed licenses 2011 onwards) or the number thereof (for ALS to 2011). Using the camera on your computer, scan the code to verify identity and data specific to the person carrying license, and you can even verify photography, ensuring 100% identity of the person.
Senador Isaias	Nuvolsoft	x		Official Application of Isaiah Gonzalez Cuevas, Senator of the Republic by the state of Baja California Sur
I Informe Gobierno	Governrnent of Veracruz & Rafael Alejandro Ballester Diez		x	Government Information
CONARTE	Government of Nuevo Leon & CONARTE		x	CONARTE is theater, cinema, music, dance, literature, photography, visual arts, popular culture and cultural heritage enjoy!

continued on following page

Table 2. Continued

App Name	Agency / Organization	Federal	State	Short Description
ZacatecasTravel	Government of Zacatecas & CEOS New Media Agency		x	This is the official app of the Ministry of Tourism of the State of Zacatecas Travel Guide that provides for everyone.
Isseg Móvil	ISSEG GUANAJUATO		x	You can view: Out, offers and promotions ISSEG Pharmacy. Location of Parking ISSEG. Balance, deposits and withdrawals of voluntary savings. Balance and movements CREDISSEG. Simulators, notices and contributions ISSEG.
Cédula Móvil SEP	Ministry of Education	x		The new application consultation on the National Register of Professionals on the Mobile version, administered by the Directorate General of Professions is of a public nature under Articles 25 and 32 of the Rules of the Regulatory Law of Article 5 concerning the constitutional exercise of professions in the Federal District and aims to expand the search criteria of professionals who register their titles and feature professional license with effect paten
Unidad del Vocero de QRoo	Governement of Quintana Roo &Rex Systems Software Developer		x	Spokesman QRoo is the official app used by government spokesperson unit of Quintana Roo, for social communication.
Traza tu ruta	Ministry of Communications and Transport	x		The official Android application for the site of the Ministry of Communications and Transport, Mexico. Trace any route between two cities - the application will provide mileage, estimated fuel consumption, toll cost and average travel time.
Cumbre Tajín	Governemnt of Veracruz		x	Cumbre Tajin is an annual festival around the spring equinox. Collect art, ceremony, music, reflection, dance, healing and multiple expressions of artistic creation and ritual. The Totonac culture is the hostess and hosts over 5000 artists from Veracruz, Mexico and the world, who for five days offer five thousand activities in three offices: Takilhsukut Park, El Tajin archaeological zone and town of Papantla.
Bus Sonora	Government of Sonora		x	The Government of the State of Sonora is the mobile application for you Bus Sonora.Monitors paths and drives known point of the city are transiting.
redarbol	Mexico Inteligente		x	Red Tree is a digital platform that allows citizens of the City of Mexico record and document the tree species in the city. As a citizen, through this application you can make reports to find trees near your home, neighborhood, work, parks, medians or any other place in Mexico City to visit.
Conaculta-Mexico Es Cultura	Conaculta	x		Meet and cultural activities located in Mexico and make your own cultural agenda. Also, you can access our toll free number 01 800 CULTURE (01 800 2 85 88 72)
Mi Policia	SSP-DF		x	Application that approaches the citizen to information interactively their respective Quadrant, providing a quick way to call in an emergency and the location of graphically Quadrants Federal District.
Taxiaviso	Soluciones en Tecnologia Devfactor	x		Taxiaviso is a service that relies on technology, citizens and government to offer security and help to distinguish good from evil taxis. Taxiaviso helps you check all taxis operating in Mexico Airports, as well as taxis in Mexico City, Sinaloa and Nuevo León.

continued on following page

Table 2. Continued

App Name	Agency / Organization	Federal	State	Short Description
Ecobici	Epic Win Solutions S.C			Allows to see the nearest station where you find. Search cicloestaciones stations and public transport within the perimeter of Ecobici.View Full map and availability cicloestaciones. Draw strokes you make on Ecobici and know the distance, time and speed of each. May. In case of an accident or event, you can use the shortcut to call the callcenter Ecobici
IEPC	Publiapps		x	Coahuila IEPC aims to improve the process of the upcoming elections in the state of Coahuila.
UTCV Móvil	CEDESOFT		x	Formal implementation of the Technological University of Central Veracruz where students can check scores, news, and information specific to each of the races that are offered in the UTCV.
Programa Compras de Gobierno	Ministry of Economy	x		Search Tenders and Business Opportunities. Simply enter the name of the product or service you want to sell, is the recruitment procedures and business opportunities.
Veracruz Comercio	Ministry of Economic Development of Veracruz		x	The application is an effort by the Ministry of Economic Development and Port of Veracruz, to provide users with information to help them physically locate the companies that make commercial infrastructure and services of the State of Veracruz.
Reporte MH	Del. Miguel Hidalgo		x	This APP you can help report problems: Moguls.Street lighting broken. Leakage of water. Garbage collection on public roads. Desilting drainage network. Tree pruning.
Chihuahua en tu móvil	Chihuahua Municipality Coordination Systems		x	Chihuahua on your mobile is an application that allows citizens and visitors of the City of Chihuahua, interact with the city. The application allows you to report problems such as potholes, street lighting and garbage collection at no cost, directly, in addition to feedback information if necessary and provide timely follow.
Quién es Quién en los Precios	Federal Consumer	x		Who's Who is a program developed by PROFECO that will help you stay informed about the prices of products and services for smart consumption decisions and improve your family finances.
Reporte Vial Puebla	Government of puebla		x	Fasting use phones and other devices while driving.Report lights, potholes and other services for the Municipality of Puebla with this application. Operates 24 hrs. 365 days a year, the pages are generated on weekdays.Road Report information module is that you can be aware of closures of roadways in the city of Puebla to help you plan your route in advance and avoid congestion.
CEIGQro	CEIGQro		x	Implementation of the State Commission on Government Information where all information concerning the CEIG available. They may check stats, news, laws and regulations. The application will also allow requests for information.
072Móvil	Government of Federal Distric & Arbiec		x	Implementation of the Federal District to request Urbana Citizen Service infrastructure (potholes, water leaks, Pruning, Urban Cleaning)
Mérida es cultura	Directorate of Culture Merida		x	Mobile Application billboard activities and events of the Cultural Department of the Municipality of Mérida 2012-2015

complish all the components in our model, with a different degree of compliance. Such apps are GuiaPemex, IEPCC; Agente Vial Movil; Jalisco Móvil; Ayuntamiento de Mérida; Congreso Estado de San Luis P.; Profeco en 30'; Diputados; Guía Pemex; Jalisco Móvil; Agente vial; IEPCC. We now analyze some of them in the case studies in the following main section.

SOME CASE STUDIES OF MEXICAN MOBILE APPS

Several government apps are created every day to match government data or activities and are used by citizens. We chose some case studies from the Mexican government to present a degree of compliance and focus on the model we developed. We decided to choose an example of each: transparency, government efficiency and accountability. Also on this section, we present a novel app – Mujer Migrante (Migrant Woman) – that was not included in our previous sample. This is an important example of this kind of software since it has a direct social impact. On the other hand, we noticed that the government and federal agencies have developed some apps but they are registered with the name of the developer. We found a kind of applications that involve apps addressing the needs of individuals to search and manage information. They even allow users to denounce irregular practices from the public and private sector. This accountability problem is directly related to corruption. Our example for the Government Efficiency is the Federal Consumer Protection Agency (PROFECO) in Mexico. PROFECO has developed "Profeco en 30" in order to solve this problem and help citizens to start claims.

Profeco en 30

The function of the Federal Consumer Protection Agency in Mexico is to allow consumers to complain about products, bad services or frauds.

This agency uses social media and communication technologies very often. This agency has a blog, a website in Facebook, Twitter and also YouTube accounts, as well as a printed magazine. A natural ecosystem to create an app that links all the related content.

The recent PROFECO's platform "Profecto en 30" is a tool that allows consumers to report the failure of compliance from service providers. It is a mobile application where consumers can make comments and complaints that will be handled by the verification and monitoring staff of PROFECO. Figure 2 shows some screen shots of this app.

This app accomplishes our idea of open government and transparency but it is also linked to government efficiency - less than 30 minutes response - and it also helps the economic development because the claims can be from the public or private sector. There were 1,000 apps downloads in July 2013, during the first week that it was launched. In October 2013, the download account reached 5,000. It is important to mention that it became one of the most downloaded apps in Mexico. An important difference of this app is the interaction among citizens and the government agency. Only one more app presented this characteristic. Most of the apps are informative.

Jalisco Móvil

The state of Jalisco is in the northern part of Mexico. A well communicated local government that developed a number one government website in recent years (Sandoval et. al, 2011). It links the content with several mobile apps. This is another example of a government efficiency component that can also be linked to openness and accountability. The government of the State of Jalisco developed an app, which links most of the government services. The consultancy provides different features such as the "click-to-call" function for a free access to procedures and related information to the government of Jalisco.

Table 3. Mexican apps and categories of the model of analysis

APPS	Transparency	Accountability	Government Efficiency	Economic Development
072Móvil		x	x	x
Adeudo Vehicuar	x			
Agente Vial Movil			x	
Aire DF		x		
App de Desarrollo Económico SLP	x			
Ayuntamiento de Merida			x	
Bus Sonora			x	
CARNET Móvil			x	
Cédula Móvil SEP			x	
CEIGQro	x	x		
Chihuahua en tu Móvil	x			
Conaculta-Mexico Es Cultura				x
CONARTE				x
Congreso Estado de San Luis P.	x			
Consulta de Licencia Federal	x			
COPAES Móvil			x	
Cuajimalpa de Morelos		x		
Cumbre Tajín				x
Descubre Puebla				x
Diputados	x			
Durango				x
Durango Turistico				x
Ecobici	x			
Fernando Mayans		x		
Gobierno Móvil	x	x		x
Huixquilucan		x		
I Informe Gobierno	x			
ICAP			x	
Info DF Movil		x		
Jalisco Móvil_		x		
JaliscoFactBook		x		
Mérida es cultura	x			
Mi Policia	x		x	
Partido Verde Ecologista		x		
Profeco en 30	x			
Programa Compras de Gobierno		x	x	
Quién es Quién en los Precios	x	x	x	
redarbol				x
Reporte MH			x	x
Reporte Vial Puebla		x	x	x
Revista Enterate				x
S. de Desarrollo Económico SLP	x	x		x
Sefiplan David				x
Senador Isaias	x			
SIAM - Act y recargos			x	
SIAM - Estatus tramites			x	
Sitios Turísticos	x	x		
SNIIM Móvil		x		
Taxiaviso			x	x
Traza tu ruta			x	x
Unidad del Vocero de QRoo	x			
UTCV Móvil				x
Veracruz Comercio			x	
ZacatecasTravel				x

Figure 2. Profeco at 30 app

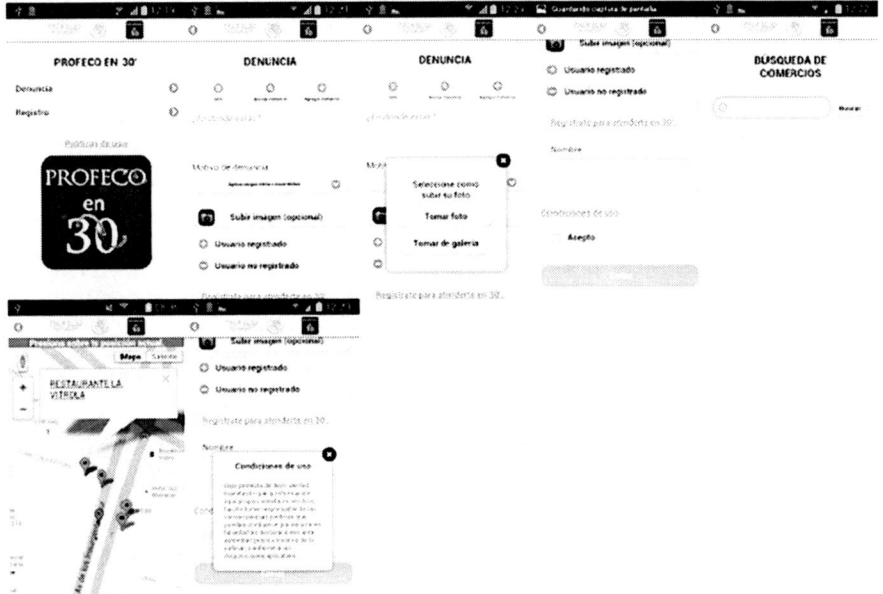

This app has become a one-stop window for government procedures. However, it is unable to start any administrative proceedings online. The only interaction that is allowed is to make claim about public services solution such as broken lamps or signals, trash on pavement, etc. You can also get information related to debt free vehicles, air quality, Fojal loans etc. This virtual window procedure includes the option of dialing from any part of the state with an 800 no cost line. This app achieved 1,000 downloads until October 2013.

A similar app in the "Jalisco Móvil" is the Get Community Payback. This free app allows users to submit suggestions for Community Payback projects to their local Probation Trust. Simply take a photo, tag the location on a map and send your suggestion. This will be routed automatically to the Probation Trust responsible for delivering Community Payback in your area. Notifications will be sent to the app when the project is assessed, completed etc., with a link to a website where updates and other Community Payback projects can be found.

Figure 3. Jalisco Movil app

Cédula Móvil SEP

The Secretariat of Federal Public Education (Secretaría de Educación Pública) also launched an app to improve citizens' services. This app has a search engine for the National Register of Professionals and provides criteria concerning the exercise of the professions in the capital city of Mexico. This solves a problem in two different perspectives. The first one is for companies and agencies that search for valid and real qualified staff. The second one is for people who want to update information about their professional register. This search can be done by entering the professional license number or by detail, entering name, first name, middle name and/or educational institution. The shown data are preliminary. In the event the user requires an official document, he/she must follow the normal procedure of "Professional Background" at the General Directorate of Professions. The information on this application is public and is continuously updated. This app has been downloaded 4,000 times at the time of writing this chapter (October 2013). See Figure 4 for an example of a web page.

A similar app to Cedula Movil SEP is Careers@Gov. This mobile application allows you to browse and apply for jobs in the Singapore Public Service. The Singapore Public Service is one of the largest employers in Singapore, employing more than 136,000 officers in 16 Ministries and more than 50 Statutory Boards.

Diputados (Congressmen)

The Congressmen App is a very rare example of accountability app. Normally, citizens develop this kind of apps or other media related companies. However, the Congress itself developed this Mexican app. The kind of information and updates users can find in this application software is remarkable. This app delivers news, events and related information of the Chamber of Deputies of the Congress of the Union, the low chamber of Mexico. The purpose of this app is intended to educate the public by providing access to information through Liaison Unit Obliges by the Federal Law of Transparency and Access to Public Government Information (LFTAIPG). These informative apps allow citizen to know and understand their congress functions and work. However, there is not a possibility of interaction. More than 1,000 downloads were register last October 2013. See Figure 5 for screen shoots.

Some features of this app are:

- Locate your MP for your GPS.
- See your congressman assessment.

Figure 4. SEP Cédula Móvil app

Figure 5. Diputados (Congressmen) app

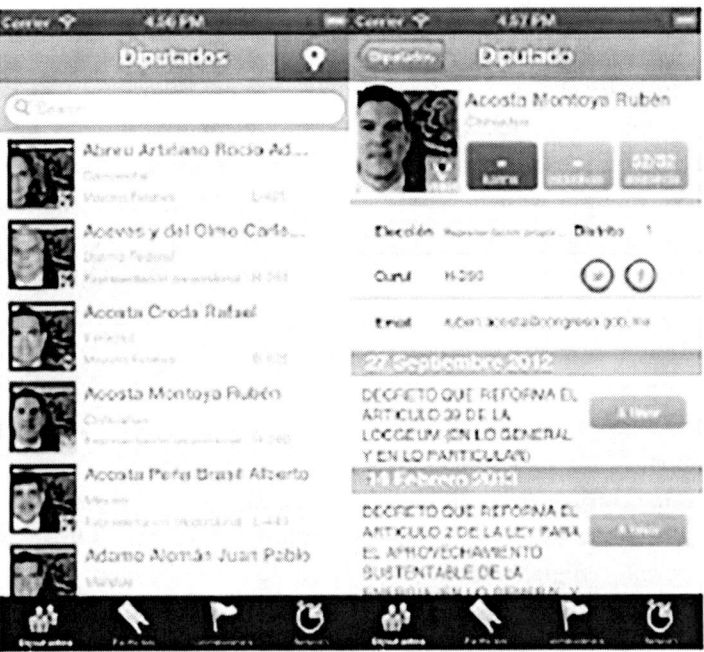

- See the agenda.
- Summary session.
- Synopsis of expert evaluations.
- Highlights of the sessions.
- List of committees.
- Voting orientation about deputies.
- View the number of initiatives.
- View the number of attendances in sessions.
- View your Twitter and Facebook.
- See the app's own assessment based on your activity.
- View your district.

A similar app is the Congress by Sunlight Foundation Follow, the latest from Washington with the free Congress app from the nonpartisan Sunlight Foundation. Learn more about your members of Congress, including their contact information and track activity on bills. Features include:

- Find lawmakers in congress, call them directly and connect with them on social media.
- Discover new and active legislation.
- Explore legislators for your location and others on an interactive map.
- Follow bills and legislators for quick access and updates.
- Review a legislator's latest votes and bill sponsorships.
- See bill activity with vote breakdowns.
- Get directions to a legislator's D. C. office.

Chat Mujer Migrante

We did not consider this app in our sample because we did not know about its existence until the World Summit of the Information Society (WSIS) awarded it a prize. The Ministry of Communications and Transport (SCT) in Mexico launched four new applications for mobile devices in favor of migrant girls and women on the International Day of the Girl in the Information Technology and Communication (ICT), celebrated on the fourth Thursday of April (annual event).

This new project introduces the technological tools in the search for options to improve the communication between migrant women, civilian authorities and organizations. It aims to offer legal and professional advice, prompt and secure, to users of these applications. Refer to Figure 6. The new related mobile apps are:

- **Chat Migrant Women:** Channels for doubts and questions of migrant women to institutions and organizations that can provide support. It also provides access to the online service portal "Migrant Women" from a mobile device either iPhone or Android.
- **Help!:** It offers choice questions about emergencies. The answers provide the functionality of direct calls to emergency phone numbers where migrants can be helped. This is only available for Android.
- **Helping Hands:** Through this app, you can access the database of the federation. There you can find telephone numbers and addresses of shelters, hospitals, location

maps and migrant support centers, among others. It is only available for the Android system.

- **Pro Women's Legal Guide:** It provides clear and simple information about laws that protect the rights of migrant women in Mexico and contact details if they require support regarding the defense of their rights. Also, it is only available for the Android platform.

Through these case studies, we found that most of the agencies are focusing their efforts to inform, publish and share information rather than creating collaboration or discussion. Nevertheless, some examples, such as Mujer Migrante and Profeco en 30 are important. These imply the first steps on real collaboration between citizen-government. The difference is that these two apps solve citizen needs – information or claims – and they try to link the information to the government or the responsible agencies in order to obtain some responses or obligations.

On the other hand, apps such as Diputados (Congressmen) or Jalisco Movil attempt to provide legitimization of their actions and results, but unfortunately still do not allow citizens to interact with the information. It is just a matter

Figure 6. Mujeres Migrantes (female-migrant app)

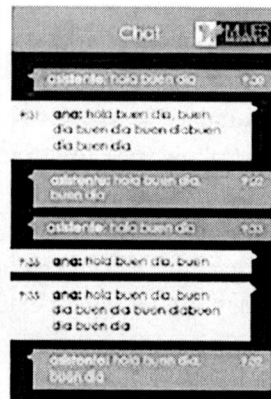

about government information related to the local government or congress actions. These two examples are different from the other government apps. Further research will provide some guidelines and the future paths for this trend.

FUTURE RESEARCH

Mobile government is now on the agenda. In recent years, many developments provided the context - smart phones technology - and the market - citizens and government - to exploit this trend. Besides, the software industry is emerging regarding this trend. The mobile marketplaces - IOS and Android - are growing exponentially and launching new apps every day. But, the question is: are all of these conditions leading to government efforts in the correct direction?

According to the present research, we see enough conditions to launch a common platform to design government apps. This platform could meet the criteria for m-government models (Antovski, 2006) to become flexible, interoperable and data sharing. Nowadays software applications have spread different conditions and software design. Even though, most of them have problems with security and privacy issues. A single platform could reduce these problems and augment citizens' downloads, because it will not depend on certain operative systems and a single government platform for developing apps. They could accomplish programming and security standards for a single country or worldwide.

The first path of research is to start developing such a trend for a common platform for developing government apps that accomplish m-government features. The second path is directly related to our model of study. We understand that this is a first attempt to test our model, unfolding many advantages and disadvantages. The main weakness is the lack of enough validity. We need to test the model in a different context, with different country apps and refine component descriptions

and variables to become more accurate and easy to understand for developers, government officials and scholars related to this topic. The model for apps classifications focuses on a preliminary idea that measurement can help to improve things. Following this idea, the purpose for classifying and marking apps in certain variable may show the lacks and strengths of those apps and could support some ideas for improvement. We need to work more on this topic in order to provide a practical tool, which improves research and practice.

A third research path, directly related to the previous one, is to focus our research on assessing apps evolution. The maturity level of the apps market and the permanent innovation of apps and software of mobile and internet devices allow us to research on future development paths. Where are the government apps going? What is the next step in the evolution of this market? Which is the operative system is going to be in order to get close to government and citizens? What are the features of the next generation of government apps? What is the future of m-government through the apps? Those questions remain unsolved and require a permanent and close observation.

The last path of research is to understand citizen apps - civic apps developed by experts, groups or single citizen that attempt to use government information resources, transforming them and exposing government failures or contradictions with its own data. Our chapter deals about government-developed apps; even if the government paid a company to develop them. We consider the use of public funding for development as part of the government. At this moment, we suggest a different but complementary area of research: citizen apps for the government. How shall we understand motivations and purposes of citizen apps? Are citizen apps more efficient than government apps? Why is the government allowing the creation of citizen apps? What types of citizen apps are in the market? How is the government influencing this market? Many questions arise from this path of research. Some focus on motivations, others to technical aspects and others to citizen services. We all understand that civic apps are as important as government apps. For the immediate future, the amount of civic apps could be bigger and faster changing than government apps. We must be aware of these changes and try to lead and understand their impact in public policy concerns.

Finally, research in the apps field, for practitioners and scholars, is wide and uncertain. According to all the new technological changes and from an everyday perspective, regarding the software and hardware industry, we think that these four trends allow us to understand key changes for the future.

CONCLUSION

The purpose of this chapter is to present a classification model for mobile government apps, supported in relation to mobile government fields and apps areas of research. In order to test the model suggested in this study, we collected data from the Mexican government apps and classified them accordingly in our given model. This initial research reveals that only nine of the 32 states of the Mexican Republic have adopted this kind of applications. There are 27 apps created by government agencies, 29 apps developed by independent agencies and 10 of these apps were created by private sector companies with resources provided by the government. Previous research reports that government portals expand their capabilities by using Web 2.0 tools, empowering citizens to collaborate with government and other NGOs to produce relevant information and solve important citizen problems (Sandoval et. al. 2011).

As a main outcome of this research, we found out that government apps are reasonably difficult to classify. Our model also lacks sophistication; the weaknesses are as follows: 1) Some government apps share variables of different components - this transposition of two or more variables makes it difficult to assess; and 2) Some components cannot be ascertained, whether these are complementary

or exclusive. We have attempted to figure this out with clearer variables. However, the government apps, that we measured, reveal something different - these are still complementary.

These lessons led us to several reflections about the process of assessment of government apps. One conclusion that we can provide is the difficulty to define boundaries on assessing apps. We therefore propose a model based on a literature review and our own experience collecting and assessing apps. Definitely, a more empirical research and a systematic approach are required to validate our model.

Another conclusion is to come up with a method of benchmarking of apps that could then promote their development. For example, we can assess PROFECO App 30 as an open government app, with very good points on government efficiency and accountability. So far, this app can be considered as an example to be followed by other apps. We consider this point as a contribution to this research, in order to highlight the good practices and to determine a valid point of view from a benchmarking perspective.

Finally, we consider that benchmarking, using a model for classifying and measuring can promote research on this field because we can divide the object of study and research about specific topics such as open government or accountability. From this perspective, we can assess if the government apps accomplish their tasks and provide real value to citizens and public officials.

Governments expand their own limits when citizens can send reports of problems with public services or safety problems, allowing government officials to respond faster and better by accumulating information (Nath, 2011). Our research provides preliminary evidence that shows the adoption of mobile apps in Mexico by government entities. This level of adoption is emerging and its use must be promoted by more government agencies. These new ways of communication, collaboration, interaction and coproduction are changing the government direction and the relationships between government, entities and citizen, improving the co-participation and the use of available open information. This chapter is, hopefully, a first step in the right direction of understanding the use of government mobile apps, their evolution and impact on governments' innovation that are largely enabled by current emerging technologies including the social media, Web 2.0 and mobile devices.

REFERENCES

Abramowicz, W., Karsenty, L., Moore, P., Peinel, O. G., Wiśniewski, M., & Tilsner, D. (2005). USE-ME.GOV (USability-drivEn open platform for MobilE GOVernment). In *Proceedings of Euro mGov 2005: The First European Mobile Government Conference* (pp. 7-16). Brighton, UK: Mobile Government Consortium International LLC.

Amaravadi, C. S. (2005). Digital repositories for e-government. *Electronic Government: An International Journal, 2*(2), 205–218. doi:10.1504/EG.2005.007095

Anthes, G. (2011). Invasion of the Mobile Apps. *Communications of the ACM, 54*(9), 16–18. doi:10.1145/1995376.1995383

Antovski, L., & Gusev, M. (2005). M-Government Framework. In *Proceedings of Euro mGov 2005: The First European Mobile Government Conference* (pp. 36-44). Brighton, UK: Mobile Government Consortium International LLC.

Antovski, L., & Gusev, M. (2006). M-GOV: The Evolution Method. In *Proceedings of Euro mGov 2006: The Second European Conference on Mobile Government* (pp. 26-35). Brighton, UK: Mobile Government Consortium International LLC.

Apple. (2013). *5 años del App. Store*. Apple. Retrieved from https://itunes.apple.com/WebObjects/MZStore.woa/wa/viewFeature?cc=mx&id=660896396&enlh=1.8321.832.83.8&mt=8&ls=1

Beath, C. M., & Walker, G. (1998). Outsourcing of Application Software: A Knowledge Management Perspective. In *Proceedings of the Thirty First Hawaii International Conference on System Sciences*, (pp. 666-674). IEEE.

Beer, D., Kunis, R., & Runger, G. (2006). A component based software architecture for e-government applications. In *Proceedings of The First International Conference on Availability, Reliability and Security (ARES 2006)* (pp. 1-8). ARES.

Berners-Lee, T. (2010, November 22). Long Live the Web: A Call for Continued Open Standards and Neutrality. *Scientific American*. Retrieved from http://www.scientificamerican.com/article.cfm?id=long-live-the-web

Borucki, C., Arat, S., & Kushchu, I. (2005). *Mobile Government and Organizational Effectiveness*. Mobile Government Consortium International LLC.

Carlson, V., & Eyler-Werve, K. (2012). *Civic Apps Competition Handbook* (Vol. 1). O'Reilly Media. Retrieved from http://shop.oreilly.com/product/0636920024484.do

Chatzinotas, S., Ntaliani, M., Karetsos, S., & Costopoulou, C. (2006). Securing m-Government Services: The Case of Agroporta. In *Proceedings of Euro mGov 2006: The Second European Conference on Mobile Government* (pp. 61-70). Brighton, UK: Mobile Government Consortium International LLC.

De Kool, D., & van Wamelen, J. (2008). Web 2.0: A New Basis for E-Government? In Proceedings of Information and Communication Technologies: From Theory to Applications, (pp. 1-7). ICTTA.

Developer, A. (2013). *Android, the world's most popular mobile platform*. Google. Retrieved from http://developer.android.com/intl/es/about/index.html

Eggers, W. D. (2013, March 13). *Getting Mobile Right: Six Steps to Success in Government*. Retrieved from http://www.governing.com/columns/mgmt-insights/col-mobile-technology-six-steps-success-implementation-government.html

Eggers, W. D., & Jaffe, J. (2013). *Gov on the go*. Deloitte. Retrieved from http://dupress.com/articles/gov-on-the-go/

Estalella, A., & Ardevol, E. (2011). E-research: Challenges and opportunities for social sciences. *Convergencia Revista de Ciencias Sociales*, *18*(55), 87–111.

Estevez, E., & Janowski, T. (2007). *Programmable Messaging for Electronic Government - Building a Foundation*. Academic Press. doi:10.1007/978-3-540-75221-9_9

Fidel, R., Scholl, H. J., Liu, S., & Unsworth, K. (2007). Mobile government fieldwork: a preliminary study of technological, organizational, and social challenges. In *Proceedings of the 8th annual international conference on Digital government research: bridging disciplines & domains* (Vol. 228, pp. 131-139). Philadelphia, PA: Digital Government Research Center.

Fu, B., Lin, J., Li, L., Faloutsos, C., Hong, J., & Sadeh, N. (2013). Why people hate your app: making sense of user feedback in a mobile app. store. In *Proceedings of the 19th ACM SIGKDD international conference on Knowledge discovery and data mining* (pp. 1276–1284). New York, NY: ACM. doi:10.1145/2487575.2488202

Garson, G. D. (2003). *Public Information Technology: Policy and Management Issues*. Hershey, PA: Idea Group Publishing.

Geiger, C. P., & von Lucke, J. (2012). Open Government and (Linked) (Open) (Government) (Data). *JeDEM - eJournal of eDemocracy and Open Government*, *4*(2), 265-278.

Goggin, G. (2011). Ubiquitous apps: politics of openness in global mobile cultures. *Digital Creativity*, *22*(3), 148–159. doi:10.1080/14626268.2011.603733

Herrick, D. R. (2009). Google this! using Google apps for collaboration and productivity. In *Proceedings of the 37th annual ACM SIGUCCS fall conference* (pp. 55–64). New York, NY: ACM. doi:10.1145/1629501.1629513

Hewson, C. (2008). Internet-mediated research as an emergent method and its potential role in facilitating mixed methods research. In *Handbook of Emergent Methods* (pp. 543–570). New York: Guilford Press.

Holwerda, T. (2011, June 24). *The History of «App» and the Demise of the Programmer*. Retrieved from http://www.osnews.com/story/24882/The_History_of_App_and_the_Demise_of_the_Programmer

Hosmer, C., Jeffcoat, C., Davis, M., & McGibbon, T. (2011). Use of mobile technology for information collection and dissemination. Data & Analysis Center for Software, 77.

Iansiti, M., & Levien, R. (2004). *The keystone advantage: What the new dynamics of business ecosystems mean for strategy, innovation, and sustainability*. Boston: Harvard Business School Press. Retrieved from http://books.google.ca/books/about/The_Keystone_Advantage.html?id=T_2QFhjzGPAC

Johnson, R. C. (2010). Apps Culture reinventing mobile Internet. *Electronic Engineering Times*, (1588), 24-26, 28, 30.

Jones, S. (1999). *Doing Internet Research*. Academic Press.

Kapogiannis, G., Touzos, M., & Kreps, D. (2006). M-Business & M-Government: Co-operation, The Greek case study. In *Proceedings of Euro mGov 2006: The Second European Conference on Mobile Government* (pp. 154-159). Brighton, UK: Mobile Government Consortium International LLC.

Karantjias, & Papastergiou. (2009). Design principles of secure federated e/m-government framework. *International Journal of Electronic Governance, 2*(4), 402-423.

Kersten, G. E., Kersten, M. A., & Rakowski, W. M. (2001). Application software and culture: Beyond the surface of software interface. *Culture (Canadian Ethnology Society)*, 1–17.

Kesavarapu, S., & Choi, M. (2012). M-government - A framework to investigate killer applications for developing countries: An Indian case study. *Electronic Government: An International Journal, 9*(2), 200–219. doi:10.1504/EG.2012.046269

King, J., Lampinen, A., & Smolen, A. (2011). *Privacy: Is there an app. for that?* Academic Press. doi:10.1145/2078827.2078843

Labs, I. (2012). *Apps for Democracy* (Vol. 2012). Academic Press.

Lerman, K. (2007). User Participation in Social Media: Digg Study. In *Proceedings of 2007 IEEEWICACM International Conferences on Web Intelligence and Intelligent Agent Technology Workshops*, (pp. 255-258). IEEE.

Lientz, B. P., & Swanson, E. B. (1981). Problems in application software maintenance. *Communications of the ACM, 24*, 763–769. doi:10.1145/358790.358796

Lientz, B. P., Swanson, E. B., & Tompkins, G. E. (1978). Characteristics of application software maintenance. *Communications of the ACM, 21*, 466–471. doi:10.1145/359511.359522

Lugano, G. (2008). Mobile social networking in theory and practice. *First Monday, 13*(11). doi:10.5210/fm.v13i11.2232

Maria Emmanouilidou, D. K. (2010). A framework for accessible m-government implementation. *Electronic Government, an International Journal, 7*(3), 252 - 269.

Maumbe, B. M., & Owei, V. (2006). Bringing M-government to South African Citizens: Policy Framework, Delivery Challenges and Opportunities. In *Proceedings of Euro mGov 2006: The Second European Conference on Mobile Government* (pp. 160-173). Brighton, UK: Mobile Government Consortium International LLC.

Milam, L., & Avery, E. J. (2012). Apps4Africa: A new State Department public diplomacy initiative. *Public Relations Review, 38*(2), 328–335. doi:10.1016/j.pubrev.2011.12.013

Murugesan, S., Rossi, G., Wilbanks, L., & Djavanshir, R. (2011a). The Future of Web Apps. *IT Professional Magazine, 13*(5), 12–14. doi:10.1109/MITP.2011.89

Murugesan, S., Rossi, G., Wilbanks, L., & Djavanshir, R. (2011b). The Future of Web Apps. *IT Professional Magazine, 13*(5), 12–14. doi:10.1109/MITP.2011.89

Ntaliani, M., Costopoulou, C., Manouselis, N., & Karetsos, S. (2009). M-government services for rural SMEs. *International Journal of Electronic Security and Digital Forensic, 2*(4), 407–423. doi:10.1504/IJESDF.2009.027672

O'Hara, K. (2012). Transparency, open data and trust in government: shaping the infosphere. In *Proceedings of the 3rd Annual ACM Web Science Conference* (pp. 223–232). New York, NY: ACM. doi:10.1145/2380718.2380747

O'Reilly, T. (2005). What Is Web 2.0. *O'Reilly Media*. Retrieved from http://oreilly.com/pub/a/oreilly/tim/news/2005/09/30/what-is-web-20.html

OECD, & ITU. (2011). *M-Government*. Paris: Organisation for Economic Co-operation and Development. Retrieved from http://www.oecd-ilibrary.org/content/book/9789264118706-en

Pocatilu, P. (2010). Developing Mobile Learning Applications for Android using Web Services. *Informatica Economica, 14*(3), 106–115.

Raths, D. (2011). Cities add citizen engagement mobile apps to their portfolios. *KM World, 20*(5), 10–11.

Sandoval-Almazan, R. (2012). Open Government in Mexico: An Assessment Preview 2007-2010. In *Proceedings of Conference for E-democracy and Open GOvernment* (Vol. 1, pp. 255-266). Krems, Austria: Austrian Institute of Technology.

Sandoval-Almazan, R., & Gil-Garcia, J. R. (2012). Are E-Government Portals Becoming Central Components for Public Information Sharing Networks? An Initial Exploration of Local Governments in Mexico. *Government Information Quarterly, 29*, S72–S81. doi:10.1016/j.giq.2011.09.004

Sandoval-Almazán, R., Gil-Garcia, J. R., Luna-Reyes, L., Luna-Reyes, D., & Murillo, G. D. (2011). The Use of Web 2.0 on Mexican State Websites: A Three-year Assessment. *Electronic. Journal of E-Government, 9*(2), 107–121.

Sandoval-Almazan, R., Gil-Garcia, J. R., Luna-Reyes, L. F., Luna, D. E., & Rojas-Romero, Y. (2012). Open government 2.0: citizen empowerment through open data, web and mobile apps. In *Proceedings of the 6th International Conference on Theory and Practice of Electronic Governance* (pp. 30–33). New York, NY: ACM. doi:10.1145/2463728.2463735

Scholl, H. J., Fidel, R., Mai, J. E., & Unsworth, K. (2006). Seattle's Mobile City Project. In *Proceedings of Euro mGov 2006: The Second European Conference on Mobile Government* (pp. 144-153). Brighton, UK: Mobile Government Consortium International LLC.

Scholl, H. J., & Klischewski, R. (2007). E-Government Integration and Interoperability: Framing the Research Agenda. *International Journal of Public Administration, 30*(8), 889–920. doi:10.1080/01900690701402668

Scholl, H. J., & Luna-Reyes, L. F. (2011). Transparency and openness in government: a system dynamics perspective. In *Proceedings of the 5th International Conference on Theory and Practice of Electronic Governance* (pp. 107-114). Tallinn, Estonia: ACM.

Sheng, H., & Trimi, S. (2006). Towards a Framework to Understand M-Government. In *Proceedings of AMCIS 2006*. Acapulco, Mexico: AIS.

Tang, L.-Y., Hsiu, P.-C., Huang, J.-L., & Chen, M.-S. (2013). iLauncher: an intelligent launcher for mobile apps based on individual usage patterns. In *Proceedings of the 28th Annual ACM Symposium on Applied Computing* (pp. 505–512). New York, NY: ACM. doi:10.1145/2480362.2480461

Thomas, J. C., & Streib, G. (2005). E-Democracy, E-Commerce, and E-Research: Examining the Electronic Ties Between Citizens and Governments. *Administration & Society, 37*(3), 259–280. doi:10.1177/0095399704273212

Townsend, A. M. (2002). Mobile and wireless technologies: emerging opportunities for digital government. In *Proceedings of the 2002 annual national conference on Digital government research* (pp. 1–5). Los Angeles, CA: Digital Government Society of North America. Retrieved from http://dl.acm.org/citation.cfm?id=1123098.1123153

Traunmüller, R. (2011). Mobile government. In *Proceedings of the Second international conference on Electronic government and the information systems perspective* (pp. 277–283). Berlin: Springer-Verlag. Retrieved from http://dl.acm.org/citation.cfm?id=2033665.2033694

Trimi, S., & Sheng, H. (2008, May). Emerging trends in M-government. *Communications of the ACM*, 53–58. doi:10.1145/1342327.1342338

van Wamelen, J., & de Kool, D. (2008). Web 2.0: a basis for the second society? In *Proceedings of the 2nd international conference on Theory and practice of electronic governance* (pp. 349-354). ACM. doi:10.1145/1509096.1509169

Walravens, N. (2012). Mobile business and the smart city: Developing a business model framework to include public design parameters for mobile city services. *J. Theor. Appl. Electron. Commer. Res.*, *7*(3), 121–135. doi:10.4067/S0718-18762012000300011

Walravens, N., & Ballon, P. (2011). The City as a Platform: Exploring the Potential Role(s) of the City in Mobile Service Provision through a Mobile Service Platform Typology. In *Proceedings of 2011 Tenth International Conference on Mobile Business (ICMB)* (pp. 60-67). doi:10.1109/ICMB.2011.22

Wei, Z., Gao, X., Jia, D., & Yang, Y. (2010). Research of mobile government based on multi-modal platform with unified engine. In *Proceedings of Intelligent Computing and Integrated Systems (ICISS), 2010 International Conference on* (pp. 786-789). ICISS. doi:10.1109/ICISS.2010.5657109

Woods, E. (2009, May 3). *Web 2.0 and the public sector - Public Sector - Breaking Business and Technology*. Academic Press.

Wu, H., Ozok, A. A., Gurses, A. P., & Wei, J. (2009). User aspects of electronic and mobile government: Results from a review of current research. *Electronic Government, an International Journal*, *6*(3), 233-251.

Xu, Q., Erman, J., Gerber, A., Mao, Z., Pang, J., & Venkataraman, S. (2011). Identifying diverse usage behaviors of smartphone apps. In *Proceedings of the 2011 ACM SIGCOMM conference on Internet measurement conference* (pp. 329-344). New York, NY: ACM. doi:10.1145/2068816.2068847

Yamakami, T. (2007). MobileWeb 2.0: Lessons from Web 2.0 and Past Mobile Internet Development. In Proceedings of Multimedia and Ubiquitous Engineering, (pp. 886-890). IEEE.

ADDITIONAL READING

Al-Khamayseh, S., & Lawrence, E. (2006). Towards citizen centric mobile government services: a roadmap. *CollECTeR Europe, 2006*, 129.

Begay, W. R. (2013). *Mobile Apps and Indigenous Language Learning: New Developments in the Field of Indigenous Language Revitalization*. The University of Arizona., Arizona EU. Retrieved from http://arizona.openrepository.com/arizona/handle/10150/293746.

Bertot, J. C., Jaeger, P. T., Munson, S., & Glaisyer, T. (2010). Social Media Technology and Government Transparency. *Computer*, *43*(11), 53–59. doi:10.1109/MC.2010.325

Brown, S. (2012). Mobile apps Which ones really matter to the information professional? *Business Information Review*, *29*(4), 231–237. doi:10.1177/0266382112465501

Fling, B. (2009). *Mobile design and development: Practical concepts and techniques for creating mobile sites and Web apps* (1st ed.). USA: O'Reilly Media, Inc. Retrieved from http://www.amazon.com/Mobile-Design-Development-Practical-Techniques/dp/0596155441.

Kabachinski, J. (2011). Mobile medical apps changing healthcare technology. *Biomedical Instrumentation & Technology, 45*(6), 482–486. doi:http://dx.doi.org/10.2345/0899-8205-45.6.482.

Moore, J. (2012). The benefits of mobile apps for patients and providers. *British Journal of Healthcare Management, 18*(9), 465–467.

Murugesan, S. (2013). Mobile Apps in Africa. *IT Professional, 15*(5), 8–11. doi:10.1109/MITP.2013.83

Näkki, P., Bäck, A., Ropponen, T., Kronqvist, J., Hintikka, K. A., & Harju, A. (2011). *Social media for citizen participation Report on the Somus project.*

Sharma, S. K., & Gupta, J. N. D. (2004). Web services architecture for m-government: issues and challenges. *Electronic Government, an International Journal, 1*(4), 462–474.

Sung, S. J. (2011). How can we use mobile apps for disaster communications in Taiwan: Problems and possible practice. Presented at the 8th Asia-Pacific Regional ITS Conference, Taipei 2011: ITS. Retrieved from http://www.econstor.eu/handle/10419/52323.

Van Jaarsveldt, L. (2012). Political engagement and government informing seeking: Increasing role of social media and mobile devices. Presented at the 23rd European Regional Conference of the International Telecommunication Society, 1-4 July 2012, Vienna, Austria, Retrieved from http://www.econstor.eu/handle/10419/60356.

Xu, Q., Erman, J., Gerber, A., Mao, Z., Pang, J., & Venkataraman, S. (2011a). Identifying diverse usage behaviors of smartphone apps. In *Proceedings of the 2011 ACM SIGCOMM conference on Internet measurement conference* (pp. 329–344). New York, NY, USA: ACM. doi:10.1145/2068816.2068847

Xu, Q., Erman, J., Gerber, A., Mao, Z., Pang, J., & Venkataraman, S. (2011b). Identifying diverse usage behaviors of smartphone apps.

Yelton, A. (2012). *Bridging the digital divide with mobile services* (1st ed., Vol. 48). Amer Library Assn Ed. Retrieved from www.amazon.com/Bridging-Digital-Services-Library-Technology/dp/0838958567.

KEY TERMS AND DEFINITIONS

Accountability: Open data needs to be tied up to an accountability "mechanism" in order to achieve accountability. (Carlson & Eyler-Werve, 2012).

E-Government: The use of information and technology to support and improve public policies and government operations, engage citizens, and provide comprehensive and timely government services" (Hans J Scholl & Klischewski, 2007: p.21).

Mobile Apps Efficiency: Mobile applications have the potential to improve service delivery, as well as efficiency and efficacy in government operations. (Trimi & Sheng, 2008).

Mobile Apps: Mobile apps create the capability of creating location-based services (LBS), such as traffic, bus or parking applications. (Trimi & Sheng, 2008).

Mobile Government: The use of interactive technologies through mobile devices – smart phones, tablets – to provide services, information and enable citizens to contact government officials.

Open Data: Public access to government data, however, remains challenging largely due to the heterogeneity and complexity of the public information ecosystem that results in high costs for locating, decoding, inter-linking and reusing existing government data to offer an open and incremental ecosystem that interconnects providers, consumers, and contributors of open government data. (N. Shadbolt et al., 2012).

Open Government: Could be understood as an integrated platform to drive government data into open and accountable information for citizens (Sandoval-Almazan, 2012).

Transparency: Open data accessible with the possibility of report flaws in data sets, allowing feedback from citizens regarding administration, collaboration among agencies, delivery of government services in a more efficient way and innovation of businesses (Carlson & Eyler-Werve, 2012).

Web 2.0: A network used as a platform, spanning all connected devices (O'Reilly, 2005).

Section 4
Social Computing and Data Modelling for Connected Services for Inclusive Government

Chapter 10
Social Computing and Cooperation Services for Connected Government and Cross-Boundary Services Delivery

Walter Castelnovo
University of Insubria, Italy

ABSTRACT

Connected Government requires different government organizations to connect seamlessly across functions, agencies, and jurisdictions in order to deliver effective and efficient services to citizens and businesses. In the countries of the European Union, this also involves the possibility of delivering cross-border services, which is an important step toward a truly united Europe. To achieve this goal, European citizens and businesses should be able to interact with different public administrations in different Member States in a seamless way to perceive them as a single entity. Interoperability, which is a key factor for Connected Government, is not enough in order to achieve this result, since it usually does not consider the social dimension of organizations. This dimension is at the basis of co-operability, which is a form of non-technical interoperability that allows different organizations to function together essentially as a single organization. In this chapter, it is argued that, due to their unique capacity of coupling several technologies and processes with interpersonal styles, awareness, communication tools, and conversational models, the integration of social computing services and tools within inter-organizational workflows can make them more efficient and effective. It can also support the "learning" process that leads different organizations to achieve co-operability.

INTRODUCTION

Under the pressure of the current global economic crisis, many governments boosted the e-Government's strategic role in supporting the economic recovery (Ubaldi, 2011). The European Union identified e-government as a fundamental element of the Digital Agenda for Europe (DAE) and considered it as one of the seven Flagship Initiatives stated in the EU's 2020 strategy for smart,

DOI: 10.4018/978-1-4666-6082-3.ch010

sustainable and inclusive growth (EC, 2010). One of the main objectives of the DAE is the use of Information and Communication Technologies (ICT) to foster the establishment of a Single Internal Market involving all the Member States since this could have a relevant impact on the EU's economy. Actually, the full development of the Single Market by 2020 is expected to increase the EU's GDP by an extra 4% that corresponds to a € 500-billion gain (EPC, 2010). The EU's strategy for attaining this objective identifies four drivers (EC, 2012):

- Developing fully integrated networks in the Single Market;
- Fostering mobility of citizens and businesses across borders;
- Supporting the digital economy across Europe;
- Strengthening social entrepreneurship, cohesion and consumer confidence

Among these, crucial for the attainment of the objective of the Single Market, is the mobility of citizens and businesses across borders. According to a European Commission's estimate (EC, 2013), there were approximately 1,790,000 immigrants and commuters between EU Member States in 2009 with an estimated growth of 22.7% by 2020 (reaching 2,196,035 individuals per annum in 2020). Besides fostering economic growth, mobility of both citizens and enterprises among the Member States can contribute substantially to strengthening the European citizens' perception of living and working in a Single Market. However, this objective can be achieved at the condition that entrepreneurs can set up and run a business anywhere in Europe independently of their original location, and that citizens are allowed to study, work, reside and retire anywhere in the EU. To guarantee these conditions to European citizens and entrepreneurs, the public administration

agencies in the Member States should be able to provide seamless, interoperable and sustainable cross-border public services.

The availability of cross-border services could have a significant impact on citizens and enterprises mobility across the EU Member States. In EC (2013), it is estimated that there would be a total demand of 1,262,887 users for cross-border services besides 140,000 branches and immigrant business start-ups between EU Member States that could utilize cross-border business services. However, to deliver cross-border services, the public administration agencies of different Member States are required to connect seamlessly across functions, agencies, and jurisdictions to deliver effective and efficient services to citizens and businesses. This would allow European public administration agencies to act as a single organization, so that citizens feel that a single (virtually integrated) organization is serving them rather than a number of different public authorities, possibly from different Member States. From this point of view, to satisfy the demand of cross-border services and to strengthen the European citizens' perception of living and working in a Single Market (which are crucial for the attainment of the objectives stated by EU's 2020 strategy), the Member States should transform their government systems toward the Connected Government model (Pallab, 2010) at both the national and the European Union level.

Connected government is usually considered as a multi-dimensional construct (Kaczorowski, 2004; Pallab, 2010), including dimensions such as:

- Citizen centricity, as the guiding principle for the public sector transformation processes, whose goal is to create greater value for citizens, not only for citizens as users/consumers or beneficiaries, but also for citizens as taxpayers, as participants in the democratic processes, as policy makers and employees in public administration agencies and as suppliers and entrepreneurs as well (Bannister, 2002; Castelnovo & Simonetta, 2007, Castelnovo, 2013).

- Back-office reorganization, to force the public administration agencies to "rethink their operations to move from being system-oriented to chain-oriented with respect to their structure, functioning, skills and capabilities, and culture and management" (UNDESA, 2008).
- Networked organizational model, to transform a fragmented system of government agencies in a networked virtual organization that operates seamlessly toward a common mission, that is to deliver more value to citizens and enterprises (Johnston, 2006).
- Standardized infrastructure and interoperability, as the condition that makes it possible to achieve the vertical integration among different levels of Government as well as the horizontal integration among government organizations belonging to the same institutional level (Microsoft, 2011).
- Public sector governance, to guarantee the consistency of the transformation processes implemented both at the Central and the Local Government level, and to assure that all the transformation processes preserve the public interest and increase the value for citizens (Castelnovo, 2012).
- Social inclusion, as a way to bridge the gap between government and citizens, to build trust in government and to assure that no citizen is left behind.

Achieving the objective of connected government at the EU level is not a simple matter at all. It involves 27 different central government systems, 89,149 municipalities, 1,126 second tier Local Government organizations and 105 third tier Local Government organizations (CEMR, 2011).

Moreover, to these it should be added a number of government agencies that in some countries are directly involved in the delivery of services to citizens and enterprises. This raises the problem of how the horizontal and vertical integration/cooperation among government bodies and agencies (that is instrumental for connected government) can be achieved within and across highly fragmented systems of (Local) Government.

This problem is particularly apparent when considered from the point of view of cross-border inter-agencies cooperation. However, the cross-border level simply magnifies problems that already affect inter-agencies cooperation at the level of the single states. Without achieving the connectedness of government agencies at the level of the single states there is no hope to achieve the level of cross-border connectedness required for the delivery of cross-border services. Based on this observation, in this chapter I will be concerned with the problem of how horizontal integration/cooperation among government agencies can be achieved in Local Government, especially in those countries whose system of Local Government is characterized by a high administrative fragmentation and by the prevalence of small municipalities. My claim is that the Social Computing and Collaboration Services and tools (SC&CSs) made available by the so-called Web 2.0 paradigm provide powerful tools for supporting inter-agencies cooperation and the development of a connected system of Local Government as a first step toward a virtually integrated/connected system of European public administrations.

This chapter is organized as follows. The next section discusses interoperability as the platform on which connected government can be based. Here, I will show how, in order to guarantee inter-agencies cooperation, a concept of non-technical interoperability is needed, similar to the concept of co-operability as defined in the Command & Control literature. In the section that follows, I will argue that the sharing of a cooperation environment and a cooperation platform among different organizations can help them to achieve co-operability. More specifically, I will suggest including in the cooperation platform the Social Computing and Cooperation Services and tools typical of the Enterprise 2.0 paradigm. Furthermore, I will discuss the possibility of using SC&CSs to support trust-building processes among the people involved in inter-agencies cooperation, the standardization of the operative processes and the establishment of a shared system of values and a shared organizational culture. Finally, with reference to the case of One Stop Shop for Production Activities in Italy, which represents a typical service of (Connected) e-Government, in the last section of the chapter I will consider how SC&CSs can be used to support the execution of inter-organizational workflows. This will exemplify the use of Social Computing and Cooperation services and tools in cases in which the delivery of services crosses organizational boundaries (i.e., cross-border, cross-administrative and cross-sectorial service delivery).

INTEROPERABILITY AS THE PLATFORM FOR CONNECTED GOVERNMENT

According to EIF (2010), a cross-border public sector service is a service supplied by public administrations (either national public administrations at any level, or bodies acting on their behalf, and/or EU public administrations) either to one another or to European businesses and citizens. The EU's approach to cross-border services as described, for instance, in EC (2013) focuses exclusively on online cross-border services; however, as explicitly recognized in EIF (2010), in order to satisfy the principles of inclusion and accessibility traditional paper-based and/or face-to-face service delivery needs (at least) to co-exist with electronic delivery. Actually, mainly due to complexity and security concerns (and to the fact that paper, mail or personal visits are still necessary to use an electronic service), despite their online availability many services are still accessed offline by European citizens and businesses, as shown in Table 1 that reports data collected through a survey involving European citizens and businesses.

Table 1. e-Government citizens and business services per usage level (EC, 2013)

Service		Usage	Accessed Online	Accessed Offline	Both
Citizens	Income tax declaration	66,6%	48,3%	35,8%	15,9%
	ID request	65,8%	65,8%	58,6%	17,8%
	Vehicle tax	56,5%	49,9%	38,8%	11,3%
	Enroll as a student	43,2%	49,5%	27,7%	22,7%
	Ordering a birth certificate	33,2%	40,4%	47,1%	12,5%
	Register for a pension	16,2%	19,4%	64,5%	16,1%
	Register for legal aid	14,5%	31,4%	60,5%	8,1%
Business	Business tax declaration	56,8%	46,9%	35,4%	17,7%
	Consult a business register	37,1%	61,4%	2,9%	22,4%
	Establish a new legal entity	27,1%	45,2%	46,8%	8,1%
	Submitting a tender	21,4%	46,9%	22,4%	28,6%

The data reported in Table 1 clearly show that in the implementation of cross-border services both online and offline (face-to-face, paper based) interactions between citizens/businesses and public administrations should be taken into the account, especially if cross-border services are to be considered as a means to strengthen the European citizens' perception of living and working in a Single Market.

The basic service scenario I will consider in this chapter is the following: *An enterprise E established in Member State X, wishing to establish in Member State Y, submits a request (either online or offline) for establishment in Member State Y.* To process this request:

- E could be required to interact (either online or offline) with a public administration A in State Y.
- Public administration A could need to exchange information with a public body in State X and then use this information in the execution of inter-organizational workflows involving other public agencies in State Y.

Under this service scenario, to let European citizens/entrepreneurs experience seamless cross-border services across Europe, it is necessary (i) to make citizens' interactions with public administrations across Europe as similar as possible; and (ii) to allow public administrations from different Member States to achieve high levels of cooperation/integration. In both cases, the problem is how to achieve this result, especially when at the Member State level the delivery of the services involves different agencies to which the institutional system in force assigns different competences.

The first and most obvious answer to this problem is interoperability; indeed, as stated in UNDESA (2008): *The key platform on which connected government is built upon is the concept of interoperability which is the ability of government organizations to share and integrate information by using common standards.*

Interoperability lies at the basis of the EU's approach to cross-border services delivery. Actually, the fundamental role of interoperability has been clearly pointed out at least since the adoption of the "eEurope Action Plan 2005" at the Seville summit in 2002 (EC, 2002). Indeed, in that Action Plan the objective was explicitly stated "to issue an agreed interoperability framework to support the delivery of pan-European e-Government services to citizens and enterprises" (EC, 2002, p. 10). This framework has actually been delivered in 2010 and the Member States have been invited to align their National Interoperability Frameworks (NIFs) with the European Interoperability Framework (EIF). The EIF has been designed to allow not only cross-border services but also cross-sectorial and cross-administrative services. This means that it should be considered as the basis for inter-organizational cooperation for the delivery of services originating at all the layers of government, i.e. at the local, the regional, the national and the EU level.

The EIF assumes the following definition of interoperability and interoperability framework (EIF, 2010, p. 2):

Interoperability, within the context of European public service delivery, is the ability of disparate and diverse organizations to interact towards mutually beneficial and agreed common goals, involving the sharing of information and knowledge between the organizations, through the business processes they support, by means of the exchange of data between their respective ICT systems.

An interoperability framework is an agreed approach to interoperability for organizations that wish to work together towards the joint delivery of public services. Within its scope of applicability, it specifies a set of common elements such as vocabulary, concepts, principles, policies, guidelines, recommendations, standards, specifications and practices.

The EIF considers four levels of interoperability, each of which must be taken into account when defining cross-border/sectorial/administrative services (*cross-boundary* services henceforth). The four levels are:

- **Technical Interoperability:** Which concerns the planning of technical issues involved in linking IT systems and services. Technical interoperability includes key aspects such as open interfaces, interconnection services, data integration and middleware, data presentation and exchange, accessibility and security services.
- **Semantic Interoperability:** Which concerns the meaning of exchanged information that has to be preserved by all parties. Semantic interoperability defines the exact format of the information to be exchanged and ensures that the meaning of exchanged information is understandable by any other application that can combine the received information with other information resources and process it in a meaningful manner.
- **Organizational Interoperability:** Which concerns the coordination of processes in which different organizations achieve a previously agreed and mutually beneficial goal. Organizational interoperability is concerned with bringing about the collaboration of administrations that wish to exchange information and may have different internal structures and processes. This includes aligning business processes and related data exchange, as well as meet the requirements of the user community by making services available, easily identifiable, accessible and user-focused.

- **Legal Interoperability:** which concerns the alignment of legislations so that exchanged data is accorded proper legal weight. Legal interoperability assures that the legal validity of the information exchanged to provide services is maintained across borders and data protection legislation in both originating and receiving countries is respected.

Moreover, besides the four level of interoperability the framework emphasizes the role of the political context to facilitate cooperation among public administrations. Indeed, the political context must be considered to achieve interoperability because "for effective cooperation, all stakeholders involved must share visions, agree on objectives and align priorities" (EIF, 2010, p. 21).

According to the EIF, all these objectives can be achieved by requiring all the organizations involved to formalize cooperation arrangements in interoperability agreements (for each level of interoperability considered in the EIF). However, as explicitly recognized in (EIF, 2010, 28):

ensuring interoperability between legal instruments, organization business processes, information exchanges, services and components that support the delivery of a European public service is a continuous task, as interoperability is disrupted by changes to the environment, i.e. to legislation, the needs of businesses or citizens, the organization of public administrations, business processes or technologies.

Although, the signing of interoperability agreements (at all layers of government) certainly is a necessary condition for cooperation among different public administrations, it is at least dubious that it is also a sufficient condition to maintain interoperability over time in a complex and changing environment. Indeed, the political, institutional and organizational heterogeneity between public administrations (both within and across Member

States) can affect interoperability (Misuraca, Alfano & Viscusi, 2011). Due to the political, institutional and organizational heterogeneity of Member States' public administration systems, in order to maintain interoperability over time some form of inter-organizational compatibility should be pursued at all layers of government as a way to improve inter-organizational (cross-boundary) cooperation. This includes the sharing of visions, objectives and priorities that can be ratified in co-operation agreements, but that cannot be achieved simply by signing those agreements.

BEYOND INTEROPERABILITY

The conditions that define the four interoperability levels (plus the political context) considered in the EIF are focused toward enabling the automatic exchange of data between the ICT systems of the public organizations that cooperate for the delivery of services. Indeed, in the service scenario considered in (EIF, 2010) citizens/businesses from one Member State X directly interact (online) with public administrations in another Member State Y that deliver online the services they required. The satisfaction of the interoperability requirements defined by the EIF is expected to allow the public administrations involved to exchange efficiently and effectively the information they need to deliver the required services.

However, the effectiveness of inter-organizational information exchange and integration heavily depends on interactions among social and technical processes both at the intra-organizational and at the inter-organizational level (Pardo, Cresswell, Dawes & Burke, 2004). From this point of view, even when it simply involves information exchanges among different organizations, the delivery of cross-boundary services through connected government should be considered as based on a complex networked socio-technical system (more specifically, a socio-technical system of systems). In such a socio-technical system, hu-

man participants and IT systems from different organizations perform processes using information and other resources to produce services for internal or external users (Alter, 2006). Scholl and Klischewski (2007) describe nine factors that constraint government integration and interoperability. These are:

- **Constitutional/Legal Constraints:** Integration and interoperation may be outright unconstitutional because the democratic constitution requires powers to be divided into separate levels and branches of government. Total integration and interoperability between and among branches and levels would virtually offset that constitutional imperative of checks and balances.
- **Jurisdictional Constraints:** Since under the constitution, governmental and non-governmental constituencies operate independently from each other, they own their information and business processes. Due to this, their collaboration, integration and efforts towards achieving various levels of interoperability are voluntary.
- **Collaborative Constraints:** Organizations are distinct in terms of their disposition and readiness for collaboration and interoperation with others. Past experience, socio-political organization, and leadership style influence the degree of proneness and adeptness of potential interoperation.
- **Organizational Constraints:** Organizational processes and resources may differ between organizations to such an extent that integration and interoperation might prove exceedingly difficult to achieve without standardizing processes, systems, and policies.
- **Informational Constraints:** While transactional information might be more readily shared, strategic and organizational information might not. In addition, information quality issues arise when integrating infor-

mation sources across various domains of control and quality standards.

- **Managerial Constraints:** Interoperation becomes inherently more complex as more parties with incongruent interests and needs become involved. As a result, the demands of the respective management task might exceed the management capacity of interoperating partners.

- **Cost Constraints:** Integration and interoperation between diverse constituencies might be limited to the lowest common denominator in terms of availability of funds. Also, unexpected budget constraints might pose serious challenges to long-term interoperation projects.

- **Technological Constraints:** The heterogeneity of e-Government information system platform and networking capabilities might limit the interoperation of systems to relatively low standards.

- **Performance Constraints:** As performance tests suggest, the higher the number of interoperating partners the lower the overall system performance in terms of response time. Yet, the focus on prioritized needs might enable fewer but more effective interoperations

Some of these constraints are considered in the EIF, as part of either one of the four levels of interoperability it defines or the political context. Some other constraints, most notably the collaborative and the managerial ones, have not been properly accounted for in the EIF. These constraints mainly depend on the social dimension of inter-organizational cooperation, which includes "human factors" such as feelings, motivation, trust, communication, culture, personal relationships, goals, values and commitment. These human factors could heavily affect the effectiveness of inter-organizational cooperation, even in case it

only amounts to the sharing of information. Thus, in order to allow the delivery of cross-boundary services besides technical issues also human factors should be taken into the account, which have not been properly considered in the EIF.

The impact of human factors on the effectiveness of the delivery of seamless cross-boundary services is even more critical when the delivery of the services also requires some offline (face-to face and/or paper based) activities to be performed, as it happens in the service scenario described above. Actually, in this case the delivery of the services does not depend only on the interchange of information among different organizations, enabled by interoperability (as defined in the EIF). In this service scenario in accessing cross-boundary services citizens and businesses could experience not only differences in legislations but also differences in the following:

- Political contexts, both at the national and at the local level (characterized by different priorities, strategies and objectives).
- Administrative traditions and management styles.
- Procedures (of different complexities and lengths).
- Work practices and public servants' skills, and, in the case of cross-border services, even in languages.

All these elements make it very unlikely for citizens/businesses to experience seamless cross-boundary services across Europe. If this will not reduce the citizens' (and businesses') mobility in the Internal Market, it certainly will not contribute to strengthen the European citizens' perception of living and working in a Single Market.

Although, they are not covered by the four levels of interoperability considered in EIF (2010), the elements mentioned above can nevertheless be accounted for in terms of some form of "interoperability," even though different from the one considered in the EIF. Indeed, to deliver seamless cross-boundary services, different public administrations have to be able not only to share information efficiently and effectively, but also to operate in a so seamless and integrated way to be perceived as parts of a single (virtual and networked) system of (European) public administrations. This requires different public administrations to share conditions that can increase inter-organizational compatibility, in terms of "standardization" at the level of operational processes, organizational architectures, management styles, culture and value systems as well as vision and strategy. By making different organizations more and more compatible, this form of organizational interoperability (that cannot be reduced to the organizational dimension of interoperability as defined in (EIF, 2010)) can strengthen inter-organizational (cross-boundary) cooperation much more than the signing of cooperation/interoperability agreements can do. This would affect the efficiency and effectiveness of the (online and offline) delivery of cross-boundary services, which represents a value for the users of those services. However, organizational compatibility would also allow European citizens to interact with public administrations in different Member States as if they were all members of a (virtually) integrated system of European public administrations, which would contribute to strengthen the European citizens' perception of living and working in a Single Market.

The concept of organizational compatibility can be described more precisely in terms of *co-operability*, as it has been defined in the context of joint and multinational military operations (Clark & Jones, 1999; Tolk, 2003; Stewart et al., 2004). Co-operability represents a form of non-technical interoperability aiming at the successful bridging of differences in doctrine, organization, concepts of operation, and culture so that different organizations can function together essentially as a single organization with no loss in effectiveness (US-CREST, 2000; Gompert & Nerlich, 2002).

Co-operability goes beyond the conditions that enable information exchange among different organizations; it considers also the willingness to interact and the desire to communicate clearly (Alberts, Huber & Moffat, 2010).

Co-operability can be defined based on four attributes that summarize different aspects characterizing inter-organizational cooperation (Clark & Jones, 1999), viz:

- **Preparedness:** This attribute describes the preparedness of both the organization and its members to interoperate. It is made up of doctrine, experience and training.
- **Understanding:** The understanding attribute measures the amount of communication and sharing of knowledge and information within the organization and explains how the information is used.
- **Command Style:** This attribute describes the management and command style of the organization - how decisions are made and how roles and responsibilities are allocated/delegated.
- **Ethos:** The ethos attribute concerns the culture and value systems of the organization, its goals and aspiration.

These attributes allow describing different levels of co-operability, characterized by a growing degree of organizational compatibility among different organizations, as defined in Table 2.

The co-operability attributes mainly concern the human factors that could influence the effectiveness of inter-organizational cooperation; from this point of view, co-operability can be considered as a complement of the four levels of interoperability defined in the EIF. The sharing of both interoperability and co-operability conditions allows different government organizations to achieve higher levels of efficiency and effectiveness in their interactions, which could mean better cross-boundary services. Moreover, it also allows to sensibly reducing the level of organizational

Table 2. Levels of organizational compatibility/co-operability (Clark & Jones, 1999)

Levels of co-operation cooperability	ATTRIBUTES			
	Preparedness	**Understanding**	**Command Style**	**Ethos**
Unified	Complete, normal day-to-day working	Shared	Homogeneous	Uniform
Combined	Detailed doctrine and experience in using it	Shared communications and shared knowledge	One chain of command and interaction with home organizations	Shared ethos but with influence from home organizations
Collaborative	General doctrine in place and some experience	Shared communications and shared knowledge about specific topics	Separate reporting lines of responsibility overlaid with a single command chain	Shared purpose; goals, value system significantly influenced by home organizations
Ad hoc	General guidelines	Electronic communications and shared information	Separate reporting lines of responsibility	Shared purpose
Independent	No preparedness	Communication via phone, etc.	No interaction	Limited shared purpose

heterogeneity among them, so that they can be perceived as part of a single and highly connected system of European government organizations. Finally, by basing inter-organizational cooperation on both interoperability and co-operability, the integration of different public sector organizations from different Member States can be achieved even without forcing them to conform to a unique organizational model. This allows safeguarding the autonomy of each organization participating in the cross-boundary services delivery, which Pardo, Gil-Garcia and Burke (2006) identify as one of the condition that can improve the effectiveness of inter-organizational cooperation.

THE WEB AS A COOPERATION PLATFORM

Social Computing and Cooperation Services for Inter-Organizational Trust Building

The conditions that define co-operability are quite difficult to satisfy, especially when the higher levels of co-operability are considered. Indeed, they

are not the kind of conditions that can be satisfied simply by signing inter-organizational agreements. Rather they require different organizations to enter complex processes of mutual adjustment that, in the end, will reduce their organizational heterogeneity and allow them to function together essentially as a single organization.

The process through which different organizations can achieve higher levels of co-operability can be considered as a learning process, in which different organizations engage in a series of iterative and interactive learning cycles over time, typically characterized by greater and greater trust and adaptive flexibility, as well as the willingness to make increasing and irreversible commitments.

By discussing the formation of strategic alliances, Doz (1996) describes such a learning process as in Figure 1.

According to this description of the evolution of inter-organizational cooperation, the cooperation can be activated through the definition of an agreement (for instance the interoperability agreements defined by the EIF) stating the initial conditions for the cooperation. Once the cooperation has been activated, different organizations can start adapting to each other by learning from

Figure 1. The evolution of inter-organizational cooperation (Doz, 1996)

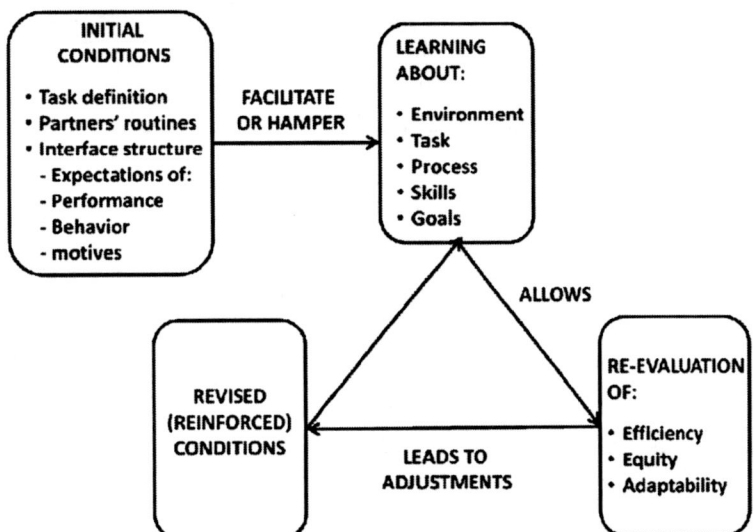

their mutual interactions. Through the iteration of the learning cycles over time, trust relationships among different organizations can be established and reinforced, which is the basis for the mutual adjustment and trust reinforcing processes that can lead to co-operability.

(Castelnovo, 2012) argues that sharing a cooperation environment and a cooperation platform can support the learning process represented in Figure 1. The cooperation environment can be conceived of as a set of conditions that, based on a shared reference model, specify the interoperability requirements different organizations should satisfy at the strategic, the organizational, the operative and the technological level in order to cooperate. These requirements are those that can be explicitly stated in the cooperation agreement. The cooperation platform can be conceived of as a set of specific tools and services that support inter-organizational cooperation (Camarinha-Matos & Afsarmanesh, 2003; Ollus, 2005; 2007). Among the services and tools that the cooperation platform should provide, particularly relevant are those that enable direct interactions among people belonging to different organizations, not necessarily limited to the interactions required for the execution of inter-organizational workflows.

Interpersonal relationships are fundamental for establishing, maintaining and strengthening trust at both the inter-personal and the inter-organizational level (Kramer, 1999). This makes the Social Computing and Cooperation Services (SC&CSs) and tools made available by the so-called Web 2.0 particularly useful to support the learning process leading to co-operability. Indeed, social computing and cooperation services can be considered as a set of applications that enable people to efficiently interact with other people, as well as to share contents and data through the Web (Parameswaran & Whinston, 2007; Young, 2007; Warr, 2008). Whereas publicly available SC&CSs are already widely diffused for private use (Pascu, 2008), their adoption and use by enterprises and government organizations is still quite limited, although it is widely expected that the adoption of SC&CSs would have a high positive impact on both enterprises and government organizations (Bughin, Chui & Miller, 2009; Doculals, 2010; Osimo, 2008).

Table 3 gives a schematic overview of some of the more common SC&CSs, grouped with respect to the four basic functionalities provided by social computing: messaging, collaboration, broadcasting and knowledge building. Some of

Table 3. Social Computing and Cooperation services (based on (Doculabs, 2010))

	Already Widely Used	**Increasingly Adopted**	**Rarely Used**
Messaging	E-mail	Instant messaging Presence detection Unified telephony	
Collaboration	File/document sharing Portal Calendaring	Discussion forum Shared whiteboard Shared workspace	
Broadcasting	Web conferencing	Audio broadcasting Blog Content syndication	Micro-blogging Video broadcasting
Knowledge building		Wiki	Polling Community building Expertise management Social filtering Knowledge market

these services are already widely used by organizations, both for internal purposes and for customer related purposes. Others are being increasingly adopted by organizations; yet others are still quite rarely used, although the greatest value is expected from their adoption.

The term "Enterprise 2.0" refers to the use of SC&CSs by enterprises, as well as to their embracing of the collaborative philosophy of Web 2.0 (McAfee, 2006; 2009). By analogy, the term Government 2.0 refers to the use of SC&CSs by government organizations (Osimo, 2008; Mergel, Schweik & Fountain, 2009; Dixon, 2010); sometimes it is also used to refer to the next step of e-Government towards more participatory arrangements (Baumgarten & Chui, 2009).

As enterprises, also government organizations can adopt SC&CSs either to support interactions among organization members (internal focus) or for facing citizens, partners and suppliers (external focus). The use of SC&CSs described in this chapter is intermediate between the two. The use of SC&CSs to establish and enforce relationships among different (public administration) organizations that cooperate for the delivery of services has an external focus (since the focus is on inter-organizational relationships). However, this use of SC&CSs concerns the strengthening of the relationships among the members of an inter-organizational network as the condition that could enable them to achieve higher levels of co-operability. Used in this sense SC&CSs have an internal focus. From this point of view it can be said that, integrated within the cooperation platform, SC&CSs are used with a "network focus."

The use of SC&CSs to establish and reinforce the relationships among the members of different organizations exploits the SC&CSs capacity of supporting the establishment and the reinforcement of different types of ties between individuals. Following McAfee's bull's eye metaphor (McAfee, 2009), the different types of ties in which an individual can be involved can be represented as in Figure 2.

Strong ties are those ties that bind people that are used to work together; people that work together only occasionally are bound by weak ties, whereas the potential ties are those ties that could be established if only there was the opportunity. McAfee (2009) makes a connection between this classification of ties and different SC&CSs. Thus, for instance, he claims that a tool like Wiki is useful for people with strong ties working closely together, whereas Blogs can be seen as a way of both nurturing and creating new weak ties, and social networking is particularly useful for creating new potential ties that can possibly evolve in weak ties (or even in strong ties).

Figure 2. Types of ties (McAfee, 2009)

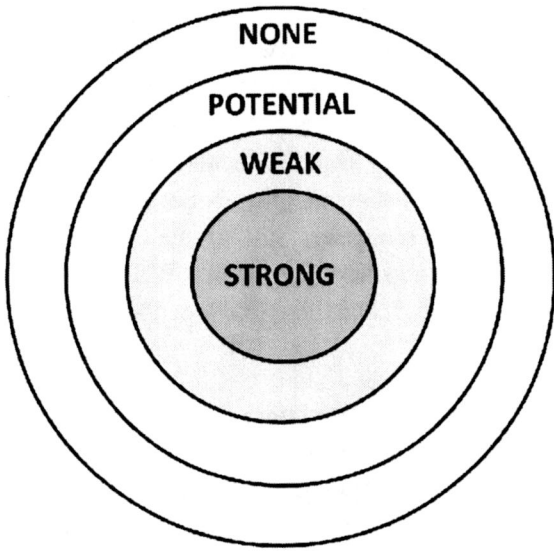

McAfee's classification of interpersonal ties can be used also to describe different types of relationships among people belonging to different organizations involved in an inter-organizational cooperation. The strong inter-organizational ties are those ties that bind people that, while belonging to different organizations, nevertheless work together on a day-to-day basis. The weak inter-organizational ties are those ties that bind people belonging to different organizations that interact only occasionally. Finally, all the employees of the organizations involved in the cooperation are linked by potential ties that could evolve into weak ties (or even into strong ties) if the cooperation will widen the range of the shared resources and activities.

The establishment of strong ties cannot be determined by the use of technological tools alone; indeed, strong ties require a level of acquaintance and trust that only working shoulder to shoulder can determine. However, the use of SC&CSs allows the systematic sharing of knowledge and the exploitation of collective intelligence; besides making the cooperation more efficient and effective, this can help the members of different organizations to achieve and maintain a shared

value system, a sharing of the goals, and a common understanding, as required by the higher levels of co-operability.

In the case of weak and potential ties, SC&CSs can play a more direct role in establishing, maintaining and strengthening interpersonal links. Tools like blogs and social networks can help establishing new interpersonal relations and allow the sharing of information and knowledge also among people loosely tied, as people belonging to very large organizations usually are (but the same holds for people belonging to different organizations as well) (Dimicco et al., 2009).

The fundamental role of weak ties to allow knowledge workers to access non-redundant information is well known. The role of weak ties for reducing organizational heterogeneity among different organizations is much less studied. However, data are available that show how the use of SC&CSs supporting the establishment of week ties can determine "social" benefits including the sharing of a common organizational culture and identity among people geographically dispersed (Jackson, Yates & Orlikowski, 2007; Richter & Riemer, 2009). This can help creating an environment conducive to trust building, which is a fundamental condition for co-operability.

SC&CSs and tools are increasingly been adopted by enterprises and government organizations and there is a growing interest in both Enterprise 2.0 and Government 2.0. However, the adoption rate of SC&CSs by enterprises and government organizations is still quite low. The 2012 AIIM's survey (AIIM, 2012) on the adoption and use of social computing tools and techniques by enterprises (including both central and local government organizations) reports that in only 50% of the organizations that responded to the survey employees are encouraged to use SC&CSs in relation to their job. Moreover, only 37% of respondents think that SC&CSs will be in regular use across the whole enterprise in the next 2 years.

The data concerning government organizations are even worse; 42% of government organizations restrict use of social tools completely (only 16% of enterprises do the same). Government organizations delivering services to citizens are three times more likely (29%) to have taken disciplinary action with staff relating to social activity than other organizations (11%). 44% of Government organizations delivering services to business actively discourage employees from using SC&CSs for their job (the average figure for this in enterprises is 19%).

The gap between the expected and the actual data concerning the adoption and use of SC&CSs in enterprises mainly depends on some well-known risks that enterprises still perceive as related to the adoption of SC&CSs. Such risks concern inappropriate behavior and content, inaccurate information, embarrassing content, non-compliance with laws, regulations, intellectual property or policies (McAfee, 2009). Besides these, Dawson (2009) identifies further risks concerning reduced staff productivity, IT security and loosing of control. However, despite all these well-known risk, AIIM (2012) reports that at least 50% of the surveyed organizations feel that integrating SC&CSs to all types of business process would be very or extremely valuable. This makes the identification of the right strategy for the introduction of SC&CSs the critical step for assuring their adoption and use.

Raeth et al. (2009) observes that despite their potential to transform workflows and organizations, SC&CSs are more likely to be implemented as tools for complementing existing work practices and structures. Moreover, they also observe that "unlike process-oriented enterprise systems for which use is often mandated, Web 2.0 systems, with their emphasis on supporting individual and group's idiosyncratic communication and collaboration activities, are more likely to be treated either officially or practically voluntary" (Raeth et al., 2009, p. 1). Instead, Chui, Miller and Roberts (2009) identify the integration of SC&CSs within existing workflows as a critical success factor for

the adoption of SC&CSs. According to them, this avoids the risk that employees perceive the use of SC&CSs just as another "to do" on an already crowded list of tasks. Bughin, Chui and Miller (2009) found that successful companies not only tightly integrate Web 2.0 technologies with the workflows of their employees but also create a "networked company," linking themselves with customers and suppliers using Web 2.0 tools (Bughin, Chui, & Miller, 2009, p. 1).

SOCIAL COMPUTING AND COOPERATION SERVICES FOR SUPPORTING INTER-ORGANIZATIONAL WORKFLOWS

In this section of the chapter, I will describe a possible use of SC&CSs that is in line with the observations above. More specifically, I will consider the possibility of using SC&CSs to support the execution of the inter-organizational workflows related to the procedures for starting, transforming or closing a business in Italy. This case is particularly interesting since it involves the activity of One Stop Shop for Production Activities in Italy (SUAP – Sportello Unico per le Attività Produttive), which is a typical example of connected e-Government. Moreover, under the Italian legislation in force, the SUAPs have competence on most of the 10 cross-border services related to business and start-up considered in (EC, 2013). From this point of view, the case that will be discussed in the following pages can be considered as related to the cross-border delivery of services through connected government.

One-Stop Shop for Production Activities (SUAP) as the single point of access to services and information for business offered by different public authorities has been established in Italy since the late '90s as part of the policies for reducing the administrative burdens on enterprises. According to the Italian law, each municipality must establish its own SUAP by choosing among establishing it individually, sharing it with other municipalities through inter-municipal cooperation or delegating it to the local Chamber of Commerce.

Under the legislation currently in force, a new business can be started up simply through a communication (Segnalazione Certificata di Inizio Attività – SCIA, Certified Communication of the Start of the Activity) submitted online to the SUAP through the national portal www.impresainungiorno.it. This portal acts as the Point of Single Contact (PSC), as required by the so-called Service Directive issued in 2006 by the European Union. The online procedure is such that it guarantees that the SCIA contains all the required information and that all the required documents have been attached (if the communication is incomplete it is rejected). A SCIA that has been checked for completeness is registered to the system that automatically sends a receipt to the applicant. Upon receiving this receipt, an entrepreneur can start a new business.

Upon registering the SCIA, the SUAP has to operate all the required controls within an interval of time defined by the law (60 days). If any of the controls detects any inadequacy, then the business can be stopped (either temporally until the inadequacy will be removed or definitively, if the inadequacy cannot be removed), otherwise it can operate without requiring any further authorizations.

The new legislation did not reduce the number of public agencies and authorities having competences concerning starting, transforming or closing a business, which besides municipalities includes Fire Departments, Public Health agencies, Regional and Provincial governments, and other public agencies depending on the type of the business. This means that the controls that have to be operated may involve many different public agencies. However, given the lack of resources affecting them, those agencies could have difficulties in promptly operating all the required ex-post controls on all the SCIAs submitted to the SUAP.

In such a situation, the only way to satisfy the requirements stated by the law is by identifying what controls really need to be operated, based on a careful evaluation of the potential risks each SCIA submitted to the SUAP could entail. Since each agency's evaluation can have consequences on the result of the processing of a SCIA, thus affecting the other authorities' activities and decisions, it is necessary for all the actors involved to share both their evaluation of the potential risks it could entail and their decisions whether or not to perform some controls on it. This requires the SUAP to implement an efficient and effective system for coordinating decisions involving many different public bodies. The decision whether or not to perform controls on a given SCIA could entail some form of risk. Hence, the SUAP should implement an inter-organizational risk management system, i.e. a system whose goal is "to more effectively allow organizations to share information and perform necessary activities with regard to risk management that may affect their collective behavior" (Meyers, 2006, p. 6).

The Team Risk Management (TRM) approach described in (Higuera et al., 1994) defines methods and tools that can be used to manage risks at the inter-organizational level. In the TRM approach, three types of risk management processes are considered:

- **The Baseline Risk Assessment:** Which is an intra-organizational process executed independently by each organization to individually identify the risks associated with their respective organizations.
- **The Team Review Process:** Which is an inter-organizational process conducted jointly by the organizations involved to share and jointly evaluate the most important risks they individually identified.
- **A Set of Continuous Processes:** Implemented at both the intra and the inter-organizational level, which comprise a cyclic set of activities by means of

which each organization involved manages risks at the intra-organizational level and, through a repeated execution of the team review process, at the inter-organizational level as well.

Figure 3 below shows how TRM can be integrated within the SUAP workflow. After registering a SCIA submitted to it through the PSC, the SUAP sends the relevant documentation to all the agencies having competences on it. Upon receiving the documents, the agencies individually perform the baseline risk assessment process, with the aim of identifying all the potential risks that SCIA could entail. Besides evaluating the potential risks involved, during this phase each partner can also plan the execution of its control activities on that SCIA. The team review allows the partners to jointly evaluate the SCIA based on the potential risks each one of them identified

Figure 3. Integration of TRM within the SUAP's workflow

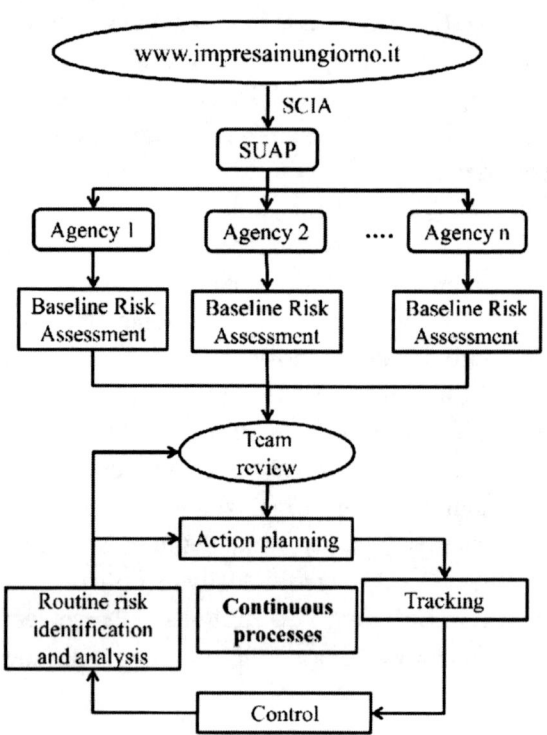

individually, thus arriving at a shared risk evaluation. Such an evaluation forms the basis for jointly deciding whether to perform the required controls on that SCIA.

As observed above, the SUAP's workflow is constrained by strict temporal requirements since all the possible controls should be made within 60 days from the online submission of the SCIA to the PSC. This raises two questions: (i) what risk management processes can be made more efficient and effective in order to conclude the controls within the terms fixed by law, and how this can be done; (ii) what technologies and tools can be used to make those processes more effective and efficient, both at the intra-organizational and at the inter-organizational level.

Castelnovo (2011) describes some requirements that should be satisfied in order to make both the Baseline Risk Assessment and the Team Review processes more efficient and effective. Table 4 summarizes such requirements.

Both the Baseline Risk Assessment and the Team Review essentially amount to group decision processes, i.e. a processes in which different people (possibly belonging to different organizations) "work together to analyze and prioritize alternative solutions to problems and choose one alternative through extensive communication, analysis, deliberation, and negotiation" (Turban, Liang & Wu, 2011, p. 138).

Based on a comparison of different frameworks for group decision making, Turban, Liang and Wu (2011) identify the following four main activities for group decision making:

- **Intelligence:** Which amounts to information gathering and sharing for the purpose of problem identification and determining its importance.
- **Design:** Which amounts to finding alternative solutions and analyzing them.
- **Choice:** Which amounts to selecting an effective course of action.

Table 4. Requirements for the Baseline Risk assessment and the Team Review

Baseline Risk Assessment	A1. Identify the people to be involved in the risk assessment, whatever organization they belong to.
	A2. Promptly inform them that their advices are needed to evaluate a SCIA.
	A3. Support their risk assessment activity by giving them access to information concerning how analogous cases have already been treated previously.
Team Review	B1. Allow all the involved organizations to decide and share their decision without requiring them to participate in face-to-face meetings.
	B2. Let all the involved organizations be promptly informed of the possible occurrence of secondary risks (risks rising as a consequence of the implementation the risk handling actions), wherever they have be detected.
	B3. Promptly inform all the managers that have to implement the risk handling actions agreed on (whatever organization they belong to) of what they are expected to do and when.
	B4. Let all the managers (whatever organization they belong to) intervene in the decision making process to contribute to the best timing for the execution of the risk handling actions agreed on, also considering the priorities already defined by the organizations they belong to.

- **Implementation:** Which amounts to managing the process and implementing the solution.

All these activities can be made more efficient and effective by using Collaboration 2.0 tools, which amounts to the use of Web 2.0-based social software services and tools (i.e. SC&CSs) to facilitate intra and inter-organizational collaboration (Coleman & Levine, 2008; Sari et al., 2008). Table 5 summarizes the use of SC&CSs to support group decision-making (Turban, Liang & Wu, 2011, p. 144).

By considering the inter-organizational risk management processes described above as group decision making processes, the same Collaboration 2.0 services and tools (SC&CSs) Turban, Liang and Wu suggest to use to support the activities reported in Table 5, can be used to support the SUAP's inter-organizational workflow as well.

The requirement A1 in table 4 concerns the identification of the people (domain experts) to involve in the baseline risk assessment (and in the routine risk identification and analysis, as well), which can be considered as a case of expertise management. A way to satisfy the requirement A1 is by requiring all the organizations involved in the SUAP's activity to associate an expertise profile to each of their knowledge workers (or at least to those directly involved in the execution of the SUAP's workflow). This would allow the manager responsible for the SUAP to find the competencies required for the processing of a given SCIA simply by browsing (either automatically or manually) the expertise profiles. This can be done by transforming the usual "directory of people" from the "yellow pages" model to a social network model, analogous to the IBM's "Blue Pages" described in (Richter & Riemer, 2009). Besides making the search of experts easier, representing expertise through social network profiles also allows the possibility of making the expertise description more reliable by means of social filtering mechanisms that can be used to further qualify the level of expertise of registered experts (Damianos et al., 2007; Braun, Kunzmann & Schmidt, 2008).

By resorting to this solution, upon receiving a SCIA the SUAP can prompt all the agencies involved to activate the concurrent execution of their baseline risk assessment processes through the following steps:

- Identify all the agencies that need to be involved in the processing of that SCIA.

Table 5. Collaboration 2.0 tools supporting group decision making

	Group Decision Making Process Tasks	**Collaboration 2.0 tools**
Intelligence	Problem Identification	BlogBlogs, IM, Polls (voting)
	Information finding/sharing	RSS feeds, blogblogs, Twitter
	Soliciting expert's opinion	Social networks answering function Enterprise social bookmarking
	Prioritize problems (importance)	Voting, blogblogs, IM, discussion groups (forums), chat room
Design	Search for alternative solutions	Search tools, expert/answering
	Idea generation—brainstorming	Discussion groups, blogblogs, IM, chat room
	Experts' opinions	Answer function, IM, Twitter
	Organize alternatives	WikiWiki
	Identify criteria of choice	WikiWiki, blog, IM, discussion groups
	Prioritize criteria of choice (importance)	Discussion groups, voting, IM, chat room
Choice	Analysis (Forecasting, risk, comparisons)	Collaborative decision making (Social networking plus Business Intelligence analysis)
	Selection of an alternative	Polling, discussion groups, IM
Implementation	Meeting management	Twitter, RSS feeds
	Project management	Presence awareness, Twitter, blog
	Report writing	WikiWiki
	Training (if needed)	Virtual worlds

- Identify what are the competences required to properly assess the potential risks involved in that SCIA.
- Search the network (by means of a social network search engine) for people with the required competences.
- Send the relevant documentation directly to the people identified in the previous step, irrespective of the organization they belong to (this step accounts for the requirement A2 in table 4 as well).

A different approach to searching the experts (that could also be integrated in the searching of profiles approach) is by means of microblogging. People normally use microblog to provide to others updates on their activities, observations and interesting content. However, based on data concerning the IBM's microblogging application BlueTwit, Ehrich and Shami (2010) found that workplace employees are mostly using the tool to post information, engage in brief directed conversation and asking help from colleagues. This suggests the possibility of using a microblog post to identify the people to involve in the baseline risk assessment; this could be done in the form of a request for help to which people with the required competences answer. By means of the same post, these people get also informed of the fact that their advices are needed, thus satisfying the requirement A2 as well. A further alternative is the use of automatically generated RSS feeds instead of microblogs posted by people allocated to the SUAP.

Microblogs can be used in all the situations in which the people involved need to be promptly informed of the occurrence of potentially critical events, as in the case of a partner's detection of a secondary risk arising during the execution of the risk handling actions he is in charge of. In this case, as he detects the secondary risk, a partner can post a microblog informing of this, almost in real time, all the other partners that will thus be prepared to define and share the most adequate measures to cope with the new situation. Hence, by resorting to microblogging also the requirements B2 and B3 can be satisfied.

Blogs represent powerful tools for sharing expertise, improving cooperation and enabling business intelligence creation (Andrus, 2005; Kolari et al. 2007). In the case we are considering, the use of blogs satisfies also the requirements A4, B1 and B4. Upon receiving a SCIA, the manager responsible for the functioning of the SUAP can initiate a blog devoted to its risk assessment. The managers responsible for risk assessment (both the baseline risk assessment and the routine risk identification and analysis) within the organizations involved in the processing of that SCIA can then post comments on that blog. In this way, all the people involved in the risk evaluation can share their evaluation of: (i) the potential risks involved (for their organizations); (ii) the need of performing the ex-post controls on that SCIA; and (iii) the timing for the execution of the controls. This allows:

- The definition of a shared risk profile for the SCIA under processing (requirement A4).
- A shared decision concerning whether to perform the controls, and when (requirement B1), possibly also by resorting to polling mechanisms that can be integrated quite easily in a blog (Rodriguez et al., 2007).
- A shared planning of the execution of all the ex-post controls (whatever organization is responsible for each control activity), compatible with the other activities each partner has to perform as part of his normal activity (requirement B4).

Blogs can be used to share information and decisions both at the inter-organizational and at the intra-organizational level. In the latter case, the use of blogs would allow the manager responsible for the internal risk assessment to collect information and advices that could be useful for a sound risk assessment from people within the organization (whatever unit they belong to).

Risk assessment, and risk treatment as well, can be more efficient and effective when information concerning previous experiences can be used to derive useful insights on how to treat the case at hand. One way to do this is by allowing the use of tools that enable groups to jointly create and share content; Wikis are typical collaborative tools that can be used to this end, even in critical contexts (Wagner, 2004; Andrus 2005). In the processing of a SCIA, blogs can be used to make a shared risk assessment and a shared risk treatment strategy emerge from the collaborative effort of people belonging to different organizations (as well as to different units within the same organization). Wikis, instead, can be used to document the whole risk assessment and risk treatment processes for that SCIA. The intrinsic collaborative nature of the Wiki enables documents to be written collectively by all the people involved in the risk management and the risk treatment processes (whatever organization they belong to), thus allowing a more structured, complete, reliable and useful documentation of the way a SCIA has been treated. Moreover, the possibility of using both content tags and meta-information tags and links to organize the content of Blogs and Wiki pages makes it particularly easier to find the information needed (Kolari et al., 2007). This allows a more efficient use of past experiences in both the risk assessment and the risk treatment processes. From this point of view, by using Wikis to document the risk management processes also the requirement A3 in table 4 above can be satisfied.

CONCLUSION

The use of Social Computing and Cooperation services and tools (SC&CSs) by organizations, as well as their embracing of the collaborative philosophy of Web 2.0, to improve both internal and external relationships is continuously increasing and there is a general expectation that this will determine a deep transformation in the way organizations will work in the interconnected world. Besides enterprises, also government organizations are increasingly adopting SC&CSs with the aim of making their processes more transparent, fostering citizens' participation and promoting collaboration and cooperation among government agencies.

Inter-agencies cooperation lies at the basis of Connected Government whose aim is to enable government organizations to connect seamlessly across functions, agencies, and jurisdictions to deliver effective and efficient services to citizens and businesses. In the countries of the European Union, this objective concerns the delivery of services within each Member State and the delivery of cross-border services as well. Indeed, the possibility for European citizens and businesses of accessing cross-border (and cross-boundary) services (either delivered completely online or through some form of face-to-face/paper based interaction) represents a fundamental step toward the strengthening of the perception of living and working in a Single Market.

Connected Government is difficult to achieve in the European Union due to the high organizational heterogeneity among government organizations, both at the national and at the cross-border level. Without reducing organizational heterogeneity among government organizations from different Member States, it is very unlikely that their interactions with European public administrations could strengthen the European citizens' perception of living and working in a Single Market.

Interoperability allows different organizations to interact through the exchange of information.

However, interoperability by itself cannot reduce organizational heterogeneity since it also depends on the social dimension of organizations, which includes "human factors" such as feelings, motivation, trust, communication, culture, personal relationships, goals, values and commitment. Besides possibly making European citizens experience striking differences when interacting with public administrations in different Member States, all the above-mentioned elements can affect the effectiveness of inter-organizational cross-boundary cooperation and the delivery of seamless cross-boundary services as well.

Organizational heterogeneity can be reduced through co-operability, which is a form of non-technical interoperability aiming at the successful bridging of differences in doctrine, organization, concepts of operation, and culture so that different organizations can function together essentially as a single organization with no loss in effectiveness. The process that can lead different organizations to achieve higher levels of co-operability can be considered as a learning process in which different organizations engage in a series of iterative and interactive learning cycles over time, typically characterized by greater and greater trust and adaptive flexibility, as well as the willingness to make increasing and irreversible commitments. In the chapter, I argued that the use of SC&CSs to support the interactions among different public administrations could make this learning process more efficient and effective. This is due to the role SC&CSs can play in the establishment, maintenance and strengthening of inter-personal relationships, both within and across the organizational boundaries, which is fundamental for every trust-building process.

Many barriers have to be overcome in order to allow an effective use of SC&CSs within organizations (Bughin, 2008). A detailed discussion of those barriers is outside the scope of this chapter; however, there is at least one point that deserves some attention here. The take up of SC&CSs by employees is not spontaneous. (AA.VV., 2010) reports that companies quote employees resistance as the single biggest barrier to adoption; McAfee (2009) has reached the same conclusion based on his interviews with lead companies. This point is crucial for the use of SC&CSs suggested in this chapter. Actually, both the use of SC&CSs to support the learning process that can lead to co-operability, as the basic condition for connected government, and the use of SC&CSs to make the SUAP's workflow more efficient and effective heavily rely on the employees' willingness to collaborate, within and across organizational boundaries.

However, this problem concerns collaboration in itself rather than the adoption and use of services and tools that can make collaboration easier, as SC&CSs are. People are the critical ingredient in any collaboration; no benefit can be realized without people being willing to trust and share both the content and themselves (Coleman & Levine, 2008). The success of any process based on collaboration depends on the ability to address people, process, and technology issues simultaneously with the heaviest focus on people. Coleman and Levine (2008, p. 24) observe that usually 80% (or more) of the effort in collaboration goes into the development of the enabling technologies instead, whereas only 20% (or less) is focused on the relationships or interactions between people. This suggests a first indication of what should be done to overcome the barriers that could limit the use of SC&CSs: invest more on people and on their relationships than on technology, which is just an enabling element whose value is in its capacity of supporting people in the way they work, without imposing further burdens on them.

This leads to another important success factor for the adoption and use of SC&CSs. The best practices show that the highest chances of success depend on the integration of the new tools within existing workflows and on their application to the core activities of the organization. The use of SC&CSs described in this chapter concerns the procedures for starting, transforming or closing a business in Italy. The efficiency and effectiveness of these procedures, implemented by the One Stop Shop for Production Activities, are critical for the reduction of the administrative burdens on enterprises, which represents a strategic policy objective for the Italian Public Administration. From this point of view, the use of SC&CSs described in the chapter satisfies the success condition above for the adoption of SC&CSs by organizations.

Under the legislation in force, the procedures for starting, transforming or closing a business in Italy require the execution of inter-organizational workflows in which different public agencies have to perform a coordinated set of activities, according to the competencies set to them by laws and regulations. All these activities are tied together through informal and formal communication processes that enhance the cooperative interactions and trust between the people and the organizations involved.

Communication (both formal and informal) represents a fundamental condition for enabling effective cooperation. On the one hand, through communication all the actors involved in the execution of inter-organizational workflows can build and reinforce a shared vision, create a sense of collective ownership and responsibility, enhance inter-organizational trust, which can make the cooperation more effective and easier to manage. On the other hand, at the operative level, communication enhances the interactions among all the actors involved, both at the intra-organizational and at the inter-organizational level, and allows information and knowledge to be shared among them, which makes the execution of inter-organizational workflows more efficient and effective.

Fundamental as they are for effective cooperation, communication and knowledge sharing are not so easy to manage at the inter-organizational level. However, as Coleman and Levin (2008, p. 9) point out, "when you couple technology and process with interpersonal styles, awareness, communication tools, and some conversational models that support 'sustainable collaboration', you have a winning combination." Social Computing and Cooperation services provide such a winning combination that, as shown in the chapter, can be exploited to enhance inter-organizational cooperation, to allow the seamless delivery of cross-administration/sectorial/border services and to pursue the objective of Connected Government.

REFERENCES

AA.VV. (2010). *Enterprise 2.0 study - D4 Final report*. European Commission.

AIIM. (2012). *Social in the Flow-transforming processes and sharing knowledge*. AIIM Industry Watch.

Alberts, D. S., Huber, R. K., & Moffat, J. (2010). *NATO NEC C2 maturity model*. U.S. Department of Defence.

Alter, S. (2006). *The Work System Method: Connecting People, Processes and IT for Business Results*. Larkspur, CA: Work System Press.

Andrus, D. C. (2005). *The Wiki and the Blog: Toward a Complex Adaptive Intelligence*. Central Intelligence Agency.

Bannister, F. (2002). Citizen Centricity: A Model of IS Value in Public Administration. *Electronic Journal of Information Systems Evaluation, 5*(2).

Braun, S., Kunzmann, C., & Schmidt, A. (2008). People Tagging & Ontology Maturing: Towards Collaborative Competence Management. In *Proceedings of 8th International Conference on the Design of Cooperative Systems COOP '08*. Carry-le-Rouet, France: COOP.

Bughin, J. (2008). The rise of enterprise 2.0. *Journal of Direct. Data and Digital Marketing Practice, 9*, 251–259. doi:10.1057/palgrave.dddmp.4350100

Bughin, J., Chui, M., & Miller, A. (2009). How companies are benefiting from Web 2.0 - McKinsey Global Survey Results. *The McKinsey Quarterly*.

Camarinha-Matos, L. M., & Afsarmanesh, H. (2003). Elements of a base VE infrastructure. *Computers in Industry, 51*, 139–163. doi:10.1016/S0166-3615(03)00033-2

Castelnovo, W. (2011). Social Computing Tools for Inter-organizational Risk Management. In *Proceedings of the 5th European Conference on Information Management Evaluation*. ECIME.

Castelnovo, W. (2012). An Architecture Driven Methodology for Transforming from Fragmented to Connected Government: A Case of a Local Government in Italy. In *Enterprise Architecture for Connected E-Government: Practices and Innovations*. Hershey, PA: IGI Global. doi:10.4018/978-1-4666-1824-4.ch015

Castelnovo, W. (2013). A Stakeholder Based Approach to Public Value. In *Proceedings of the 13th European Conference on E-Government*. ACPI.

Castelnovo, W., & Simonetta, M. (2007). The Evaluation of E-Government projects for Small Local Government Organisations. *Electronic Journal of E-Government, 5*(1).

CEMR. (2011). *EU Subnational Government – 2010 Key Figures, 2011/2012 Ed*. Council of European Municipalities and Regions. Retrieved June 22, 2013, from http://www.ccre.org/docs/Nuancier2011Web.EN.pdf

Chui, M., Miller, A., & Roberts, R. P. (2009). Six ways to make Web 2.0 work. *The McKinsey Quarterly.*

Clark, T., & Jones, R. (1999). Organisational Interoperability Maturity Model for C2. In *Proceedings of the Command and Control Research and Tecnology Symposium.* Retrieved June 22, 2013, from http://www.dodccrp.org/events/1999_CCRTS/pdf_files/track_5/049clark.pdf

Coleman, D., & Levine, S. (2008). Collaboration 2.0: technology and best practices for successful collaboration in a Web 2.0 world. Cupertino, CA: Happy About.info Publishing Company.

Damianos, E. L., Cuomo, D., Griffith, J., Hirst, D. M., & Smallwood, J. (2007). Exploring the Adoption, Utility, and Social Influences of Social Bookmarking in a Corporate Environment. In *Proceedings of the 40th Hawaii International Conference on System Sciences.* IEEE.

Dawson, R. (2009). *Implementing Enterprise 2.0: A Practical Guide To Creating Business Value Inside Organizations With Web Technologies.* Advanced Human Technologies.

DiMicco, J. M., Geyer, W., Dugan, C., Brownholtz, B., & Millen, D. R. (2009). People Sensemaking and Relationship Building on an Enterprise Social Networking. In *Proceedings of HICSS 2009.* IEEE.

Dixon, B. E. (2010). Towards E-Government 2.0: An Assessment of Where E-Government 2.0 is and Where It Is Headed. *Public Administration & Management, 15*(2), 418–454.

Doz, Y. L. (1996). The Evolution of Cooperation in Strategic Alliances: Initial Conditions or Learning Processes? *Strategic Management Journal, 17,* 55–83. doi:10.1002/smj.4250171006

EC. (2002). *eEurope 2005: An information society for all.* Communication from the Commission - COM(2002) 263 final. Retrieved June 22, 2013, from http://eur-lex.europa.eu/LexUriServ/LexUriServ.do?uri=COM:2002:0263:FIN:EN:PDF

EC. (2010). *EUROPE 2020 - A strategy for smart, sustainable and inclusive growth.* Communication from the Commission - COM(2010) 2020. Retrieved June 22, 2013, from http://ec.europa.eu/eu2020/pdf/COMPLET%20EN%20BARROSO%20%20%20007%20-%20Europe%202020%20-%20EN%20version.pdf

EC. (2012). *Single Market Act II - Together for new growth.* Communication from the Commission - COM(2012) 573. Retrieved June 22, 2013, from http://ec.europa.eu/internal_market/smact/docs/single-market-act2_en.pdf

EC. (2013). *Study on Analysis of the Needs for Cross-Border Services and Assessment of the organisational, Legal, Technical and Semantic Barriers.* European Commission. Retrieved June 22, 2013, from http://ec.europa.eu/digital-agenda/en/news/final-report-study-analysis-needs-cross-border-services-and-assessment-organisational-legal

Ehrlich, K., & Shami, N. S. (2010). Microblogging inside and outside the workplace. In *Proceedings of 4th International AAAI Conference on Weblogs and Social Media.* Washington, DC: AAAI.

EIF. (2010). *European Interoperability Framework (EIF) for European public services.* Retrieved June, 22, 2013 from http://eur-lex.europa.eu/LexUriServ/LexUriServ.do?uri=COM:2010:0744:FIN:EN:PDF

EPC. (2010). *Digital Single Market Newsletter.* Retrieved June 22, 2013, from http://www.epc.eu/dsm/1/Digital_Single_Market.pdf

Gompert, D. C., & Nerlich, U. (2002). *Shoulder to Shoulder - The Road to U.S. European Military Co-operability: A German American Analysis*. RAND Corporation. Retrieved June, 22, 2013, from http://www.rand.org/content/dam/rand/pubs/monograph_reports/2005/MR1575.pdf

Higuera, R. P., Gluch, D. P., Dorofee, A. J., Murphy, R. L., Walker, J. A., & Williams, R. C. (1994). *An Introduction to Team Risk Management*. (Version 1.0), Special Report CMU/SEI-94-SR-1, Carnegie Mellon University.

Jackson, A., Yates, J., & Orlikowski, W. (2007). Corporate Blogging: Building community through persistent digital talk. In *Proceedings of the 40th Hawaii International Conference on System Sciences*. IEEE.

Johnston, P. (2006). *21st Century Networked Local Government*. Cisco Systems. Retrieved June, 22, 2013 from http://www.cisco.com/web/about/ac79/docs/wp/21st_Century_Networked_Local_Government.pdf

Kaczorowski, W. (Ed.). (2004). *Connected Government*. London: Premium Publishing.

Kolari, P., Finin, T., Lyons, K., Yesha, Y., Perelgut, S., & Hawkins, J. (2007). On the structure, properties and utility of internal corporate blogs. In *Proceedings of the International Conference on Weblogs and Social Media ICWSM 2007*. ICWSM.

Kramer, R. M. (1999). Trust and Distrust in Organizations: Emerging Perspectives, Enduring Questions. *Annual Review of Psychology, 50*, 569–598. doi:10.1146/annurev.psych.50.1.569 PMID:15012464

McAfee, P. A. (2006). Enterprise 2.0: The Dawn of the Emergent Collaboration. *MIT Sloan Management Review, 47*(3), 21–28.

McAfee, P. A. (2009). *Enterprise 2.0: New Collaborative Tools for Your Organization's Toughest Challenges*. Cambridge, MA: Harvard Business Publishing.

Mergel, I., Schweik, C., & Fountain, J. (2009). *The Transformational Effect of Web 2.0 Technologies on Government*. Retrieved June, 22, 2013 from http://papers.ssrn.com/sol3/papers.cfm?abstract_id=1412796

Meyers, B. C. (2006). *Risk Management Considerations for Interoperable Acquisition, TECHNICAL NOTE, CMU/SEI-2006-TN-032*. Carnegie Mellon University.

Microsoft. (2011). *Connected Government in a Connected World*. Microsoft Corp. Retrieved June, 22, 2013 from http://www.microsoft.com/download/en/details.aspx?id=8295

Misuraca, G., Alfano, G., & Viscusi, G. (2011). Interoperability Challenges for ICT-enabled Governance: Towards a pan-European Conceptual Framework. *Journal of Theoretical and Applied Electronic Commerce Research, 6*(1), 95–111. doi:10.4067/S0718-18762011000100007

NUS Institute of Systems Science, National University of Singapore. (2010). Retrieved June 22, 2013, from http://unpan1.un.org/intradoc/groups/public/documents/unpan/unpan039390.pdf

Ollus, M. (2005). A Holistic Approach towards Collaborative Networked Organizations. In *Innovation and the Knowledge Economy: Issues, Applications, Case Studies*. IOS Press.

Ollus, M. (2007). Approaches and solutions supporting collaboration in networks. In *Proceedings of the International Cluster Conference*. Venice, Italy: Academic Press.

Osimo, D. (2008). *Web 2.0 in Government: Why and How?* JRC Scientific and Technical Reports, European Communities.

Pallab, S. (2010). *Understanding the Impact of Enterprise Architecture on Connected Government*. Academic Press.

Parameswaran, M., & Whinston, A. B. (2007). Social Computing: An Overview. *Communications of the Association for Information Systems, 19*, 762–780.

Pardo, T. A., Cresswell, M. A., Dawes, S. S., & Burke, G. B. (2004). Modeling the Social & Technical Processes of Interorganizational Information Integration. In *Proceedings of the 37th Hawaii International Conference on System Sciences.* IEEE.

Pardo, T. A., Gil-Garcia, R., & Burke, G. B. (2006). Building Response Capacity through Cross-boundary Information Sharing: The Critical Role of Trust. In *Exploiting the Knowledge Economy: Issues, Applications, Case Studies.* Amsterdam: IOS Press.

Pascu, C. (2008). *An Empirical Analysis of the Creation, Use and Adoption of Social Computing Applications.* JRC Scientific and Technical Reports. European Communities.

Raeth, P., Smolnik, S., Urbach, N., & Butler, B. (2009). Corporate Adoption of Web 2.0: Challenges, Success, and Impact. In *Proceedings of the pre-ICIS 2009 SIM Academic Workshop Enterprise and Industry Applications of Web 2.0.* Phoenix, AZ: SIM.

Richter, A., & Riemer, K. (2009). Corporate Social Networking Sites - Modes of Use and Appropriation through Co-Evolution. In *Proceedings of 20th Australasian Conference on Information Systems Corporate Social Networking Sites.* Melbourne, Australia: Academic Press.

Rodriguez, M., Steinbock, D., Watkins, J., Gershenson, C., Bollen, J., Grey, V., & Degraf, B. (2007). Smartocracy: Social Networks for Collective Decision Making. In *Proceeding of the 40th Annual Hawaii International Conference on System Sciences* (pp. 90-100). IEEE.

Sari, B., Schaffers, H., Kristensen, K., Loh, H., & Slagter, R. (2008). Collaborative knowledge workers: Web tools and workplace paradigms enabling enterprise collaboration 2.0. In *ECOSPACE IP-eProfessional Collaborative Workspace.* Dienstag.

Scholl, H. J., & Klischewski, R. (2007). E-Government Integration and Interoperability: Framing the Research Agenda. *International Journal of Public Administration, 30*(8), 889–920. doi:10.1080/01900690701402668

Stewart, K., Clarke, H., Goillau, P., Verrall, N., & Widdowson, W. (2004). Non-technical Interoperability in Multinational Forces. In *Proceedings of 9th International Command and Control Research and Technology Symposium.* Copenhagen, Denmark: Academic Press.

Tolk, A. (2003). Beyond Technical Interoperability - Introducing a Reference Model for Measures of Merit for Coalition Interoperability. In *Proceedings of the 8th International Command and Control Research and Technology Symposium (ICCRTS).* Washington, DC: ICCRTS.

Turban, E., Liang, T. P., & Wu, S. P. J. (2011). A Framework for Adopting Collaboration 2.0 Tools for Virtual Group Decision Making. *Group Decision and Negotiation, 20*, 137–154. doi:10.1007/s10726-010-9215-5

Ubaldi, B. (2011). The impact of the Economic and Financial crisis on e-Government in OECD Member Countries. *European Journal of ePractice, 11*, 5-18.

UNDESA. (2008). e-Government Survey 2008: From e-Government to Connected Governance. New York: United Nations Department of Economic and Social Affairs.

US-CREST. (2000). *Coalition Military Operations - The Way Ahead Through Co-operability.* U.S. Center for Research and Education on Strategy and Technology. Retrieved June 22, 2013, from http://www.uscrest.org/CMOfinalReport.pdf

Wagner, C. (2004). Wiki: A Technology For Conversational Knowledge Management And Group Collaboration. *Communications of the Association for Information Systems, 14*, 265–289.

Warr, A. W. (2008). Social software: fun and games, or business tools? *Journal of Information Science, 34*(4), 591–604. doi:10.1177/0165551508092259

Young, O. (2007). *Topic Overview: Web 2.0.* Forrester Research.

ADDITIONAL READING

AA.VV. (2004). *Current Perspectives on Interoperability. Technical Report, CMU/SEI-2004-TR-009 - ESC-TR-2004-009.* Carnegie Mellon Software Engineering Institute.

Castelnovo, W. (2008). From Cooperation to Cooperability, *Electronic Government, Proceedings of the 7th International Conference* (EGOV 2008). Lecture Notes in Computer Science, Vol. 5184, Springer Berlin/Heidelberg. pp. 352-363.

Castelnovo, W. (2011). Risk Management in a Cooperation Context. *Proceedings of The 11th European Conference on E-Government*, ECEG 2011.

de Kool, D., & van Wamelen, J. (2008). Web 2.0: A new basis for e-government? *Proceedings of the Third International Conference on Information and Communication Technologies: From Theory to Applications.* pp. 1-7.

Guijarro, L. (2007). Interoperability frameworks and enterprise architectures in E-Government initiatives in Europe and the United States. *Government Information Quarterly, 24*, 89–101. doi:10.1016/j.giq.2006.05.003

Husin, M.H., Deegan, G., & Evans, N. (2012). Social Twins Enterprise 2.0 and Government 2.0. *European Journal of ePractice, 17*, 51-67.

Janssen, M., Charalabidis, J., Kuk, G., & Cresswell, T. (2011). E-government Interoperability, Infrastructure and Architecture: State of-the-art and Challenges. *Journal of Theoretical and Applied Electronic Commerce Research, 6*(1), I–VIII. doi:10.4067/S0718-18762011000100001

Luna-Reyes, L., Zhang, J., Gil-Garcıa, R. J., & Cresswell, A. M. (2005). Information Systems Development as Emergent Socio-Technical Change: A Practice Approach. *European Journal of Information Systems, 14*, 93–105. doi:10.1057/palgrave.ejis.3000524

Mathiassen, L., & Sørensen, C. (2008). Towards A Theory of Organizational Information Services. *Journal of Information Technology, 23*(4), 313–329. doi:10.1057/jit.2008.10

Nilsson, A. (2008). Management of Technochange in an Interorganizational e-Government Project. *Proceedings of the 41st Hawaii International Conference on System Sciences.*

Porta, M., House, B., Buckley, L., & Blitz, A. (2008). *Value 2.0 - Eight new rules for creating and capturing value from innovative technologies.* IBM Institute for Business Value.

Punie, Y., Lusoli, W., Centeno, C., Misuraca, G., & Broster, D. (2009). *The Impact of Social Computing on the EU Information Society and Economy.* JRC Scientific and Technical Reports, European Communities.

Richter, A., & Koch, M. (2008). Functions of Social Networking Services. *The 8th International Conference on the Design of Cooperative Systems* (COOP '08).

Wattiez, A. (2011). *An Evaluation of the Critical Success Factors to Implement Enterprise 2.0 Solutions Within Belgian Organizations.* Henley – University of Reading.

KEY TERMS AND DEFINITIONS

Connected Government: The result of a transformational process that leads Government organizations to achieve a level of integration such that citizens and enterprises can interact with government as with a single entity rather than with a number of different public authorities.

Co-Operability: A form of non-technical interoperability aiming at the successful bridging of differences in doctrine, organization, concepts of operation, and culture so that different organizations can function together essentially as a single organization with no loss in effectiveness.

Cooperation Platform: A set of specific ICT tools and services that support inter-organizational cooperation.

Cross-Border/Sectorial/Administrative Services: Services delivered to citizens and businesses through inter-organizational cooperation and that can be freely accessed irrespective of the agency, department, administration or country that deliver them.

Government 2.0: The use of Web 2.0 services and tools within government organizations as well as their embracing of the collaborative philosophy of Web 2.0.

Interoperability: Ability of disparate and diverse organizations to interact towards mutually beneficial and agreed common goals, involving the sharing of information and knowledge between the organizations, through the business processes they support, by means of the exchange of data between their respective ICT systems.

Social Computing and Cooperation Services and Tools (SC&CS): ICT applications, based on the Web 2.0 paradigm, that enable people to interact efficiently with other people, as well as to share contents and data through the Web.

Chapter 11
Emerging and Traditional ICT as Critical Success Factors for Local Governments:
A Longitudinal Analysis

Enrique Claver-Cortes
Universidad de Alicante, Spain

Susana de Juana-Espinosa
Universidad de Alicante, Spain

Jorge Valdés-Conca
Universidad de Alicante, Spain

ABSTRACT

It was not long ago when Information and Communication Technologies (ICT) were not ubiquitous and Web 2.0 was the stuff of science fiction. However, these technologies are now here to stay, and local governments should learn how to make the most of them. In this chapter, the situation of emerging ICT in Spain in general and for Spanish e-government in particular is described. Next, the results of an empirical study based on a longitudinal quantitative survey are shown. The survey was carried out in 2005, before the advent of Web 2.0, and again in 2012. In the survey, the Chief Information Officers (CIO) of Spanish municipalities express their opinions on critical success factors that may enhance or hinder the effectiveness, connectivity, and transparency of their strategies for a connected government (c-government). The comparative findings reveal that political issues set off, then and now, local e-government success and failure, whereas ICT-based issues, once very important for these CIOs, have been downgraded in their minds. Therefore, the emergence of social media, mobile technologies, Web 2.0, and connected government has not had a truly significant role in the quest for e-government success on their own, but in combination with other factors. The chapter also discusses the related factors.

DOI: 10.4018/978-1-4666-6082-3.ch011

INTRODUCTION

Modern public administrations, especially local governments, need to keep up with the fast changes in demands and behaviors of the society they serve. The increased connectivity of citizens and businesses leads to new expectations in regards to the quality, transparency and efficiency of public services as well as access to public figures and institutions. Public administrations face the challenge of rebuilding their capacity to finance themselves, attracting and retaining a competent labor force, and engaging citizens in designing innovative solutions to address public issues. To do so, they require: a) rethinking local revenue sources, b) renegotiate labor relations and c) rebuild citizens' view of society (Warner, 2010). E-government strategies that help in connecting citizens, public employees and political boards, otherwise known as c-government or connected government, and those aiming to improve public performance, or e-administration, are two ways to address these challenges.

E-government refers to the provision of internal administration services to its external environment, which is related directly to the need for internal transparency of a public organization. The need to implement e-government policies has resulted in the adoption of many visions and strategic agendas. However, each vision is driven by its own unique set of social, political, and economic factors and requirements, which are known as critical success factors (CSF). These visions will be reflected in their use of Information and Communication Technologies (ICT), applications and mechanisms that governments employ to consider citizens as customers, and thus provide them with best price and quality services (Osborne & Gabler, 1992).

As part of ICT, social networks and other Web 2.0 technologies play a key role in the evolution of e-government, especially in the c-Government dimension which brings in a focus for citizens empowerment (Al-Rababah & Abu-Shanabad,

2010). However, it was not until 2005 that Tim O'Reilly coined the term Web 2.0, and therefore public administrations have not had the chance to employ these tools until recently. It is interesting to see if the advent of such technologies has made a significant impact on how public administrations perceive they should face the development and implementation of their e-government plans.

According to the extant literature, there are many factors that affect the performance of public administrations. These factors might help a public organization to achieve success in designing and implementing an e-government strategy, but they can also create difficulties. ICT is one of these factors, as well as the social aspect of e-services and how governmental and organizational decisions affect on the development of e-government strategies. Public administrations therefore must learn to recognize the effect and inter-effect of such factors in order to provide effective governance, increased transparency, effective processes and efficient services through the use of the Internet and ICT.

Both the European e-government Action Plan (2011-15) and the Malmö Ministerial Declaration on e-government, support the use of ICT in civic life. Furthermore, the Europe for Citizens Program (2007-2013) promotes initiatives that facilitate the active participation in the civic and democratic life of the European Union. In particular, Spain is, generally speaking, a fairly advanced country in the information society, notably in the area of e-government services and availability of broadband networks. Therefore, it is a well-developed country in terms of e-government, and its practitioners have a certain degree of knowledge of the topic.

The majority of the existing research on e-government, particularly local e-government, consists of the description of individual, limited initiatives, and avoiding theoretical frameworks that may provide them with a solid foundation (Becker et al., 2003). It is essential then to contrast what academics and practitioners have to say in the matter of local e-government, in order to bridge the gap between what has been written and what is done. With this framework in mind, the contribution of this chapter is twofold: Firstly, to discuss how the latest commonly used technologies and tools are being utilized by the Spanish government to enhance the effectiveness in the provision of e-services; and secondly, to discuss how these technologies and tools can be used by local governments to encourage citizens to effectively access e-services and fully participate in the affairs of the municipality.

The aim of this chapter is to reveal whether the emergence of the Web 2.0 and other emerging technologies have had a significant impact on the perception of ICT as an e-government CSF in local government. To do so, the first part of the chapter describes the situation of emerging ICT in Spain in general and for c-government in particular. Subsequently, it will illustrate empirically whether the emergence of mobile and Web 2.0 technologies, as well as other circumstances, have made an effect on ICT's role as a CSF for local c-government. Many other studies have also considered this topic, such as Agostino in Italy (2013) and Bonsón et al. (2012) in Spain, but they tend to offer a more participatory perspective instead of being efficiency oriented, whereas Claver-Cortes *et al.* (2013) only considers the barriers towards an efficient e-government, an analysis that we aim to complete with that of the balancing success factors.

The dynamic nature of this research provides a comparative analysis of the difference in the position of ICT as a positive and negative CSF before and after the advent of the current economic crisis. Moreover, since the technologies at the disposal of local councils in 2012, namely the Web 2.0 and social media, were not available in 2005, it will help local governments to find out the role emerging technologies play in their search for modernization. Understanding the implications of these changes may help other public administrations in formulating more efficient e-government communication strategies.

STATE OF THE ART

E-Government Critical Success Factors

The last decade has seen e-government implementation endeavors move from cataloguing basic government information to providing interactive e-services to citizens; this is, from e-administration to c-government. Given that the public sector is often described as bureaucratic, inefficient and less-technologically savvy (Kamal & Alsudairi, 2009), e-government may arrive as a necessary revolution, since it could offer many benefits that were not foreseen (Irani et al., 2008). Particularly, the emergence of the Web 2.0 and other connecting technologies has brought up a change in the way public administration relate to their stakeholders, bringing them closer. Social media has helped governments to become more transparent, collaborative and participatory (Mergel, 2013); especially local governments since they are closer to citizens (Cotterill & King, 2007; Agostino, 2013). However, there are some challenges that still need to be addressed by these organizations in their quest to improve the e-provision of public services (Bertot et al., 2011; Bonsón et al., 2012).

In order to map and understand the factors shaping e-government, an analytical research has been developed which attempts to identify the key variables, or CSF for e-government success, distinguishing between factors which act as drivers and those acting as barriers (Smith, Macintosh and Millard, 2011). CSF are defined by Boynton and Zmud (1984) as "those few things that must go well to ensure success." The role that each factor plays in e-government performance will depend on how public administrations approach them (Joia, 2004). Hence, we feel it is necessary to address the barriers and drivers in parallel since gaining access to a missing resource may have more negative effects than having it available but not using it.

Drivers for a Connected Government

First, a certain level of interest and commitment on the part of the political board regarding the project may provide the opportunity to rethink how government services are offered, enhancing transparency, openness, accountability, participation, and other benefits (Criado & Ramilo, 2003; Torres et al., 2005). In addition, adequate legal support may help governments to obtain funding, and even force them to deploy an e-government strategy, thus provoking major changes in policies and legislation, as well as institutional infrastructures (Yoon & Chae, 2009). Finally, good strategic planning takes into account the actual needs of the municipality. Since citizens expect better Government services, this would be a way to press for the adoption of more efficient public services (Tan et al., 2005; Torres et al., 2005).

There are some authors who support the introduction of private sector practices. This philosophy is known as New Public Management (NPM) (Hood, 1995; Pollitt, 2000). Two NPM elements relate directly to the success of e-government: human resources management policies and quality management programs. Regarding human resources management policies, public sector workers may respond positively to an initiative they perceive as contributing to the organizational mission. Human resources managers may thus encourage the deployment of a wide knowledge management program associated with the organizational goals and missions of the e-government strategy (Yao et al.., 2007).

The purpose of public quality management programs is to achieve excellence in the provision of on-line services (Dewhurst et al., 1999; Teicher et al., 2002). These quality management programs are usually based on the ISO 9000 standard, or on quality models such as the EFQM model, and may take the shape of public service charters or excellence awards. These programs have a greater effect when impelled from the inside of the public organization, since they act as a self-motivating force.

Finally, the role of ICT as purveyors of organizational, strategic and operational benefits should be noted, as Gil-Garcia & Pardo (2005) have pointed out. One of the most relevant facilitators is the design of a customer-oriented website (Janssen et al., 2006).This includes a usable interface design, self-explanatory and friendly, so that users maximize their efficiency in terms of time and money (Pieterson et al., 2005). Furthermore, the provision of an appropriate technical infrastructure and the redesign of local and interagency processes are other critical elements that will determine future success (Ebrahim & Irani, 2005).

Also, there are a few acknowledged advantages to the deployment of the Web 2.0 and social networks for e-government, such as (Arizmendi, 2012):

- Allowing citizens to actively participate in the production of web contents;
- Letting information and contents to not only be generated but also to be continuously assessed and evaluated;
- Disintermediating the connection between citizens and administration, preventing the interference of interested third parties (liberal professionals, businesses and/or political parties);
- And making communication through social networks faster and more effective, sometimes even more than the face-to-face channels.

To sum up, table 1 compiles a list of the most common drivers found in literature and the main literature supporting it.

Barriers to a Connected Government

Conversely, there are several hindering factors to the development of a successful e-government strategy. The actual endowment of technological, financial and human resources to a local e-government project will largely depend on their considered political value, so that their importance

Table 1. Literature review on e-government drivers

	Barret & Green (2001)	Caffrey (1998)	Criado & Ramilo (2003)	Duwes & Pardo (2002)	Dewhurst et al. (1999)	Ebrahim & Irani (2005)	Garson (2003)	Jannsen et al. (2006)	Melits-ki (2003)	Pieters on et al. (2005)	Scott et al. (2004)	Tan et al. (2005)	Teicher et al. (2002)	Torre s et al. (2005)	Yao et al. (2007)	Yoon & Chae (2009)
A capable council project manager			X	X			X									
A workflows and process improvement			X	X		X								X		
A well-designed information system						X	X									
A Customer-oriented portal							X			X						
Financial support	X	X	X	X												
Legislative support		X	X													X
Political awareness of the importance of IT projects			X	X								X		X		
Political commitment and leadership			X											X		
Qualified human resources	X				x		X		X		X				X	
Quality management programmes		X			X								X			
Support from higher institutions			X											X		
Strategic consideration of e-Government policies	X		X			X	X		X							
Usability of council's website							X	X		X						

as barriers is associated to political leadership, or lack thereof (Eyob, 2004). A good example of these political barriers would be the Spanish Administrative Procedure and Legal System Act, passed in 1958 and still in force. Indeed, some political leaders may be afraid of the changes that e-government entails, even of losing power, consequently becoming a real threat for its success (Ebrahim & Irani, 2005).

Other authors have found that public administrations need to change their cultural values and principles towards those of the information society instead of keeping to the traditional bureaucratic model (Riege & Lindsay, 2006). The concepts of innovation and bureaucracy seem to be almost mutually exclusive: bureaucracy does not seem to suffer innovation gladly. This barrier is in the mindset of public officials, as policy makers, and citizens, as end-users. Competition, which is the leading motivator for innovation in the private sector and in a free-market society, is perceived as less significant for old-style bureaucracies

(Vigoda-Gadot et al., 2005). As a consequence, changes in the internal workflow may be necessary to make the local government's administration run smoothly (Waisanen, 2002). Sundberg and Sandberg (2006) observe that the major handicaps in e-government implementation processes stem from the pervasive command and control traditional of bureaucratic structures. Therefore, they propose that this process of change should be continuous and strategic rather than have local governments embarking on huge radical projects.

Li (2003) and Norris et al.. (2001) agree in identifying the human factor-based barriers as the most relevant ones; as opposed to technology-based barriers that could be easily overcome with money. The shortage of ICT skills affects both the provision and the implementation of e-government services, from the supply and demand sides respectively and therefore should be tackled from both points of view. The shortage of skilled public workers has led many public sector organizations to offer ICT training courses for citizens and public sector employees, and even to recruit qualified personnel from the private sector (Criado, 2012).

Finally, Ebrahim and Irani (2005) consider that the most important barriers are rooted in the deployment of adequate ICT infrastructures for a user's experience of easy and reliable electronic access to governmental services. Overall, they agree with Bonham et al. (2001) on the fact that governments consider the lack of technical infrastructures as a significant barrier to the development of the public sector organizations' capability to provide online services and transactions. Among these barriers, the most outstanding are the high cost of ICT professionals and consultancies, and the cost of installation, operation and maintenance of e-government systems.

In addition, between the ICT–related barriers, one of the most frequently cited is the security and privacy of the transactions. Indeed, an e-government strategy might be considered successful only when all the stakeholders are comfortable using the applications and have the means to carry out electronic transactions with their council. There-

fore, investing in security matters is a necessary but not a sufficient condition for the success of the e-government strategy and the confidentiality of personal data and other sensitive information (Bonham et al., 2001; Ebrahim & Irani, 2005). Hassan et al. (2011) point out that there has not been found a way to overcome the barrier of data protection and privacy laws. Following this train of thought, Web 2.0 and social networks have been known to be easily used to disseminate false or biased information that could damage the image of the public organization behind it.

As before, table 2 summarizes the most common barriers found in literature and a selection of the main literature supporting them.

The Role of ICT in Spanish E-Government

Before describing the state of the art in regards to e-government in Spain, it is necessary to look at the broader picture of ICT. Indeed, ICT are the key to an efficient and effective development of

Table 2. Literature review on e-government barriers

	Bonham & al (2001)	Criado (2012)	Criado & Ramilo (2003)	Ebbers et al. (2008)	Ebrahim & Irani (2005)	EU (2007)	Eyob (2004)	Hassan et al. (2011)	ICMA (2002)	Janssen et al. (2006)	Li (2003)	Norris & al (2001)	Riege & Lindsay (2006)	Sundberg & Sandberg (2006)	Vassilakis et al. (2005)	Waisanen (2002)
Citizens' lack of demand for e-services						X			X						X	
Complexity of work processes and bureaucracy		X			X								X	X	X	X
Cost of IT infrastructure	X				X					X						
Digital illiteracy				X	X											
Disregard for e-Government advantages					X						X	X				
Employees' resistance to change		X	X		X										X	
Lack of political commitment						X			X							
Politicians' lack of knowledge about benefits						X	X		X							
Scarcity of financial resources						X	X									
Scarcity and unsuitability of IT				X	X	X	X			X					X	
Scarcity of qualified personnel		X					X				X	X				
Security, privacy and trust issues						X	X	X							X	
Unsuitable legal framework			X												X	

e-government for local, regional and state public organizations (Criado, 2012), especially in regards to c-government plans.

Therefore, it should be noted as the first important fact that Spain's expenditure on IT in 2011 was significantly lower than the European average. However, it was slightly higher in terms of communication technology. In addition, the contribution of the Information and Communications industry in Spain is quite low, since it does not quite reach 4% of the country's GDP. In any case, these data should be subjected to a more realistic analysis. Considering its small contribution to the country's GDP, it is striking how unusually high the investment activity is for ICT, and the relevance it has on the national innovation system. In fact, nearly 19% of investments in innovation taking place in the country belong to the ICT industry.

It is also noteworthy that the Spanish market has clearly opted in recent years towards promoting a mobile Internet-like model. In fact, Spain holds the second position in Europe's ranking of intelligent terminal penetration, which undoubtedly affects the rate of citizen accessibility to e-government services. In addition, 63% of Spanish citizens have a Smartphone, leading the ranking of the five biggest European economies. 84% of the Spanish households have an Internet connection, although only 67% are using an ASDL connection (Fundación Telefónica, 2012). However, not everybody seems to be so eager to use these ICT: Spanish 65+ elders show a below average level of integration in the European information society, not very high to begin with.

The nature of social networks and the Web 2.0 as promoters of open government and e-democracy cannot be ignored by public administrations either: 83% of Spanish citizens are registered as users of one or more social networks. According to the Spanish National Observatory for Telecommunications and Information Society (ONTSI, 2012), the most popular social media in 2012 were Facebook (15 million users), Tuenti (11 million users) and Twitter (3 million users), the latter being

the one experimenting a more dramatic increase in the realm of public administrations.

Particularly considering local administration levels, the social networks' level of diffusion in 2010 for the 75 largest municipalities within the EU (5 from each EU-15 original members) was around 30% (Bonsón et al., 2012). Specifically in Spain, in 2012, 54 out of the largest 62 municipalities (over 100,000 inhabitants) had employed at least one the principal social networks, and over 60% of them were present in three or more of them.

Taking a closer look at e-government in Spain, sophistication of online public services is above the EU average, although there is an uneven development and quality as well a lack of integration among the services offered by different administrations and departments. As a consequence of this prevalence, the EU chose a number of Spanish infrastructures and services as exemplary best practices in e-government within the EU, such as the @Firma platform for e-signature (Arizmendi, 2012). Spain is also the European leader in terms of public administration web accessibility (United Nations, 2012). As such, in 2012 the country received an award from United Nations for its full accessibility to electronic public services, since currently over 90% of procedures and services provided by the State are electronically accessible for the citizen (Arizmendi, 2012).

Moreover, it was already mentioned the importance of ICT expenditure in the Spanish national innovation system. In this sense, the most innovative e-government projects in Spain aim to contribute to increase interoperability between supra-state, state and regional public organizations, as well as to ease the migration to IPv6 systems. In fact, in 2011, 45% of Spanish municipalities (and 90% of the population) were connected through the SARA (*Security Auditor's Research Assistant*) network.

From a social point of view, it is noteworthy that Spain is one of the 14 countries (6 of which are European) where, besides promoting m-government (to facilitate public access, e-government plans cater to diversity, particularly that of vulnerable social groups: e.g. less-fortunate citizens, blind users, illiterate citizens, elderly and young people and women. It has been found that social networks offer a quite adequate platform for several of these groups to obtain information on public service provision (United Nations, 2012). As a consequence, the number of potential e-government users in Spain has grown steadily with Internet availability.

Narrowing the scope to local e-government, there are more than 8100 municipalities in Spain, ranging from less than 5000 inhabitants (75% of them) to over 4 million people (the two major cities). A research carried out in 2011 by the Orange Foundation (Fundación Orange, 2012) shows that about two thirds of them are undertaking first steps and they expect continuation and further dissemination of performance-oriented e-government strategies.

This research also found that the average availability of e-service provision in Spanish municipalities is 69%. This is a clear improvement over previous years since the same report in 2004 (Fundación Orange, 2005) did not quite reach 30% of the same services. Whereas census consultation and modification was the most common e-service in 2004, available in almost 90% of the municipalities, online tax-collection is one of the most popular e-services in 2012. On the other hand, there are a number of local e-services with low availability levels still in 2012, such as "Assistance to Disability," "Rehabilitation and Improvement Subsidies" or "Home Care" where the vast majority of municipalities are still in a low stage of development. Moreover, for the latter two services the development trend was negative between 2010 and 2011 (Fundación Orange, 2012).

Currently, only 70% of the downloaded e-forms (above EU average) were also sent back online to the administration (below EU average). This ratio is higher for firms with more than 50 employees (Fundación Orange, 2012), basically because of the need for an electronic signature for processing

forms and people's concerns regarding network security. Even more, many Spanish municipalities did not have a catalog of procedures still in 2012; while different websites can be found to provide the same service. Generally speaking, Municipal website contents are heterogeneous and few of them meet the international web accessibility standards W3C (Fundación Orange, 2012). These figures show Spanish citizens have to discover the full potential for e-government yet.

All in all, there has been a considerable decline in Spain's position in the United Nations e-government ranking. Whereas in 2011 the country ranked 9th in the world e-government development index by country and 3rd in the e-Participation index (United Nations, 2012), the following year Spain dropped 14 positions within said ranking to become 23rd (Fundación Orange, 2012). Still, if we were to consider only the top 25 emerging countries in e-government, Spain is in third place, behind Austria and Iceland, and even in first position if we limited the geographical area to Southern Europe (United Nations, 2012). Nevertheless, progress has been substantial since 2005 when Spain could be found in the 39th position worldwide and 26th in Europe in terms of e-government readiness (United Nations, 2005).

PAST AND PRESENT: AN EMPIRICAL ANALYSIS

To understand what has happened during this period, an empirical study based on a quantitative survey was carried out in 2005 and then in 2012, in which the Chief Information Officers (CIO) of Spanish municipalities' expressed their impressions on the impact of ICT, social and organizational CSF in order to enhance the effectiveness and transparency of their e-government strategies. Carrizales et al. (2006) and Heeks and Bailur (2007) have highlighted the need for longitudinal studies in the e-government field, in order to better understand the impact of the Internet on public entities.

This time gap was chosen because of four contextual factors:

- First, and most important: social networks have made their presence known in Spain during this period: Facebook in 2007, Twitter in 2009, Tuenti in 2006, LinkedIn in 2009, etc. Blogging in Spain started to become popular in 2006, and the first Facebook profiles for public administrations are from 2009. In light of the current global recession, Web 2.0 and social networks have become affordable resources, since they are mostly access-free, that provide public administration with innovative solutions in public service provision, communication, transparency and social image.
- Second, the general environment has changed notably from an expansion situation in 2005 to that of the current economic, social and political crisis.
- In addition, both general and local elections have been held over twice this time span, resulting in a political change at all levels, from a more conservative government to a more "left wing" position and back.
- And last, but not least, legislation has evolved greatly on this topic since Spanish local governments are currently under pressure to make services available via the Internet, because of legal requirements stat-

ed by the central government. Indeed, the National ICT strategic plan, implemented in 2007, comprises a series of specific programs focused on citizens, companies, the digital context, digital public services and local e-administration (European Union, 2010) to overcome this dislocation. A direct consequence of this plan is the Act for Citizens' Electronic Access to Public Services, also known as "Act 11/2007." Its purpose was to force Spain's local administrations to provide electronic access to their most important services by 2010. Therefore, the existing legal framework has forced a number of changes in the local administration arena; among them, a compulsion for developing a website as an interface for stakeholders to interact with their local councils, which didn't exist in 2005.

Table 3 illustrates a comparative evolution of Spain's e-government data between 2005 and 2012 (United Nations, 2005, United Nations 2012). The first column refers to the global e-government ranking where it can be seen that, as it was mentioned before, Spain escalated from position 39th in 2005 with an index of 0.5874 to 23rd in 2012 with a notably higher index: 0.7770.

Table 3 also shows at which stage of maturity the current on-line service provision is, in Spain, as per Layne and Lee's categorization (2001). The percentage of municipalities offering e-services has increased for every maturity stages from 2005 until now, especially the supply of higher

Table 3. E-Government in Spain: 2005 and 2012

	Global position		On-line service						e-Participation		ICT Infrastructure Index	Human Capital Index
	Ranking	Index	Stages %					Index	Ranking	Index		
			I	II	III	IV	Total					
2005	39	0.5847	66	42	0	4	37	0.3723	34	0.0794	0.3919	0.9760
2012	23	0.7770	92	67	71	58	66	0.7582	14	0.5000	0.6318	0.9409

[Source: Based on data from United Nations (2005) and United Nations (2012)]

level services (III- Transactional and IV-One stop shop), which were almost non-extant in 2005. The most popular e-services are those from stages I (Informative/Catalogue) and II (Interactive). Indeed, almost every municipality offers stage I online services (92%) in 2012, and 66% of the municipalities provide some kind of e–services in every maturity stage.

With regards to e-participation, there has also been a notable increase during this period. While in 2005 Spain was 34 in the e-participation ranking, it has climbed up to the 14th position in a seven-year span. The above mentioned development of local e-service provision in stages II, III, and IV is at the root of said increase, along with the emergence of social networks, Web 2.0 and m-government policies. All of these favor two way communication flows with citizens as well as public generation of contents and assessment of public policies. This phenomenon is supported by a considerable investment in ICT infrastructures, whose index has doubled from 2005 until 2012. However, if we take a look at the e-garticipation index for 2012, it shows that it is still at only 50%. This is motivated in part by a poor process interoperability and part by public budget constraints. In fact, in 2011 the budget for investment in modernization of public services e-government decreased to 2007 levels.

Finally, the e-government human capital index has barely changed in this period, over 90% in both years. This index is measured from the citizens' alphabetization degree and their education level, reflecting the population's ability to demand, evaluate and consider e-government activities. Therefore, a high human capital index indicates that Spanish citizens are ready to actively participate in online public services. This supports the dramatic rise in numbers of III- and IV-stage municipalities within the last seven years.

Methodology: Data Collection and Questionnaire Design

In 2005, an electronic questionnaire was sent to the Spanish municipalities that both had an official website and a population over 5000 inhabitants. Both the papers of Layne and Lee (2001) and Baum and Maio (2000) establish that the less developed stage of e-government is that of web presence; therefore, our first selection criterion is simply having a website. Even though this is not a instrument required for e-government policies, it is believed that it is a must because, although almost every city council had an e-mail address, not all of them had a website at the time, so that only a certain percentage have consciously taken a step forward towards their integration into the new economy. As for the second criterion, municipalities with population under 5000 usually depended on supramunicipal institutions (called "*Diputaciones*") for their IT management, therefore lacking autonomy in the adoption of new management practices in order to better satisfy citizens' needs (Hood, 1995; Pollitt, 2000).

The questionnaire was addressed to the Chief Information Officer (CIO) of the local council or, should this position not exist, to the person in charge of IT issues. In fact, employee's perceptions of e-government involve direct recognition of the strategic potential and its contribution to the success of the organization (Koh et al., 2006). Other researchers have also relied on the practical knowledge that CIOs possess regarding the implementation of e-government strategies, like Ward (2006), and their role as key players in transformation of public services has been highlighted by the SOCITM (UK Society of Information Technology Management) (SOCITM, 2006).

The same questionnaire was re-sent in the last semester of 2011 and first months of 2012 to the 165 councils that participated in the 2005 stage (see table 4). Surveys are an increasingly important source of information on the characteristics of public organizations (Walker & Enticott, 2004). This research method is especially helpful when the aim is to track the attributes of a large number of organizations over time. Survey data may, however, be problematic as representations of the "reality" of organizational change. In particular, if measures were taken from a single snapshot, then managerial recall of the organizational attributes in the past may be poor. Our survey method avoids such problems because we have data from both years, occupants of the same organizational roles were surveyed both years, and we asked the same questions in both periods (Casadesus & Karapetrovic, 2005; Ashworth et al., 2007). All respondents were guaranteed (and received) complete anonymity, so were able to express their actual views.

CIOs were presented with the two lists of factors shown in tables 1 and 2, on a Likert scale, and they were asked to value their impact on e-government success from 1 (none) to 5 (very strong), with regard to their particular municipalities. For the following steps, only those items which were valued over 3 in the pre-test were taken into account. Since a research instrument has content validity if researchers agree that the instrument is made up of a group of items covering the issues to be measured, i.e. that it represents a specific thematic universe, we consider our list of items suitable because they have been obtained from a review of the literature and a qualitative test of the questionnaire among experts on the e-government field.

A factorial analysis was deployed to ensure construct validity. Both factors were validated in 2005 and again in 2012 with Cronbach's alpha test for reliability (Cronbach, 1951), as seen in table 3; this is the most widely used reliability estimate in empirical research for measuring the internal consistency of multidimensional scales (Peterson, 1994).

Finally, criterion-related validity is measured through the analysis of the correlation coefficients of both factors with other measures, in such a way that it is useful to measure a present or future behavior (Conca, Llopis, & Tari, 2004). These two factors have a high criterion-related validity when, taken together, they are highly and positively related with a local council's e-government measure of performance. In this case, there is a significant correlation between both factors and a perceived increase in efficiency and transparency, both for 2005 and 2012.

Results and Discussion

Local E-Government Drivers

Table 5 shows a summary of the final items deemed as positive influences for local e-government strategies, their value (mean) and their ranking in 2012. The parameter "Item value" refers to the degree of effect that such a factor has on e-government success, according to the perceptions of the municipal CIOs (1: none at all; 5: very strong). It can be seen that not only the ranking has changed, but also the relative values of certain items have varied dramatically.

Table 4. Research universe and rate of response

Year	Municipalities	Responses	SEM	α test facilitators	α test barriers
2005	960	16 (16.7%)	7.78	0.84	0.8
2012	165	89 (54%)	10.6	0.732	0.825

Table 5. Ranking of the most relevant drivers

Drivers	Ranking 2012	Item value 2012 (SD)	Ranking 2005	Item value 2005 (SD)	Change
Public Human Capital	1 (↑)	4.45	2	4.15	+7.2%
Political leadership and commitment	2 (↓)	4.39	1	4.36	+0.7%
ICT infrastructure availability	3 (↑)	4.31	7	3.99	+8.0%
Availability of financial resources (from non-municipality sources)	3 (↑)	4.31	4	4.07	+5.9%
E-Government explicit strategic planning	5 (↑)	4.19	6	3.99	+5.0%
Website usability degree	6 (↓)	4.1	3	4.08	+0.5%
Acknowledgment of e-Government benefits by political statements	7 (↓)	4.06	5	4.03	+0.7%
Designation of a formal project manager	8 (=)	4.01	8	3.94	+1.8%
Quality management models application	9 (=)	3.7	9	3.61	+2.5%
Changes in decision making processes towards less bureaucracy	10 (=)	3.55	10	3.43	+3.5%

For every item, it is shown that mean values are higher in 2012 than in 2005, although some of them only slightly so. This evidences that CIOs are more conscious of the effects that these factors may have on their e-government strategies, especially of the role played by the political board and public employees and have recognized the role of ICT in their success, as posited by Bertot et al. (2012).

The top position in 2012 is that of Public Human Capital, which held a close second in 2005. This reflects that, as part of the literate human capital that Spain has as was seen in table 3, public employees who work in knowledge management and public policy formulation may improve their performance by using social networks since, according to McKinsey (2012), knowledge–based employees' performance increases up to 20-25% when using social networks.

In 2005 the third position went to a usable and friendly website, but this item has dropped down in the scale in 2012 considerably, presumably because it has become a somewhat cheap and known commodity. However, ICT infrastructures availability have been pushed up four places, being the item that has increased the most in absolute terms between 2005 and 2012 (8%), meaning that CIOs are aware of the boost that an intelligent expenditure in ICT may bring to their councils, which concurs with table 3. Connectivity, then, leans more on hardware developments than on software advances. This could be attributed to the advent of social media and the need to be in the Web 2.0, since they will be regardless of it being official or not (Agostino, 2013). Municipalities consider that they to control their presence emerging technologies in order for e-government to progress.

The lower positions are consistently held by New Public Management related elements: quality management programs and changes in workflow and decision-making processes. The former may owe its place to the scarcity of these programs among local councils, even 6 years after the first launch of this study; therefore, those CIOs that have not implemented any of these practices still do not have a full understanding of the effect of quality management strategies. The latter may respond to the pervasive red tape in Spanish public organizations, strictly ruled by administrative laws that clash more often than not with leaner workflows. This has hardly changed even with the political changes and the economic situation, although it might have to vary if Spain aims for a more efficient public sector. The power of emerging technologies may override this lack of acknowledgment, thus stripping it of its importance as a driver in 2012 since the political board has the means to receive feedback on their decisions in real time.

In sum, those public organizations that wish to formulate and implement a successful e-government strategy, must still exploit strengths based on human capital, political commitment, adequate infrastructures and a set of organizational behaviors consistent with the current legislative and economic situation. Political changes and a challenging environment have only reinforced the effects that these factors have on local e-government accomplishments.

Local E-Government Barriers

We proceed now to discuss the evolution of the CSF that act as barriers for a connected e-government. Table 6 follows the same layout as Table 5. Every item's mean values are considerably higher in 2012 than in 2005. Even more, six years ago these items presented quite low values when compared to those of the facilitators, but today they are within a similar range; pessimism is the norm now in Spanish municipalities. This means that CIOs are warier of the negative effects of these barriers, particularly the scarcity of financial resources and the resistance of employees to changes, which is not surprising considering the current economic and social environment.

The barriers to e-government that have experimented a greater increase are "Complexity of public work processes and bureaucracy" and "Employees' resistance to changes," rising their perceived value by 17.2% and 18.2% respectively. Starting with the former, it seems that local government managers may want to ponder the use and implementation of social networks and Web 2.0 to overcome this barrier, by means of sharing information between stakeholders (employees, citizens, businesses). It should be noted that the item "useful and friendly website" is not considered as important as a driver in 2012, mainly because it is the content what matters, rather than the channel supporting it.

As for the latter barrier, it can be overcome with appropriate human resource policies, such as training and change management procedures. This is related to the rise in one position of the need for qualified human resources in Spanish local governments. Actually, it is even stronger in Spain since most civil servants cannot be laid off easily, and therefore technophobes must be either relocated or left as a lost cause. Social networks and Web 2.0 technologies may help in sharing information and questions, as well as platforms for online training.

Table 6. Ranking of the most relevant barriers

Barriers	Ranking 2012	Item value 2012 (SD)	Ranking 2005	Item value 2005 (SD)	Change
Lack of political commitmment	1 (=)	4.33	1	4.12	5.1%
Scarcity of financial resources	2 (↑)	4.2	3	3.93	6.9%
Complexity of public work processes and bureaucracy	3 (↑)	4.02	6	3.43	17.2%
Scarcity of qualified human resources	3 (↑)	4.02	4	3.90	3.1%
Employees' resistance to changes	5 (↑)	3.97	8	3.36	18.2%
Politicians' lack of knowledge about e-Government's issues	5 (↓)	3.97	2	3.93	1.0%
Cost of ICT infrastructure availability	7 (↓)	3.94	5	3.62	8.8%
Decision making processes lie on politicians instead on IT professionals	8 (↓)	3.78	7	3.42	10.5%
Citizens do not demand e-services	9 (=)	3.62	9	3.33	8.7%
Unsuitable legislation	10 (=)	3.17	10	3.09	2.6%

The most important barrier is the same in the two surveys: the political commitment, or lack thereof in this case, which confirms the importance of the behavior of the political board in the final performance of e-government. Such hierarchical structures need the approval of top management for every stage in the work process, which may lead to service inefficiencies and, more importantly, denial of funds. The allocation of resources within local institutions belongs mostly with the political levels, although they are not the ones to put these modernization strategies into practice, which leads to many negative implications for the expected performance. A reason for this could be found in the Theory of Planned Behavior (Ajzen, 1991), which states that people's willingness to act is the best predictor for their behavior. There remains, however, a certain gap between the willingness to perform a specific behavior and to actually do so. Such a gap stems usually from the influence of contextual and situational factors, such as the availability of alternative behaviors and time. In this case, the most important factor would be that of availability of resources, specifically finance, and lack of political culture in Spain (Parrado, 2008).

Considering the rest of barriers there are a number of significant changes in the ranking of their perceived importance. Information on e-government is commonplace in Spanish municipalities today, so politicians are now e-government savvy; and public employees have the necessary skills, even if they are more resistant to change because of the uncertainty and mistrust that transpires all over their organizations. The cost of ICT infrastructures holds a 7th position, instead of a 5th, because ICTs are cheaper today, the difficulty residing in the allocation of the necessary budget for it.

Taking a look at the last ranked barrier, Spain's unsuitable legislation on e-government, it remains in the last position because of the rapid advances that Spain has been making in this area since 2001, which have been reinforced in 2007 and 2009.

In brief, the weaknesses of the municipalities' e-government strategies come essentially from political and bureaucratic aspects and the availability of monetary resources, instead of human resources as in 2005, as illustrated by table 6. It seems that political and situational changes are perceived to affect the municipalities' strategies, making them less attuned towards the achievement of a successful e-government performance.

To sum up, the results show that the most relevant CSF, both positively and negatively, are those related to political leadership and commitment, which have increased their perceived effect over the considered time gap. Within the public sector, management commitment is therefore critical. However, financial issues, ICT and human resources are viewed differently as drivers (increasing their relative importance) than as barriers: they are still important, but bureaucracy has climbed a few positions once the difficulties of implementing an e-government strategy have arisen. Following the works of Cotterill and King (2007), social networks do have some effect in e-government delivery, but they do not it on their own but in combination with other political factors, such as policy making and democracy enhancing.

It was stated that there were four issues that may have had an impact on the situation of local e-government in Spain. From last to first, the change in legislation, mentioned in the last position, has been dismissed by the surveyed CIOs since the perceived effect of this barrier has not suffered many alterations. Another cause for change that has not had a powerful effect is that of political swaps in local and state government. Politicians are assumed to be more ICT savvy and conscious of the usefulness of e-government, as is society, but practitioners see no significant differences in political will and leadership between 2005 and 2012. The second issue is that of the advent of the economic crisis in Spain. Indeed, the time for endless spending and strong investment is gone, and the lack of resources is a stronger barrier that it was seven years ago. It is also reflected in the higher position of ICT infrastructures between e-government drivers. As a matter of fact, the negative effects of this recessive environment can be somewhat lessened by the implementation of emerging technologies, the first listed issue. Their ubiquity, easiness of use and low cost may hold back the need for expensive and sophisticated hardware and software, because what is important for their success is connectivity (reach), contents (richness) and safety.

CONCLUSION AND PRACTICAL IMPLICATIONS

This paper contributes to the current body of knowledge on e-government and c-government by reflecting the effects of formulating an e-government strategy in times of growth, but implementing it in times of recession. Understanding the implications of this duality may help other local governments formulate more efficient strategies by revealing which are the most important CSFs and how they have changed over time. The advent of emerging technologies and the Web 2.0 is a cornerstone in the development of ICT as a CSF during this period and therefore the way local governments face their e-government challenges before and after this fact is a topic that must be addressed.

Both the literature review and the perceptions of the CIOs demonstrate that the driver and barrier factors are very similar, while the differences lie on the perceived specific influence of each factor. Indeed, the empirical analysis has found that Spanish local councils regard a number of requirements for the satisfactory implementation of e-government strategies, which are consistent through time in nature but not so much in perceived effect. Therefore, it has been proved that public managers need to consider these factors in both directions, because, depending on how they are deployed and the consideration they are given, something that was viewed as a handicap could be neutered or even developed into a positive

asset. That is the case of human resources, for instance: with the proper training and management a reluctant public employee could become a potential ICT leader, and so a negative effect is overridden by a much more positive influence.

The most relevant CSF, both positively and negatively, are those issues related to political leadership and commitment, which have increased their perceived effect over the considered period of time. The most outstanding change has been that of the perception of bureaucratic barriers, which were not so obvious at the formulation stage as six years prior. This is particularly true in the case of public sector organizations, where hierarchy rules over need in many cases. Social networks and emerging technologies help visualize and overcome these barriers, by letting communication among public employees and citizens flow easier and cheaper.

Public sector hierarchies tend to change as politicians do, which frequently results in the changing of management structures, responsibilities and roles. As a collateral effect, this has altered the CIOs' perceptions regarding the role that the availability of resources plays in the success of local e-government, and portrays both hardware and human capital as the most valuable assets and monetary resources as the most constricting, whereas soft ICT issues, once very important in the minds of the CIOs, have been relegated in importance, alongside with legal support.

As a result, ICT, and social media in particular, could and should be considered as significant means to help overcome these barriers and boost these drivers. However, local governments must keep in mind that these are mere tools, necessary but not sufficient, as was demonstrated by the fact that their presence improves the effects of the complexity of public work processes and bureaucracy, whereas it has no effect on the lack of political will. Its absence, then, might have a negative effect on the relationship between municipalities and their stakeholders, while its presence does not imply a positive effect by itself.

This research reveals a consistent emphasis on political issues as positive and negative triggers for local e-government success across time, whereas ICT-based issues, once very important for the CIOs, occupy now a second or third row seat. Emerging technologies have therefore become a necessary commodity for local e-government in Spain, but the content they provide as well as their control is still in the hands of the political establishment, which in turn will determine their final degree of success.

RESEARCH LIMITATIONS AND FUTURE DIRECTIONS

Several limitations should be noted. First of all, this study focuses on one group of stakeholders, CIOs (public employees with a high ICT proficiency level), and one type of municipalities (medium-large). Future research might want to consider this topic from other complementary perspectives. Particularly, if we were to consider politicians, then it would be interesting to uproot why they are so skeptical of social media and e-government.

Also, there is a potential sampling bias due to the fact that the research was carried out on a random sampling of a number of pre-selected subjects. Besides, the measures used here were perceptual rather than objective. An opportunity for research would be to revise this classification employing other taxonomies or statistical methodologies. Nevertheless, the results shown in this paper are reasonable as they come from the perspective of the practitioners themselves.

Finally, considering that this paper reflects the first stages of the analysis, the interpretation of these relationships between these CSF and the presence of Web 2.0 technologies, social networks

and other e-government tools might benefit greatly from a deeper quantitative analysis. A linear regression and a structural equations model will be adequate methodologies to explore this avenue of thought.

REFERENCES

Agostino, D. (2013). Using social media to engage citizens: a study of Italian municipalities. *Public Relations Review*, *39*, 232–234. doi:10.1016/j.pubrev.2013.02.009

Ajzen, I. (1991). The theory of planned behavior. *Organizational Behavior and Human Decision Processes*, *50*(2), 179–211. doi:10.1016/0749-5978(91)90020-T

Al-Rababah, B., & Abu-Shanab, E. (2010). E-Government and gender digital divide: The case of Jordan. *International Journal of Electronic Business Management*, *8*(1), 1–6.

Arizmendi. (2012). Redes Administrativas. *Redes Sociales*. Retrieved on May 20, 2013, from http://www.gobiernolocal.org/docs/publicaciones/RDGL_18_19_baja.pdf

Ashworth, R., Boyne, G., & Delbridge, R. (2007). Escape from the Iron Cage? Organizational Change and Isomorphic Pressures in the public sector. *Journal of Public Administration: Research and Theory*, *19*, 165–187. doi:10.1093/jopart/mum038

Barret, K., & Green, R. (2001). *Powering Up. How Public Managers Can Take Control of Information Technology*. Washington, DC: CQ Press.

Baum, C. H., & Di Maio, A. (2000). *Gartner's Four Phases of E-Government Model*. Retrieved on June 6, 2013, from http://www.gartner.com/DisplayDocument?id=317292

Becker, J., Algermissen, L., & Niehaves, B. (2003). *Implementing e-Government Strategies: A procedural model for process oriented e-Government projects*. Paper presented at the meeting of the International Business Information Management Conference (IBIMA). Cairo, Egypt.

Bertot, J. C., Jaeger, P. T., & Hansen, D. (2012). The impact of polices on government social media usage: Issues, challenges and recommendations. *Government Information Quarterly, 29*, 30–40. doi:10.1016/j.giq.2011.04.004

Bonham, G., Seifert, J., & Thorson, S. (2001). The transformational potential of e-Government: the role of political leadership. In *Proceedings of the 4th Pan European International Relations Conference*. Kent, UK: Academic Press.

Bonsón, E., Torres, L., Royo, S., & Flores, F. (2012). Local e-Government 2.0: Social media and corporate transparency in municipalities. *Government Information Quarterly, 29*(2), 123–132. doi:10.1016/j.giq.2011.10.001

Boynton, A. C., & Zmud, R. W. (1984, Summer). An Assessment of Critical Success Factors. *Sloan Management Review*, 17–27.

Caffrey, L. (1998). *Information Sharing between and within Governments: A Study Group Report*. London: Commonwealth Secretariat.

Carrizales, T., Holzer, M., Kim, S. T., & Kim, C. G. (2006). Digital governance worldwide: A longitudinal assessment of municipal web sites. *International Journal of Electronic Government Research, 2*(4), 1–23. doi:10.4018/jegr.2006100101

Casadesus, M., & Karapetrovic, S. (2005). Has ISO 9000 lost some of its lustre? A longitudinal impact study. *International Journal of Operations & Production Management, 25*(6), 580–596. doi:10.1108/01443570510599737

Claver-Cortes, E., de Juana-Espinosa, S., & Valdes-Conca, J. (2013). The effect of emerging technologies on local e-Government barriers in Spain: a longitudinal perspective. In *Proceedings of the 22nd IBIMA Conference*. Rome, Italy: IBIMA.

Conca, F. J., Llopis, J., & Tarí, J. J. (2004). Development of a Measure to Assess Quality Management in Certified Firms. *European Journal of Operational Research, 156*, 683–697. doi:10.1016/S0377-2217(03)00145-0

Cotterill, S., & King, S. (2007). Public sector Partnerships to deliver local e-government: a social network study. In *Proceedings of EGOV 2007* (LNCS), (vol. 4656, pp. 240-251). Berlin: Springer-Verlag.

Criado, J. I., & Ramilo, M. C. (2003). E-Government in practice: An analysis of web site orientation to the citizens in Spanish municipalities. *International Journal of Public Sector Management, 126*(3), 191–218. doi:10.1108/09513550310472320

Criado Grando, J. I. (2012). *Redes Sociales y Open Government*. Retrieved on June 2, 2013, from http://www.gobiernolocal.org/docs/publicaciones/RDGL_18_19_baja.pdf

Cronbach, L. J. (1951). Coefficient alpha and the internal structure of tests. *Psychometrika, 16*(3), 297–334. doi:10.1007/BF02310555

Dawes, S. S., & Pardo, T. (2002). Building collaborative digital government systems. In W. J. McIver, & A. K. Elmagarmid (Eds.), *Advances in Digital Government: Technology, Human Factors, and Policy* (pp. 259–273). Norwell, MA: Kluwer Academic Publishers. doi:10.1007/0-306-47374-7_16

Dewhurst, F., Martínez Lorente, A. R., & Dale, B. G. (1999). TQM in public organisations: an examination of the issues. *Managing Service Quality*, 9(4), 265–273. doi:10.1108/09604529910273210

Ebbers, W. E., Pieterson, W. J., & Noordman, H. N. (2008). Electronic government: rethinking channel management strategies. *Government Information Quarterly*, 25(2), 181–201. doi:10.1016/j.giq.2006.11.003

Ebrahim, Z., & Irani, Z. (2005). E-Government adoption: architecture and barriers. *Business Process Management Journal*, 11(5), 589–611. doi:10.1108/14637150510619902

European Union. (2007). Retrieved on October 17, 2010, from www.egovbarriers.org/downloads/deliverables/solutions_report/Solutions_for_eGovernment.pdf

Eyob, E. (2004). E-Government: breaking the frontiers of inefficiencies in the public sector. *Electronic Government*, 1(1), 107–114. doi:10.1504/EG.2004.004140

Fundación Orange. (2005). *Informe Anual sobre el Desarrollo de la Sociedad de la Información en España 2005*. Retrieved on June 3, 2013, from http://fundacionorange.es/areas/25_publicaciones/EESPA_A2005_COMPLETO_V3.pdf

Fundación Orange. (2012). *Informe Anual sobre el Desarrollo de la Sociedad de la Información en España 2012*. Retrieved on June 3, 2013, from http://www.proyectosfundacionorange.es/docs/eE2012.pdf

Fundación Telefónica. (2012). *La sociedad de la información es España 2011*. Retrieved on April 15, 2013, from http://e-libros.fundacion.telefonica.com/sie11/aplicacion_sie/ParteA/pdf/SiE_2011.pdf

Garson, G. D. (2003). Toward an information technology research agenda for public administration. In G. D. Garson (Ed.), *Public Information Technology: Policy and Management Issues* (pp. 331–357). Hershey, PA: Idea Group Publishing.

Hassan, H. S., Shehab, E., & Peppard, J. (2011). Recent advances in e-service in the public sector: state-of-the-art and future trends. *Business Process Management Journal*, 17(3), 526–545. doi:10.1108/14637151111136405

Heeks, R., & Bailur, S. (2007). Analyzing e-Government research: Perspectives, philosophies, theories, methods, and practice. *Government Information Quarterly*, 24(2), 243–265. doi:10.1016/j.giq.2006.06.005

Hood, C. (1995). The New Public Management in the 1980s: variations on a theme. *Accounting, Organizations and Society*, 20(2/3), 93–109. doi:10.1016/0361-3682(93)E0001-W

International City/County Management Association and Public Technology (ICMA). (2002). *Digital government survey*. Washington, DC: Author.

Irani, Z., Love, P. E. D., & Jones, S. (2008). Learning lessons from evaluating eGovernment: reflective case experiences that support transformational government. *The Journal of Strategic Information Systems*, 17(2), 155–164. doi:10.1016/j.jsis.2007.12.005

Janssen, M., Gortmaker, J., & Wagenaar, R. W. (2006). Web service orchestration in public administration: challenges, roles and growth stages. *Information Systems Management*, 23(2), 44–55. doi:10.1201/1078.10580530/45925.23.2.20060301/92673.6

Joia, L. A. (2004). Developing Government-to-Government enterprises in Brazil: a heuristic model drawn from multiple case studies. *International Journal of Information Management*, 24(2), 147–166. doi:10.1016/j.ijinfomgt.2003.12.013

Kamal, M. M., & Alsudairi, M. (2009). Investigating the importance of factors influencing integration technologies adoption in local government authorities. *Transforming Government: People. Process and Policy*, *3*(3), 302–331.

Koh, C. E., Prybutok, V. R., Ryan, S., & Ibragimova, B. (2006). The importance of strategic readiness in an emerging e-Government environment. *Business Process Management Journal*, *12*(1), 22–33. doi:10.1108/14637150610643733

Layne, K., & Lee, J. (2001). Developing fully functional E-Government: A four stage model. *Government Information Quarterly*, *18*(2), 122–136. doi:10.1016/S0740-624X(01)00066-1

Li, F. (2003). Implementing E-Government strategy in Scotland: current situation and emerging issues. *Journal of Electronic Commerce in Organizations*, *1*(2), 44–65. doi:10.4018/jeco.2003040104

McKinsey Global Institute. (2012). *The Social Economy: Unlocking value and productivity through social technologies*. Retrieved on May 26, 2013, from http://www.mckinsey.com/insights/high_tech_telecoms_internet/the_social_economy

Melitski, J. (2003). Capacity and e-government performance: An analysis based on early adopters of Internet technologies in New Jersey. *Public Performance and ManagementReview*, *26*(4), 376–390. doi:10.1177/1530957603026004005

Mergel, I. (2013). Social media adoption and resulting tactics in the US federal government. *Government Information Quarterly*, *30*, 123–130. doi:10.1016/j.giq.2012.12.004

Norris, D. F., Fletcher, P. D., & Holden, S. H. (2001). *Is your local government plugged in? Highlights of the 2000 electronic government survey. Paper prepared for the International City/County Management Association (ICMA) and Public Technology, Inc*. Baltimore, MD: PTI.

O'Reilly, T. (2005). *What is Web 2.0: design patterns and business models for the next generation of software*. Retrieved on Sept 30th, 2013, from http://oreilly.com/web2/archive/what-is-web-20.html

ONTSI. (2011). *Informe Anual de los Contenidos Digitales en España 2011*. Retrieved on May 20, 2013, from www.ontsi.red.es/ontsi/es/estudios-informes/informe-anual-de-los-contenidos-digitales-en-espa%C3%B1a-2011

Osborne, D., & Gabler, T. A. (1992). *Reinventing Government: How the Entrepreneurial Spirit is Transforming the Public Sector*. Reading, MA: Addison-Wesley.

Parrado, S. (2008). Failed policies but institutional innovation through layering and diffusion in Spanish central administration. *International Journal of Public Sector Management*, *21*(2), 230–252. doi:10.1108/09513550810855672

Peterson, R. A. (1994). A Meta-analysis of Cronbach's Coefficient Alpha. *The Journal of Consumer Research*, *21*(2), 381–391. doi:10.1086/209405

Pieterson, W., Ebbers, W., & Van Dijk, J. (2005). The opportunities and barriers of user profiling in the public sector. In *Proceedings of the 2005 EGOV*. Copenhagen, Denmark: EGOV.

Pollitt, C. (2000). Is the emperor in his underwear? An analysis of the impacts of public management reform. *Public Management*, *2*(2), 181–199. doi:10.1080/14616670041229

Riege, A., & Lindsay, N. (2006). Knowledge management in the public sector: stakeholder partnerships in the public policy development. *Journal of Knowledge Management*, *10*(3), 24–39. doi:10.1108/13673270610670830

Scott, M., Golden, W., & Hughes, M. (2004). Implementation strategies for e-Government: a stakeholder analysis approach. In *Proceedings of ECIS: The European Information Systems Profession in the Global Networking Environment.* Turku, Finland: ECIS.

Smith, S., Macintosh, A., & Millard, J. (2009). *Major factors shaping the development of eParticipation.* Retrieved on May 15, 2013, from http://is-lab.uom.gr/eP/index2.php?option=com_docman

SOCITM. (2006). *Modern public services: a role for change: The CIO as agent of transformation.* Retrieved on January 19, 2010, from http://www.socitm.gov.uk/socitm/Services/Socitm+Insight/News/Role+of+CIO.htm

Sundberg, H. K., & Sandberg, K. W. (2006). Towards e-Government: a survey of problems in organizational processes. *Business Process Management Journal, 21*(2), 146–161.

Tan, C., Pan, S., & Lim, E. (2005). Managing Stakeholder Interests in e-Government Implementation: lessons learned from a Singapore e-Government Project. *Journal of Global Information Management, 13*(1), 31–53. doi:10.4018/jgim.2005010102

Technosite. (2011). *Monitoring eAccessibility in Europe 2011: Annual Report.* Retrieved on June 6, 2013, from http://www.eaccessibility-monitoring.eu/researchResult.aspx

Teicher, J., Hughes, O., & Dow, N. (2002). E-Government: a new route to public sector quality. *Managing Service Quality, 12*(6), 384–393. doi:10.1108/09604520210451867

Torres, L., Pina, V., & Royo, S. (2005). E-Government and the transformation of public administrations in EU countries – beyond NPM or just a second wave of reforms? *Online Information Review, 29*(5), 531–553. doi:10.1108/14684520510628918

United Nations. (2005). *Global e-Government Readiness Report 2005.* Retrieved on June 4, 2013, from http://unpan1.un.org/intradoc/groups/public/documents/un/unpan021888.pdf

United Nations. (2010). *United Nations e-Government Survey 2010.* Retrieved on November 25, 2011, from http://unpan1.un.org/intradoc/groups/public/documents/un/unpan038851.pdf

United Nations. (2012). *United Nations e-Government Survey 2012.* Retrieved on June 4, 2013, from http://unpan1.un.org/intradoc/groups/public/documents/un/unpan048065.pdf

Vassilakis, C., Lepouras, G., Fraser, J., Haston, S., & Georgiadis, P. (2005). Barriers to electronic service development. *E-service Journal, 4*(1), 41–63. doi:10.2979/ESJ.2005.4.1.41

Vigoda-Gadot, E., Shoham, A., Schwabsky, N., & Ruvio, A. (2005). Public Sector Innovation for the Managerial and the Post-Managerial Era: Promises and Realities in a Globalizing Public Administration. *International Public Management Journal, 8*(1), 57–81.

Waisanen, B. (2002). The future of E-Government: Technology fuelled management toll. *Public Management, 84*(5), 6–9.

Walker, R. M., & Enticott, G. (2004). Using Multiple Informants in Public Administration: Revisiting the Managerial Values and Actions Debate. *Journal of Public Administration: Research and Theory, 14*(3), 417–434. doi:10.1093/jopart/muh022

Ward, M. (2006). Information Systems Technologies: a public-private sector comparison. *Journal of Computer Information Systems, 46*(3), 50–56.

Warner, M. (2010). The future of local government: twenty-first-century challenges. *Public Administration Review, 70,* S145–S147. doi:10.1111/j.1540-6210.2010.02257.x

Yao, L. J., Kam, T. H. Y., & Chan, S. H. (2007). Knowledge sharing in Asian public administration sector: the case of Hong Kong. *Journal of Enterprise Information Management, 20*(1), 51–69. doi:10.1108/17410390710717138

Yoon, J., & Chae, M. (2009). Varying criticality of key critical success factors national e-strategy along the status of economic development of nations. *Government Information Quarterly, 26*, 5–34. doi:10.1016/j.giq.2008.08.006

ADDITIONAL READING

Al-Sebie, M., & Irani, Z. (2003). E-Government: Defining Boundaries and Lifecycle maturity. In Proceedings of the 3rd European Conference on e-Government, Ireland: Trinity College of Dublin, 2003, pp. 19-29.

Almarabeh, T., & Abuali, A. (2010). A General Framework for E-Government: Definition, Maturity Challenges, Opportunities, and Success. *European Journal of Scientific Research, 39*(1), 29–42.

Andersen, K. V. (2006). E-Government: five key challenges for management. Electronic. *Journal of E-Government, 1*(4), 1–18.

Angelopoulos, S., Kitsios, F., & Papadopoulos, T. (2010). New service development in e-Government: identifying critical success factors. Transforming Government: People. *Process and Policy, 4*(1), 95–118.

Archer, N. P. (2005). An overview of the change management process in eGovernment. *International Journal of Electronic Business, 3*(1), 68–87. doi:10.1504/IJEB.2005.006389

Eynon, R., & Margetts, H. (2007), Organisational Solutions for Overcoming Barriers to eGovernment. European Journal of ePractice, 1, 1-13.

Heeks, R., & Bailur, S. (2007). Analyzing e-Government research: perspectives, philosophies, theories, methods and practice. *Government Information Quarterly, 2*(24), 243–265. doi:10.1016/j.giq.2006.06.005

Janssen, M., Chun, S. A., & Gil-Garcia, J. R. (2009). Building the next genereation of digital information infrastructures. *Government Information Quarterly, 2*(26), 233–237. doi:10.1016/j.giq.2008.12.006

McAdam, R., & Walker, T. (2004). Evaluating the best value framework in UK local government services. *Public Administration and Development, 24*(3), 183–196. doi:10.1002/pad.275

Schwester, R. (2009). Examining the barriers to e-government adoption. Electronic. *Journal of E-Government, 1*(7), 118–127.

Sharif, A. M., & Irani, Z. (2010). The logistics of information management within an eGovernment context. *Journal of Enterprise Information Management, 23*(6), 694–723. doi:10.1108/17410391011088600

Stamoulis, D., Gouscos, D., Georgiadis, P., & Martakos, D. (2001). Revisiting public information management for effective e-Government services. *Information Management & Computer Security, 9*(4), 146–153. doi:10.1108/09685220110400327

KEY TERMS AND DEFINITIONS

Critical Success Factor: An element that is necessary for an organization or project to achieve its mission. It is a critical factor or activity required for ensuring the success of a company or an organization. The term was initially used in the world of data analysis, and business analysis. CSF are those aspects of a strategy that must be achieved to successfully meet objectives and, if possible, to secure competitive advantage. Therefore, they represent those managerial or enterprise area, that must be given special and continual attention to bring about high performance. They include issues vital to an organization's current operating activities and to its future success.

E-Government: (Short for electronic government, also known as e-gov, Internet government, digital government, online government, or connected government) refers to the provision of internal administration services to its external environment, which is related directly with the need for internal transparency of the public organization. This term describes the use of technologies to facilitate the operation of government and the diffusion of government information and services. Simply stated, e-Government is a chance for public organizations to detect and fulfill the needs of their stakeholders more efficiently, and a means to promote a conscience of goodness regarding the development of Information Technologies.

E-Participation: It refers to the use of information and communication technologies to broaden and deepen political participation by enabling citizens to connect with one another and with their elected representatives. e-Participation can be seen as part of e-Democracy.

M-Government: Also known as Mobile government, it is the extension of e-Government to mobile platforms, as well as the strategic use of government services and applications which are only possible using smartphones, laptop computers, personal digital assistants (PDAs) and wireless internet infrastructure.

New Public Management: It is a new perspective which evolved in early 1990s, which seeks to merge the core values of business administration in the domain of public administration. It signifies the adoption of major principles of business administration in the domain of public sector, aiming at efficiency, effectiveness and economy in performance of public sector by employing modern managerial tools such as performance appraisal, cost cutting, functional autonomy, financial incentives, output targets, innovation, market orientations, responsiveness and accountability.

Open Government: It is the governing doctrine which holds that citizens have the right to access the documents and proceedings of the government to allow for effective public oversight. The narrow definition of Open Government consists of transparency, participation and collaboration of the state towards third actors like the economy or the citizenship. In its broadest construction it opposes reason of state and other considerations, which have tended to legitimize extensive state secrecy. Actually, it isn't necessarily grounded in technology although it relates to it.

Social Media: It is a social structure made up of individuals (or organizations) called "nodes," which are tied (connected) by one or more specific types of interdependency, such as friendship, kinship, common interest, financial exchange, dislike, sexual relationships, or relationships of beliefs, knowledge or prestige. It is applied to websites for social networking, electronic communication and microblogging, through which users create online communities to share information, ideas, personal messages, and other multimedia content.

Web 2.0: It is the term given to describe a second generation of the World Wide Web that is focused on the ability for people to collaborate and share information online. It refers to those websites that use technology beyond the static pages of earlier web sites. A Web 2.0 site may allow users to interact and collaborate with each other in a social media dialogue as creators of user-generated content in a virtual community, in contrast to websites where people are limited to the passive viewing of content. Over time Web 2.0 has been used more as a marketing term than a computer-science-based term. Blogs, wikis, and web services are all seen as components of Web 2.0.

Chapter 12
Data Modelling of a Multifaceted Electronic Card–Based Secure E–Governance System

Abhishek Roy
The University of Burdwan, India

Sunil Karforma
The University of Burdwan, India

ABSTRACT

In the current climate of global economic decline, the developing countries are facing severe challenges in maintaining an efficient administration within an affordable budget. If this economic slowdown continues, there will be serious difficulties which will hamper the socio-economic development of the entire region. To respond to the situation, the governments must reduce budget expenses and still maintain efficiency and openness. To do so, the administration must deploy ICT-based mechanisms to fulfil the desired objectives. In this chapter, the authors present the development of a multifaceted electronic card-based secured e-governance mechanism to attempt to redress the inherent issues and explore new dimensions of interdisciplinary research. The proposed system will also act as the all-purpose electronic identity of the Citizen and hopefully replace the existing identity instruments such as Voter Card, Permanent Account Number Card, Driving License, Ration Card, Below Poverty-Line Card, Employment Card, Health Card, Insurance Card, etc. Moreover, this electronic instrument will also enable Citizen to perform financial transactions. Clearly, the authentication procedure of the proposed mechanism must also exist otherwise the intruders will be able to breach the system and execute their ill intentions. To ensure appropriateness of security features of the mechanism, the authors have also implemented a user authentication technique using object-oriented modelling of RSA digital signature algorithm for a Government-Citizen (G2C) type of e-governance. For better management of such a huge amount of sensitive information, the authors also discuss data modelling techniques used during user authentication of the proposed model.

DOI: 10.4018/978-1-4666-6082-3.ch012

INTRODUCTION

In the current global economic meltdown, the governments of the developing countries like India are facing severe challenges in maintaining an efficient administration throughout its jurisdiction within an affordable budget. In our country the government is facing real hardship in maintaining this costly administration especially when the Indian Rupee (INR) is facing new highest of inflation on regular basis due to natural outcome of this global recession phenomenon. If this economic meltdown process continues in future, ongoing developmental projects may stop abruptly and the socio-economic development of the entire region may reach to a state of permanent coma. There can be a long drawn debate among the economists about the cause and cure of such economic issues; however no one will disagree that, to respond to this global recession the government must reduce their budget expenses thereby maintaining its efficiency. To do so, the administration must deploy Information and Communication Technology based mechanism which will fulfil the desired objectives. This generates the necessity of electronic mode of governance called e-governance, which will help to reduce the budget expenses of the government with the application of advanced technologies. The desired electronic mechanism must provide efficient administration in all respect, especially in terms of money, time, target output, etc, by bridging the gap during intra and inter departmental communications among various governmental agencies. Realizing the necessity of e-governance in Indian perspective (Roy & Karforma, 2011, 2012; Hoda, et al., 2012; Roy, Banik, Karforma & Pattanayak, 2010; Sur, Roy, & Banik, 2010), we have proposed a Citizen centric multifaceted smart card based e-governance mechanism (Roy & Karforma, 2013a, 2013b) which will solve our problems and explore new dimensions of interdisciplinary research works. This electronic instrument will replace all the existing identity instruments of the Citizen like,

Voter Card, Permanent Account Number (PAN) Card, Driving License, Ration Card, BPL (Below Poverty Line) Card, Employment Card, Health Card, Insurance Card, Tour & Travels Record, etc and will perform their functions under one head. As a value addition, using this electronic instrument the Citizen can perform various financial transactions, which they usually perform with the help of various debit cards, credit cards, etc. as the service server of the proposed mechanism will connect to the server of multiple banks of the Citizen. Since, it is an ICT (Information and Communication Technology) based electronic instrument, it will perform the intra and inter departmental communications among various governmental agencies instantly. Moreover, the Government will be able to communicate with the citizen in a timely manner at a nominal cost and citizen will also be able to access the resultant benefits very easily. Hence, it is clear that this proposed instrument will need to manage huge amounts of sensitive information about the Government and the Citizen, which may be tampered if an intruder successfully breaches the security parameters of this electronic mechanism and escalates its privileges in subsequent phases. To guard against this weakness, focus must be placed over the proper identification of the original users by deploying standard verification (Sarkar & Roy, 2012, 2013; Roy, et al., 2013; Roy, Sarkar, et al., 2012) procedures. Hence, to prevent this menace, we have already implemented the user authentication mechanism using an object oriented approach of RSA digital signature algorithm (Roy & Karforma, 2012a, 2012b; Roy, Banik, & Karforma, 2011) in a G2C type of e-governance transactions.

The organisation of this chapter is as follows. The importance of e-governance in terms of efficient administration is discussed in the next section. Then, the vital security parameters of standard e-governance mechanism are discussed. Here, we have presented a literature review on the relevant topic for better understanding of the scenarios. Our proposed e-governance model is explained in the following section. Here, we have discussed the initial Data Flow Diagram (DFD), Entity Relationship Diagram (ERD) and Table Structure of our proposed e-governance mechanism which have scope for further modifications during practical implementation. The chapter also explores the future scope of research work for our proposed model. Towards the end of the chapter, conclusions are presented and references are provided.

NECESSITY OF E-GOVERNANCE

Before proceeding further, first we should point out the sectors where the existing conventional form of governance is suffering from huge operational overheads. The main points may be summarized and listed as follows.

- Offline mode of governance.
- Huge man power requirement for deployment of administration.
- Huge operational expenses for maintenance of present form of administration.
- Lack of intra and inter-departmental communications among the governmental agencies.
- Execution delay of government orders.
- Inadequate mechanism to trace the development process of the government orders from start to finish in terms of time, money and desired output.
- Deep rooted corruption within and outside the administration.
- The habit of forced adjustments gradually implanted within the Citizenry due to under-estimation of their grievances and problems.
- Lack of surveillance over the supreme authorities of the society regarding their economic and correlated developments.

- Cross border infiltration and insurgent activities hampering the national development in regular basis.
- Inadequate mechanism for early detection and subsequent neutralization of the cross border infiltrations.
- Firm execution of rule of law equally at each and every level of society, at least for the sovereignty and prosperity of the state.

Apart from the above mentioned aspects, there may also be other aspects which will be considered gradually as the requirement generates. However, even from the above stated points, it is clear that conventional paper-based form of governance is an obsolete one in the present days of complex human lives. As mentioned in "Vulgaria" by Horman (1440), the headmaster of Winchester and Eton, the solution of this problem can also be sought from the problem itself. This generates the necessity of advanced technology based electronic form of governance called as e-governance. E-Governance or Electronic Governance is a technology-based electronic mechanism which is used by the governmental agencies to deploy administration under its jurisdiction. The primary participants of this system are:

- The Government.
- The Citizen.
- The Business.

Since, it is an electronic mechanism, the entire message transmission between the Government and the Citizen are done through the publicly available communication medium i.e. Internet. That means that all the classified information that was supposed to be communicated in offline mode will now be communicated online. Thus, Internet becomes the primary mode of data transmission in this procedure. This becomes the main point of concern for the information scientist, as they need to depend solely on such a communication channel which is very much susceptible to security infringement attempts made by the eavesdropper. Before proceeding further, the information scientist must design the blue print of an electronic mechanism which should be free from all security pit falls. To make the entire mechanism fully secure and efficient, the security scientists need to concentrate on the following aspects:

- The e-governance mechanism must be made highly secure.
- Only the authenticated user of the system will be allowed to access the e-governance mechanism. Use of various standard cryptographic check post like digital signature algorithms, complex Elliptic Curve Cryptosystems (ECC) like Elliptic Curve Digital Signature Algorithm (ECDSA), Elliptic Curve Diffie Hellman (ECDH), Elliptic Curve Discrete Logarithm Problem (ECDLP), Elliptic Curve ElGamal (ECElGamal), ECRSA, biometric applications, etc, should be used to achieve this objective.
- The e-governance mechanism should be capable enough for early detection of the unauthenticated access and their subsequent neutralization so that the system can be saved from being compromised. Combination of cryptographic algorithms, stenographic algorithms along with the biometric standards will be beneficial in this scenario.
- The authenticated user of the e-governance mechanism should get the information about any unauthenticated attempts made by the intruder, so that security features can be restored at immediate basis. In this case the use of – Internet based technologies, wireless mobile communication technologies, cloud computing, etc, will explore new horizon of research works.

This means that the information scientists must ensure that the above mentioned features are present irrespective of the operational procedure of the system. Now, we should consider the various operational procedures over which the proposed electronic mechanism must perform efficiently. The operational procedure mainly varies based on the role of the participants of the e-governance mechanism. The standard operational models of the e-governance mechanism may be categorized as follows:

- Government to Citizen (G2C) e-governance model.
- Citizen to Government (C2G) e-governance model.
- Government to Government (G2G) e-governance model.
- Government to Business (G2B) e-governance model.
- Business to Government (B2G) e-governance model.
- Business to Business (B2B) e-governance model.

Further explanation of these categories is now provided in the following sub-sections.

Government to Citizen (G2C) E-Governance Model

In this type of e-governance model, the electronic transaction is initiated by the Government which is further continued by both the participants i.e. the Government and the Citizen. Online submission of various form, taxes, bills, etc may be considered as the examples of G2C type of e-governance model. During this transaction both the parties must ensure that they are dealing with the authenticated user only. The schematic diagram of the process is demonstrated in Figure 1.

Figure 1 shows that the Government and the Citizen are connected to each other through Internet for the electronic transactions. The output of the electronic transactions is stored for future reference in the database connected with the Government.

Citizen to Government (C2G) E-Governance Model

In this model, the electronic transaction is initiated by the Citizen which is further continued by both the participants i.e. the Government and the Citizen. Accessing various governmental facilities like, health facility, employment facility, ration facility, insurance facility, etc may be considered as the examples of C2G type of e-governance model. During this transaction both the parties must ensure that they are dealing with the authenticated user only. The schematic diagram of the process is demonstrated in Figure 2.

Figure 2 shows that the Citizen and the Government are connected to each other through Internet. The outcomes of the electronic transactions are stored for future reference in the database connected with the Government.

Government to Government (G2G) E-Governance Model

In this type of e-governance model, the electronic transaction is executed between two different governmental agencies which may be either inter-departmental or intra-departmental in nature. The online official communications between various governmental agencies may be shown as the example of this type of electronic transactions. During this transaction both the parties must ensure that they are dealing with the authenticated user only. The schematic diagram of the process is demonstrated in Figure 3.

Figure 3 shows that the Government G1 and the Government G2 are connected to each other through Internet. The outcome of the electronic transactions carried out between these two participants is stored for future reference in the databases connected with both the governments.

Figure 1. Government to Citizen (G2C) type of E-Governance model [Participant – 1: Government, Participant – 2: Citizen]

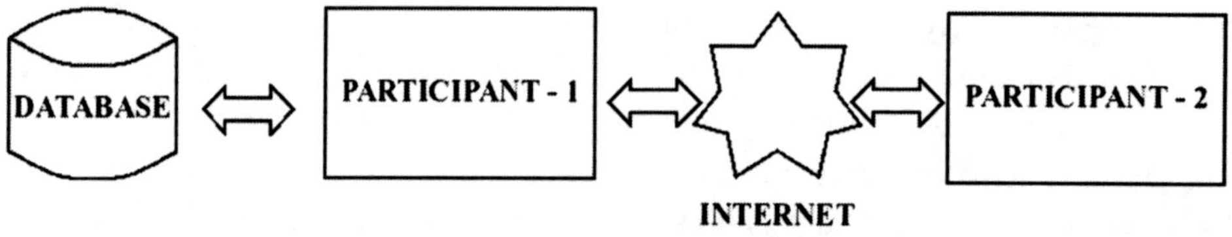

Figure 2. Citizen to Government (C2G) type of E-Governance model [Participant – 1: Citizen, Participant – 2: Government]

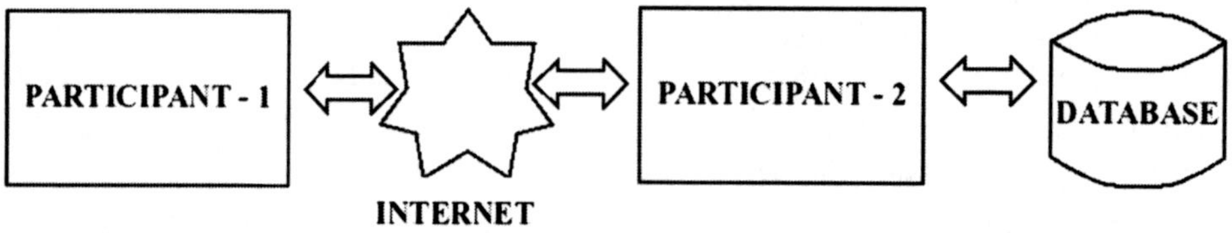

Figure 3. Government to Government (G2G) type of E-Governance model [Participant – 1: Government G1, Participant – 2: Government G2, where G1 and G2 are different]

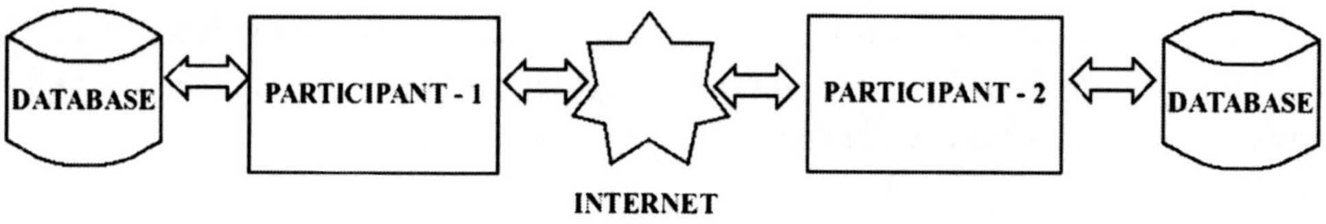

Government to Business (G2B) E-Governance Model

In this model, the electronic transaction is executed between the Government and the Business entity. The online official communications like intimation of audit operations, various government verifications, etc, carried out between the governmental agencies and the business entities may be shown as the example of this type of electronic transactions. During this transaction both the parties must ensure that they are dealing with the authenticated user only. The schematic diagram of the process is demonstrated in Figure 4.

Figure 4 shows that the Government and the Business are connected to each other through Internet. The outcome of the electronic transactions carried out between these two participants are stored for future reference in the databases connected with both the parties.

Business to Government (B2G) E-Governance Model

In this type of e-governance model, the electronic transaction is executed between the Business entity and the Government. The online official commu-

nications like submission of audit reports, assets declarations, payment of business related taxes, etc, carried out between the business entities and the governmental agencies may be considered as the example of this type of electronic transactions. During this transaction both the parties must ensure that they are dealing with the authenticated user only. The schematic diagram of the process is demonstrated in Figure 5.

Figure 5 shows that the Business and the Government are connected to each other through Internet. The outcome of the electronic transactions carried out between these two participants are stored for future reference in the databases connected with both the parties.

Business to Business (B2B) E-Governance Model

In this type of e-governance model, the electronic transaction is executed between two different business entities. The various online business transactions carried out between two different business houses may be cited as the example of these types of transactions. During this transaction both the parties must ensure that they are dealing

Figure 4. Government to Business (G2B) type of E-Governance model [Participant – 1: Government, Participant – 2: Business]

Figure 5. Business to Government (B2G) type of E-Governance model [Participant – 1: Business, Participant – 2: Government]

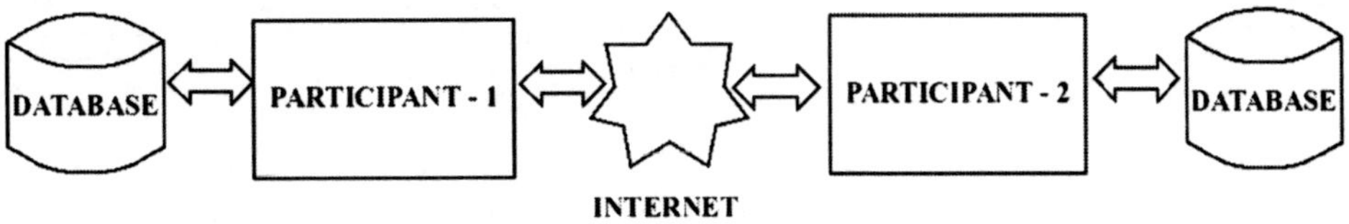

285

with the authenticated user only. The schematic diagram of the process is demonstrated in Figure 6.

Figure 6 shows that the business houses are connected to each other through Internet. The outcome of the electronic transactions carried out between these two participants is stored for future reference at the databases connected with both the parties. Since in this case both parties involved are from the business world, this model may be cited as the example of Electronic Commerce or E-Commerce also. Even the electronic communication between the Citizen and the Business via Citizen to Business (C2B) and Business to Citizen (C2B) models are considered as appropriate examples of E-Commerce transactions.

Summary

The above models show that, although e-governance and E-Commerce are completely two different domains, there are inter-connections between these two domains. These inter-connections among various domains increase as we intend to develop these electronic transactions in more realistic manner. The involvement of mobile technologies with the electronic governance and electronic commerce will explore the new horizons of research in Mobile Governance or M-Governance and Mobile Commerce or M-Commerce. Similarly the involvement of Cloud Computing with electronic governance and electronic commerce will explore the new domain of Cloud Governance or C-Governance and Cloud Commerce or C-Commerce. Whatever may be the technical approach, it is no more a hidden fact that electronic

mode of governance i.e. e-governance will help to provide an efficient administration within an affordable budget.

SECURITY PARAMETERS OF E-GOVERNANCE

Once, we understand the necessity of electronic governance, we should concentrate on the security parameters of the mechanism. The primary objective of the security system may be noted as Privacy, Integrity, Non-Repudiation, Authentication (PINA) as explained below:

- **Privacy:** The privacy of the information must be maintained from the intruders.
- **Integrity:** The information must reach from source to destination in unaltered manner.
- **Non-Repudiation:** The sender of the information must not be able to deny the act of sending the information to the receiver.
- **Authentication:** The electronic mechanism must be accessible only to its authenticated users. This will help to defend the system from being compromised by the intruders.

To increase the privacy of the information, secure encryption techniques like RSA, Elliptic Curve Cryptography (ECC), etc may be used. Use of secure communication channels will make the task easier in this case. To provide the integrity, non-repudiation and authentication feature of the

Figure 6. Business to Business (B2B) type of E-Governance model [Participant – 1: Business B1, Participant – 2: Business B2, where B1 and B2 are different]

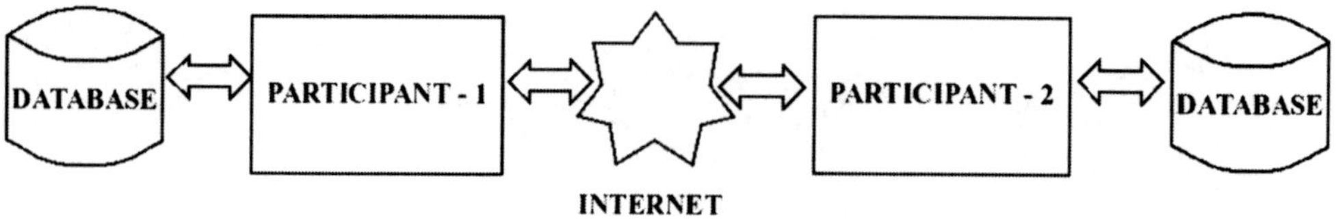

electronic mechanism, the use of complex digital signature algorithms like RSA digital signature algorithm, Elliptic Curve Digital Signature Algorithm (ECDSA), biometric technique, etc will be beneficial. Ultimately, the identity of the user must be verified through a proper verification procedure so that only the authenticated user are able to access the mechanism, else the entire electronic mechanism will become very much susceptible to security threats. The following literature survey will reveal the scenario in a broader sense.

Researchers are using the Information and Communication Technology (ICT) based applications for better decision making at various level of governance (Potekar & Giragaonkar, 2004) so that e-services may be delivered at the door steps of the rural populace (Raja, Ramana, et al., 2012). It is better to deliver these E-Services to the Citizen through multiple local language based customized interfaces (Mittal, Kumar, et al., 2004) using various advanced technologies like cloud computing (Charalabidis, Koussouris & Ramfos, 2011), etc. Even, the Mobile Data Management (MDM) with the help of low-end phones (Manjunath, Muthanna, et al., 2012) may be used for this purpose. While doing so, the privacy of the Electronic Identity (eID) (Nimalaprakasan, Ramanan, et al., 2009) in context of data access and data management needs to be maintained with utmost priority. This task becomes more challenging in the context of developing countries which are suffering from extreme shortage of resources in various allied sectors (Rahman & Rajon, 2011; Rajon & Rahman, 2012). In such a situation, the entire process must be designed in a very systematic manner focusing over the main principles of efficient public management during implementation of e-governance (Alguliev & Yusifov, 2009). Then, the existing mechanisms in the international platform (Alguliev, Imamverdiyev, & Yusifov, 2011) should be studied carefully for efficient practice (Janowski, Estevez, & Ojo, 2012)

and crisis management (Da-li, Hua-lin, & Changnan, 2008) of the society using e-governance. The conceptual models like Software Architectural Design Model for e-governance Systems (SADM-EGS) (Murthy & Kumar, 2003), etc using software engineering tools like UML Use Cases (McKenzie, Crompton, & Wallis, 2008) and others should be documented in proper manner for better digital record maintenance of the system (Pappel, Pappel & Saarmann, 2012). Furthermore, password based user authentication protocols (Zhai & He, 2010) need to be deployed for defending various type of fatal attacks like dictionary attack, replay attack (Bredin, Miguel, et al., 2006) that will be mounted by the intruders over the mechanism for fulfilment of ill-intentions. The application of biometric and graphical parameters (Wang, Li, et al., 2011) in this case will increase the strength of the security system. As the e-governance mechanism have to handle large number of keys, researchers are also working to reduce the key management load (Lee, Li, & Kim, 2009) with the help of identity based secret signature scheme. Even, scientist are trying to enhance the security parameters of the similar type of applications by performing comparison of standard cryptographic algorithms such as ECC (Mohammed, Emarah, & El-Shennawy, 2001) and RSA (Savari, Montazerolzohour, & Thiam, 2012). Encouraged by the ongoing research works on multi-application cryptographic smart card based e-governance mechanism (Liu, Wei, Siu, Chan, & Choi, 2001; Naccache & M'Raihi, 1996), we have proposed our own electronic mechanism which will find solution to our problems while providing maximum data security (Mohandes, 2010). Using our proposed electronic instrument whose detailed description is provided further, Citizen will be able to perform various electronic transactions including financial transactions like payment of electricity bill (Simpson, 1996), telephone bills, water tax, land tax, etc.

PROPOSED E-GOVERNANCE MODEL

From the above mentioned literature survey, it is clear that the researchers throughout the globe are working for implementation of security features for developing a secure electronic mechanism in customized manner for finding solutions to their specific problems. For this purpose, they are using various cryptographic algorithms like digital signature algorithms, Elliptic Curve Cryptosystem (ECC), biometric techniques, etc. over smart card applications. Even the combination of standard cryptographic applications with biometric techniques, have been figured out in this connection. In this scenario, we have only two options open to us – to watch and appreciate the success achieved by others, or start building our own electronic mechanism which will cater to the needs of our problems. The first option is very easy and effortless, whereas the second one is very tough and laborious. In spite of various hurdles, with the urge of nation building we have opted for the second option.

We consider designing a Citizen centric electronic instrument which will help the governmental agencies to deploy Electronic Governance or e-governance throughout its jurisdiction. Like other standard electronic mechanism, our proposed instrument will use Information and Communication Technology (ICT), i.e. Internet for data transmission process. The use of public communication medium i.e. Internet makes the security of this entire mechanism very much susceptible to the attempts of the hackers made initially to breach the system and escalate its privileges in subsequent phases. To avoid this menace, we have tried to design the e-governance mechanism in a realistic manner such that, it will successfully prevent the access of the unauthorised users. At least it should be capable enough to detect the unauthorised access at the initial stage so that precautionary measures may be initiated and the integrity of the e-governance mechanism may

be restored with utmost priority. Hence, we have proposed an e-governance mechanism using a multifaceted electronic card called Multipurpose Electronic Card (MEC) which will defend the sovereignty of the state along with its socio-economic developments in various ways. This electronic card will authenticate the identity of its owner during the e-governance transactions using standard cryptographic check post like digital signature algorithms like RSA, Elliptic Curve Digital Signature Algorithm (ECDSA), etc. We will use the complex Elliptic Curve Cryptosystem (ECC) like Elliptic Curve Diffie Hellman (ECDH), Elliptic Curve Discrete Logarithm Problem (ECDLP), Elliptic Curve ElGamal (ECElGamal), ECRSA, etc for improvement of the security and authentication parameters of the proposed mechanism. Even we are also considering the incorporation of biometric authentication techniques within our proposed mechanism in subsequent phases for execution of secure e-governance transactions. These electronic transactions may be precisely mentioned as payment of telephone bill, income tax, house rent, road tax, land tax, electric bill, insurance and other similar types of premium, etc, accessing various facilities like, voting facility, ration facility, health facility, employment facility, etc. This electronic card will ensure that these e-governance transactions are applicable only for the original inhabitants of the nation, which means, it will detect the identity of the intruders by preventing them from the access of these services. Hence, this multifaceted smart card will act as the ultimate alternative for Birth Certificate, Ration Card, Educational Record, PAN (Permanent Account Number), Employment Card, Voter Card, Driving License, Insurance Card, Bank Record (i.e. Debit Card, etc), Tour & Travels Record, Death Certificate, etc. We have successfully implemented the user authentication using digital signature algorithms in the context of object oriented software engineering approach using our proposed multifaceted electronic card in Citizen to Government (C2G) and Government to Citizen (G2C) type of e-governance models. The conceptual model of the proposed system is depicted in Figure 7.

Figure 7 shows that the Citizen will use the proposed instrument to interact with the Government electronically and the Government will respond in the similar fashion. In practical scenario, the Citizen will use this smart card through the service outlets which will be installed throughout the state. They will operate just like ATM counters with necessary alterations for operations. These service outlets will contain web based client interaction system installed within themselves, which will act as the interface between the Citizen and the Government during the entire transaction process. These outlets will be connected to the e-governance mechanism using wireless mobile communication system. The user of this system will be allowed to access the e-governance mechanism only after proper identity verification through standard authentication procedures. This model will explore new dimensions of research work in the domain of wireless mobile communication system and web based applications. However, in this chapter, we will concentrate only on the modelling of the user authentication system of our proposed mechanism.

The block diagram of the proposed Multipurpose Electronic Card (MEC) is shown in Figure 8. This is the initial block diagram of the proposed smart card which will have options for further enhancements. Since this mechanism will be performing electronic transactions between the Government and the Citizen, it is obvious that

huge data load will be mounted over the proposed system which needs to be managed in a secured manner. To do so, the user authentication technique must be implemented with utmost efficiency. Hence, the Figure 9 shows the secured database connectivity of the proposed e-governance model.

Figure 9 depicts the following aspects of the proposed e-governance mechanism:

- The Citizen will communicate with the government with the Multipurpose Electronic Card (MEC).
- This electronic instrument will further interact with the e-governance mechanism.
- The electronic mechanism will contain the following segments for handling the sensitive data in a secured manner:
- **Firewall:** Firewall will prevent the entry of spam ware, malware and other malicious elements within the mechanism. This will act as the strong guard which will perform the database management very securely.
- **Web Server:** After passing through the firewall of the system, data will enter the Web Server of the electronic mechanism. Here the alteration of the data will be performed as per the requirement of the transaction. Here various web based technologies will be installed with the web server for the smooth progress of the operation.
- **Application Server:** In the next phase the data will perform the necessary interaction with the Application Server of the system.

Figure 7. Conceptual diagram of the proposed smart card based E-Governance model

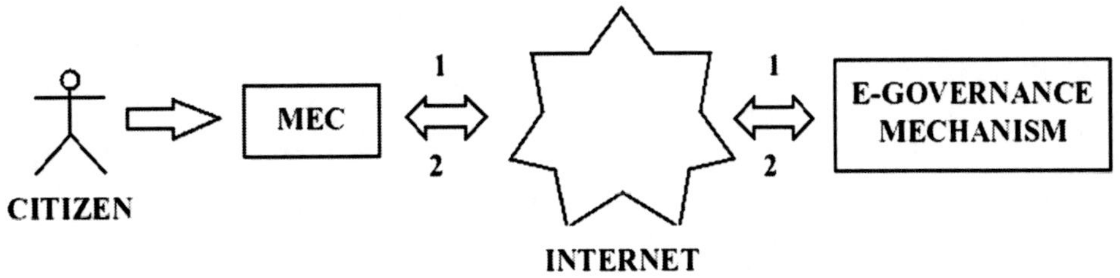

- **Database:** Finally the resultant data set of these entire operations will be stored in the database for further reference. Hence, it is very clear that huge load will be mounted on this database as it will be containing all the vital information of the citizenry so that governmental agencies can access them instantly. We have shown this database using a single database structure just to get a simple view of the complex structure. However in practical scenario to avoid the database server failure the distributed database management system will be exercised in this phase.

- The Service System of the proposed e-governance mechanism comprises of the Firewall, Web Server and the Application Server.

- The Server side comprises of the Web Server, Application Server and the Database storage.

- The entire data transmission through this mechanism will proceed in bidirectional manner which includes request from the Citizen and its corresponding reply from the Government.

From Figure 9, it is clear that the user authentication system must be efficiently installed within the e-governance mechanism so that it can be saved from being compromised by the intruders. Thus, we have implemented the user authentication scheme by the object oriented modelling of RSA digital signature algorithm using G2C type of e-governance in our proposed application. The following Data Flow Diagram (DFD), Entity Relationship Diagram (ERD) and Table

Figure 8. Block diagram of the Multipurpose Electronic Card (MEC)

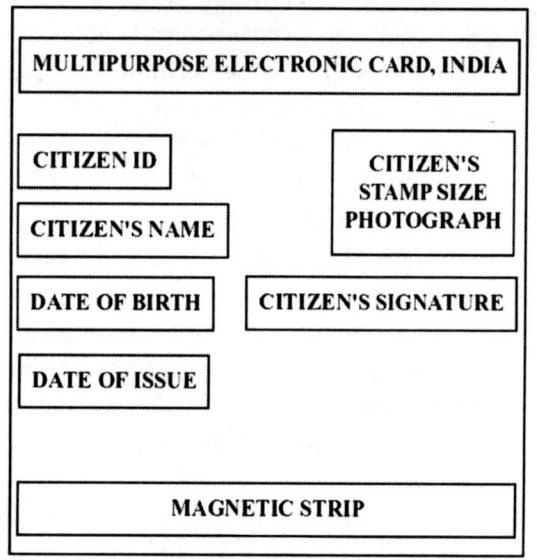

Figure 9. Secured database connectivity of the proposed E-Governance mechanism

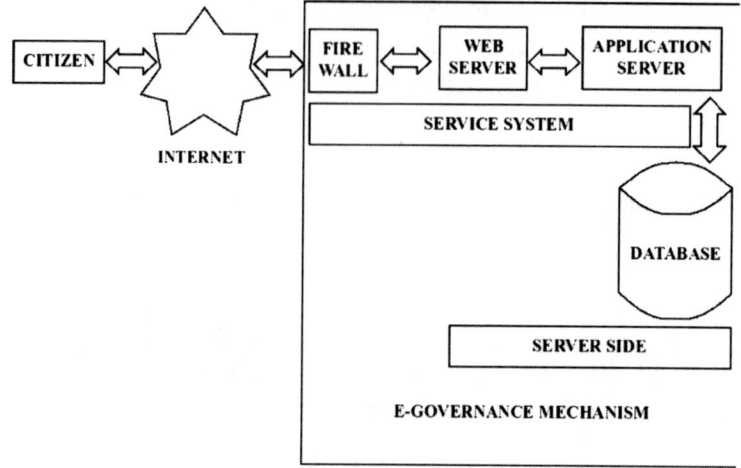

structure explains the data modelling of the said implementation.

Data Flow Diagram (DFD) of the Proposed Model

This segment will demonstrate the Context Level DFD, Level 1 DFD and Level 2 DFD. Figure 10 shows the Context Level DFD of our proposed model. Figures 11 and 12 present the Levels 1 and 2 DFD diagrams of our proposed model. Finally, the logistics management for grant of facilities by the Government to the Citizen are demonstrated in Figure 13.

Context Level DFD

Figure 10 shows the following sequence of actions:

1. The Citizen enters the information which also includes the prime numbers based on which the Government will calculate the secret key.
2. Government performs the following operations to finally generate the signature:
    a. Select the publicly available public key.

b. Calculate the secret key based on public key and the user inputs.
c. Perform hashing of the message by selecting publicly available hash functions like MD5, etc.
d. Generate the signature based on above calculations.
3. The message and the signature is delivered to the Citizen. Here the Citizen verifies the arrived hash message with the expected hash message.
4. The Government grant benefits and facilities to the Citizen.
5. The Citizen access the benefits granted by the Government.

Levels 1 and 2 DFD

The sequential description of Level 1 DFD which is shown in Figure 11 is as follows:

1. Enrolment of information in Multipurpose Electronic Card (MEC).

Figure 11. Level 1 DFD of user authentication system in our proposed model

Figure 10. Context Level DFD of user authentication system in our proposed model

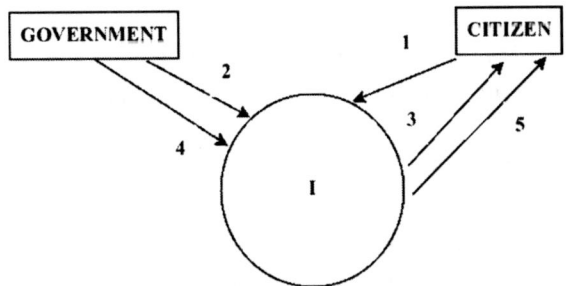

I - ENROLLMENT & AUTHENTICATION SYSTEM, where
1 : USER INPUT.
2 : SIGNATURE GENERATION.
3 : DELIVERY OF MESSAGE & SIGNATURE.
4 : GRANT OF BENEFITS.
5 : ACCESS OF BENEFITS.

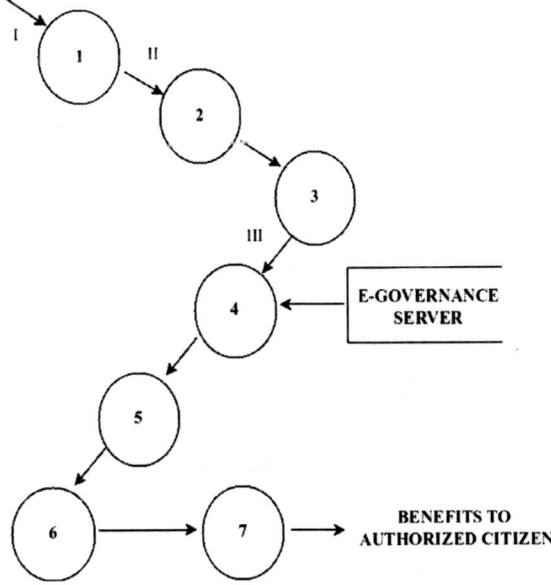

291

2. Transmission of information using Information and Communication Technology (ICT).
3. Verification of information and signature for grant of facilities.
4. Encryption of E-Services.
5. Transmission of information using Information and Communication Technology (ICT).
6. Verification of information and signature.
7. Decryption of E-Services.
   I. The information of the Citizen enters the system for enrolment purpose.
   II. The enrolled information seeks verification.
   III. The mechanism proceeds further only after successful validation.

The summary of Figure 12 is as follows:

1.1 The system accepts input from the Citizen.
1.2 Here the information of the Citizen is converted to its binary equivalent form.
1.3 Creation of signature is done.
1.4 The signature which is created recently is incorporated with the information of the Citizen.
1.5 The information of the Citizen is stored for further reference in encrypted format with the help of Multipurpose Electronic Card (MEC).
   I. Here the information of the Citizen is processed for entry into the system.
   II. Here the Citizen enters two prime numbers which are required by the Government for key generation purpose according to our object oriented modelling of RSA digital signature algorithm.

The summary of Figure 13 is as follows:

3.1 Requisite parameters are verified for granting Voter Card (VC) facility to the Citizen.

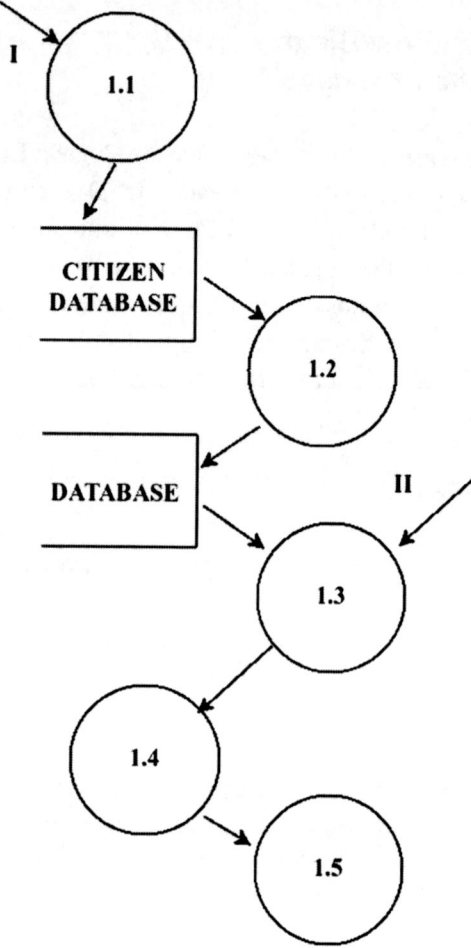

Figure 12. Level 2 DFD of user enrolment system in our proposed model

3.2 Requisite parameters are verified for granting Permanent Account Number (PAN) facility to the Citizen.
3.3 Requisite parameters are verified for granting Employment Card (EC) facility to the Citizen.
3.4 Requisite parameters are verified for granting Driving Licence (DL) and Insurance Card (IC) facility to the Citizen.

The above mentioned verification depends on the following parameters of the Citizen: 1) Date of Birth (DOB) of the Citizen; 2) Educational and Professional parameters of the Citizen; and 3) Educational parameters of the Citizen.

Figure 13. Verification of information and signature for grant of facilities

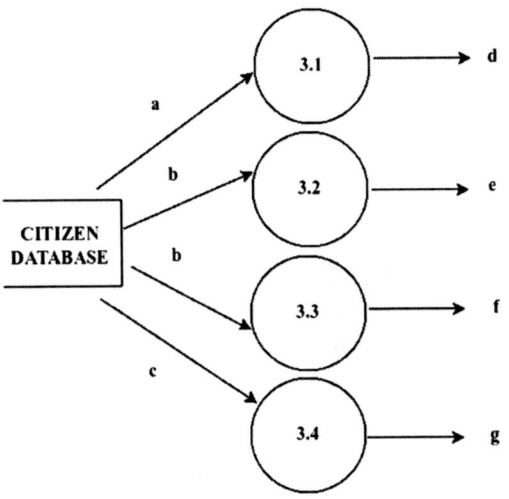

The following facilities are processed for the Citizen only after successful verification:

d.   Voter Card (VC) facility.
e.   Permanent Account Number (PAN) card facility.
f.   Employment Card (EC)
g.   Driving Licence (DL) and Insurance Card (IC) facility.

ER Diagram of the Proposed Model

Entity Relationship Diagram (ERD) of the proposed authentication mechanism is shown in Figure 14. The descriptions of the relations presented in Figure 14 are as follows:

- **Input Data:** Citizen provides the essential information to the Government through Multipurpose Electronic Card (MEC).
- **User Authentication:** Citizen authenticates its identity for access of benefits granted by the Government. The entire transaction is done with the help of Multipurpose Electronic Card (MEC).
- **Store Service:** Government stores the services provided to the Citizen in a E-Service server.
- **Benefits:** E-Service server allows the access of the services to the Citizen after proper authentication.

Though, we have tried to demonstrate the relations in a simpler form, in practical scenario the situation will definitely change as the distributed database management should be incorporated in our proposed model for successful handling of huge databases.

Figure 14. Multipurpose Electronic Card (MEC) based user authentication system using RSA in our proposed mechanism

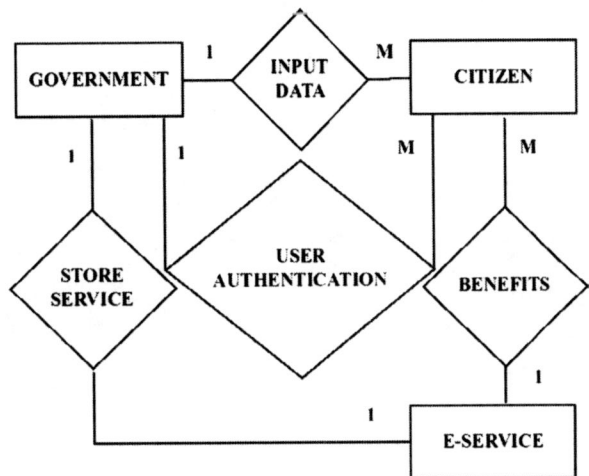

Table Structure

Based on our model depicted at the initial stage of operations, the sample database may be designed as follows:

The Structured Query Language (SQL) for creation of GOVERNMENT, CITIZEN, and SERVICE tables are shown in Box 1.

We need to enhance the table structure as the practical scenario changes and the requirement generates based on the changed scenario. However, we have tried to depict the simple table structure at the initial stage of our model.

FUTURE SCOPE

So far, we have technically validated our proposed Citizen centric smart card based E– Governance model using object oriented software engineering approach which is itself a conceptual mechanism. Hence, our proposed model has immense scope for further research work in conceptual as well as practical platforms. The interconnections among the various electronic services of the Government need more discussions. The installation of standard security parameters over the proposed model in the context of enhanced status becomes the next

Box 1.

```
CREATE TABLE GOVERNMENT
(
        ID         VARCHAR2 (30)        PRIMARY KEY,
        SERVICEID       VARCHAR2 (30),
        CITIZENID       VARCHAR2 (30),
        VC       BOOLEAN,
        PAN       BOOLEAN,
        EC       BOOLEAN,
        DL       BOOLEAN,
        IC       BOOLEAN,
        PUBLICKEY       VARCHAR2 (30),
        SERVICEID       REFERENCES SERVICE,
        CITIZENID       REFERENCES CITIZEN
);
CREATE TABLE CITIZEN
(
        CITIZENID               VARCHAR2 (30)        PRIMARY KEY,
        NAME                 VARCHAR2 (30),
        FNAME             VARCHAR2 (30),
        MNAME             VARCHAR2 (30),
        RESIDENCE             VARCHAR2 (30),
        SEX                 VARCHAR2 (10),
        DOB                 DATE,
        AGE                 VARCHAR2 (30),
        BLOOD_GR           VARCHAR2 (30),
        X                 NUMERIC (7,3),
        XII                 NUMERIC (7,3),
        UG                 NUMERIC (7,3),
        PG                 NUMERIC (7,3),
        OCCUPATION         VARCHAR2 (30),
        SALARY             NUMBER (10)
);
CREATE TABLE SERVICE
(
        SERVICEID             VARCHAR2       (30)       PRIMARY KEY,
        SERVICENAME         VARCHAR2       (30)
);
```

focus area of work. The practical implementation includes the combination of hardware domain, mathematical as well as statistical analysis and social sciences to spread the benefits of the model to the common populace.

CONCLUSION

In the above discussions, neither do we claim that we have achieved ultimate security for the proposed e-governance model, nor do we claim that our database is completely normalized. Those claims will become very much unscientific in nature as we believe that none of the electronic mechanism is fully secured in this world and the database design can be fully normalized only during its practical implementation. We refrain from the above mentioned claims because whatever we have discussed it is till date in conceptual level only. However, we firmly claim that since our proposed model is conceptually feasible, its practical implementation is also possible. Moreover, the literature survey reveals that similar type of research works are being carried out throughout the globe. We have just tried to design an electronic mechanism which will find solutions to our problems. Hence, as a future scope of research work, we will try to enhance the security features of our proposed multifaceted electronic card based e-governance model using complex Elliptic Curve Cryptosystems (ECC) like Elliptic Curve Digital Signature Algorithm (ECDSA), Elliptic Curve Discrete Logarithm Problem (ECDLP), Elliptic Curve Diffie Hellman (ECDH), Elliptic Curve ElGamal (ECElGamal), ECRSA, etc, and its corresponding data modelling.

REFERENCES

Alguliev, R. M., Imamverdiyev, Y. N., & Yusifov, F. F. (2011). *Some conceptual views on information security of the society.* Retrieved October 17, 2013, from http://ieeexplore.ieee.org/xpl/articleDetails. jsp?arnumber=6110909

Alguliev, R. M., & Yusifov, F. F. (2009). *Electronic governance as transformation technology of public management.* Retrieved October 17, 2013, from http://ieeexplore.ieee.org/xpl/articleDetails. jsp?arnumber=5372618

Bredin, H., Miguel, A., Witten, I. H., & Chollet, G. (2006). *Detecting Replay Attacks in Audiovisual Identity Verification.* Retrieved October 17, 2013, from http://ieeexplore.ieee.org/xpl/articleDetails. jsp?arnumber=1660097

Charalabidis, Y., Koussouris, S., & Ramfos, A. (2011). *A Cloud Infrastructure for Collaborative Digital Public Services.* Retrieved October 17, 2013, from http://ieeexplore.ieee.org/xpl/article-Details.jsp?arnumber=6133162

Da-li, H., Hua-lin, W., & Chang-nan, W. (2008). *Research on Framework of Public Crisis Management System under the Circumstance of E-Governance.* Retrieved October 17, 2013, from http://ieeexplore.ieee.org/xpl/articleDetails. jsp?arnumber=4680352

Hoda, A., Roy, A., & Karforma, S. (2012). Application of ECDSA for security of transaction in E-Governance. In *Proceedings of Second National Conference on Computing and Systems (NaCCS)* (pp. 281-286). The University of Burdwan.

Horman, W. (1440). *William_Horman.* Retrieved October 14, 2013, from http://en.wikipedia.org/wiki/William_Horman

Janowski, T., Estevez, E., & Ojo, A. (2012). *Conceptualizing Electronic Governance Education*. Retrieved October 17, 2013, from http://ieeexplore.ieee.org/xpl/articleDetails. jsp?arnumber=6149289

Lee, B., Li, J., & Kim, K. (2009). *Identity-Based Secret Signature Scheme*. Retrieved October 17, 2013, from http://ieeexplore.ieee.org/xpl/article-Details.jsp?arnumber=5369964

Liu, J. K., Wei, V. K., Siu, C., Chan, R. L., & Choi, T. (2001). *Multi-application smart card with elliptic curve cryptosystem certificate*. Retrieved October 17, 2013, from http://ieeexplore.ieee.org/ xpl/articleDetails.jsp?arnumber=938143

Manjunath, G., Muthanna, R., Thara, S., Dhanasekaran, H., Tiwari, P., & Das, P. (2012). *Delivering Mobile e-Governance on Low-End-Phones*. Retrieved October 17, 2013, from http://ieeexplore.ieee.org/xpl/articleDetails. jsp?arnumber=6341417

McKenzie, R., Crompton, M., & Wallis, C. (2008). *Use Cases for Identity Management in E-Government*. Retrieved October 17, 2013, from http://ieeexplore.ieee.org/xpl/articleDetails. jsp?arnumber=4489850

Mittal, P. A., Kumar, M., Mohania, M. K., Nair, M., Batra, N., Roy, P., et al. (2004). *A framework for e-Governance solutions*. Retrieved October 17, 2013 from http://ieeexplore.ieee.org/xpl/ articleDetails.jsp?arnumber=5388839

Mohammed, E., Emarah, A. E., & El-Shennawy, K. (2001). *Elliptic curve cryptosystems on smart cards*. Retrieved October 17, 2013, from http://ieeexplore.ieee.org/xpl/articleDetails. jsp?arnumber=962835

Mohandes, M. (2010). *A smart card management and application system*. Retrieved October 17, 2013, from http://ieeexplore.ieee.org/xpl/article-Details.jsp?arnumber=5687971

Murthy, D. N., & Kumar, R. V. P. (2003). *Software architectural design model for e-governance systems*. Retrieved October 17, 2013, from http://ieeexplore.ieee.org/xpl/articleDetails. jsp?arnumber=1273310

Naccache, D., & M'Raihi, D. (1996). *Cryptographic smart cards*. Retrieved October 17, 2013, from http://ieeexplore.ieee.org/xpl/articleDetails. jsp?arnumber=502402

Nimalaprakasan, S., Ramanan, S., Malalasena, B. A., Shayanthan, K., Gamage, C., & Fernando, M. S. D. (2009). *Privacy enhanced data management for an electronic identity system*. Retrieved October 17, 2013, from http://ieeexplore.ieee.org/ xpl/articleDetails.jsp?arnumber=5224184

Pappel, I., Pappel, I., & Saarmann, M. (2012). Digital records keeping to information governance in Estonian local governments. In *Proceedings of 2012 International Conference* (pp. 199 – 204). London: Information Society (i-Society).

Potekar, S. D., & Giragaonkar, K. P. (2004). *Application of ICT for better decision making in e-governance*. Retrieved October 17, 2013, from http://ieeexplore.ieee.org/xpl/abstractAuthors. jsp?arnumber=1307597

Rahman, M. M., & Rajon, S. A. A. (2011). *An effective framework for implementing electronic governance in developing countries: Bangladesh perspective*. Retrieved October 17, 2013, from http://ieeexplore.ieee.org/xpl/articleDetails. jsp?arnumber=6164814

Raja, T. N., Ramana, V. V. V., & Damodharam, A. (2012). *A framework for CSC oriented eGovernance applications*. Retrieved October 17, 2013, from http://ieeexplore.ieee.org/xpl/articleDetails. jsp?arnumber=6414944

Rajon, S. A. A., & Rahman, M. M. (2012). *Issues in implementing electronic governance: Bangladesh perspective*. Retrieved October 17, 2013, from http://ieeexplore.ieee.org/xpl/articleDetails.jsp?arnumber=6509710

Roy, A., Banik, S., & Karforma, S. (2011). Object Oriented Modelling of RSA Digital Signature in E-Governance Security. [IJCEIT]. *International Journal of Computer Engineering and Information Technology*, 26(1).

Roy, A., Banik, S., Karforma, S., & Pattanayak, J. (2010). Object Oriented Modelling of IDEA for E-Governance Security. In *Proceedings of International Conference on Computing and Systems 2010 (ICCS 2010)* (pp. 263-269). The University of Burdwan.

Roy, A., & Karforma, S. (2011a). Risk and Remedies of E-Governance Systems. *Oriental Journal of Computer Science & Technology (OJCST)*, 4(2).

Roy, A., & Karforma, S. (2011b). A Survey on E-Governance Security. [IJCECA]. *International Journal of Computer Engineering and Computer Applications*, 8(1).

Roy, A., & Karforma, S. (2012a). Object Oriented approach of Digital certificate based E-Governance mechanism. In *Proceedings of Computational Intelligence and Communication Engineering, International Joint Conference on CIIT, CENT, CSPE and CIITCom 2012* (pp. 360-366). Chennai, India: Springer.

Roy, A., & Karforma, S. (2012b). A Survey on digital signatures and its applications. *Journal of Computer and Information Technology*, 3(1&2).

Roy, A., & Karforma, S. (2013a). Object oriented metrics analysis for implementation of authentication in smart card based E-Governance mechanism. *Researchers World – Journal of Arts. Science and Commerce*, 4(2), 103–109.

Roy, A., & Karforma, S. (2013b). UML based modelling of ECDSA for secured and smart E-Governance system. In *Proceedings of National Conference on Advancement of Computing in Engineering Research (ACER13)* (pp. 207 - 222). WB India.

Roy, A., Karforma, S., & Banik, S. (2013). *Implementation of authentication in E-Governance – An UML Based Approach*. LAP Lambert Academic Publishing.

Roy, A., Sarkar, S., Mukherjee, J., & Mukherjee, A. (2012). Biometrics as an authentication technique in E-Governance security. In *Proceedings of UGC sponsored National Conference on Research And Higher Education In Computer Science And Information Technology, RHECSIT-2012* (pp. 153-160). WB India.

Sarkar, S., & Roy, A. (2012). A Study on Biometric based Authentication. In *Proceedings of Second National Conference on Computing and Systems (NaCCS)* (pp. 263-268). WB India.

Sarkar, S., & Roy, A. (2013). Survey on Biometric applications for implementation of authentication in smart Governance. *Researchers World – Journal of Arts. Science and Commerce*, 4(1), 103–114.

Savari, M., Montazerolzohour, M., & Thiam, Y. E. (2012). *Comparison of ECC and RSA algorithm in multipurpose smart card application*. Retrieved October 17, 2013, from http://ieeexplore.ieee.org/xpl/articleDetails.jsp?arnumber=6246121

Simpson, M. C. S. (1996). Smart Power, a smart card electricity payment system. In *Proceedings of IEE Colloquium on UK Electricity Prepayment Systems*. IEE.

Sur, C., Roy, A., & Banik, S. (2010). A Study of the State of E-Governance in India. In *Proceedings of National Conference on Computing and Systems (NACCS)*. WB India.

Wang, S., Li, C., Liu, J., & Wang, Z. (2011). Design of identity verification unit and management system. In *Proceedings of IET International Conference on Communication Technology and Application (ICCTA 2011)* (pp. 792–795). IET.

Zhai, S., & He, T. (2010). *Design and implementation of password-based identity authentication system*. Retrieved October 17, 2013, from http://ieeexplore.ieee.org/xpl/articleDetails. jsp?arnumber=5623039

ADDITIONAL READING

Al-Hakim, L. (2006). *Global E-Government: Theory, Applications and Benchmarking*. IGI Global. doi:10.4018/978-1-59904-027-1

Bhattacharya J (2012), e-Gov 2.0 Policies, Processes and Technologies, Tata McGraw Hill.

Griffin, D., Trevorrow, P., & Halpin, E. (2007). *Developments in e-Government A Critical Analysis*. IOS Press.

Gronlund, A. (2002). *Electronic Government: Design, Applications & Management*. Idea Group Publishing.

Obi, T. (2007). *E-Governance: A Global Perspective on a New Paradigm*. IOS Press.

Prabhu, C. S. R. (2004). *E-Governance: Concepts and Case Studies*. PHI.

Reddick, C. G. (2010). *Comparative E-Government*. Springer. doi:10.1007/978-1-4419-6536-3

Scholl, H. J. (2010). *E-Government Information, Technology, And Transformation*. M.E. Sharpe.

Shareef, M. A., Archer, N., & Dutta, S. (2012). *E–Government Service Maturity and Development: Cultural, Organizational and Technological Perspectives*. IGI Global.

Sharma, P. (2004). *E-Governance The New Age Governance*. APH Pub.

KEY TERMS AND DEFINITIONS

Authentication: It is an operational procedure which is mainly used to verify the identity of the user of an electronic mechanism before allowing the access of the electronic mechanism. This process should be exercised very carefully because it acts as the security check post of the electronic mechanism to prevent the entry of the intruders to system and defend their ill-intentions.

Citizen: Citizen refers to the individuals of the state who actively participates in the governmental transactions, in this case which is done through electronic mode of communication.

Data Flow Diagram (DFD): DFD is a graphical tool used to demonstrate the flow of data within the proposed e-governance mechanism. This data mainly contains the information of the Citizen and their facilities provided by the Government either as their birth right or acquired by dint of their credibility.

Entity Relationship Diagram (ERD): ERD is another graphical tool used to demonstrate the inter-relations and mode of message passing among the entities of the proposed e-governance mechanism.

E-Service Server: E-Service Server defines the central server of the proposed e-governance mechanism which will serve mainly two purposes – (1) It will store the information of the Citizen based on which Government and Citizen will perform e-governance transactions, and (2) It will connect to other servers (if required) connected to this electronic mechanism. For example a Citizen may have multiple bank accounts in same or different banks. In this case the E-Service Server will communicate with the servers of those banks for performing financial transactions through the respective bank accounts of the Citizen. Since this server need to handle bulk amount of data, though we have shown it as a single server in this chapter for better understanding purpose, in actual scenario the concept of distributed database management system will be applicable here.

Government: Government denotes the authority of the state who is solely responsible for deployment of governance throughout its jurisdiction which also includes its primary role of public administration.

Mechanism: Mechanism denotes our proposed E-Governance system which is based on the design and implementation of Citizen centric smart card based model. This smart card will be provided by the Government to the Citizenry for various E-Governance transactions.

MEC: Multipurpose Electronic Card (MEC) is our proposed smart card which will be used as the ICT based interface between the Government and the Citizen. This smart card will act as the all purpose electronic identity of the Citizen thereby having facilities for financial transactions using our proposed E-Governance model.

PINA: Privacy, Integrity, Non–repudiation and Authentication (PINA) are the four prime parameters based on which the security standard of the e-governance mechanism are verified. We have started to implement these features in our proposed model in a sequential manner.

Compilation of References

AA.VV. (2010). *Enterprise 2.0 study - D4 Final report.* European Commission.

ABC4Trust Project. (2009). *ABC4Trust.* Retrieved March 23rd, 2011, from http://abc4trust.de/

Abramowicz, W., Karsenty, L., Moore, P., Peinel, O. G., Wiśniewski, M., & Tilsner, D. (2005). USE-ME.GOV (USability-drivEn open platform for MobilE GOVernment). In *Proceedings of Euro mGov 2005: The First European Mobile Government Conference* (pp. 7-16). Brighton, UK: Mobile Government Consortium International LLC.

Adenike, A. O., & Oyesoji, A. A. (2010). The relationship among predictors of child, family, school, society and the government and academic achievement of senior secondary school students in Ibadan. *Nigeria Procedia Social and Behavioural Sciences, 5*, 842–849. doi:10.1016/j.sbspro.2010.07.196

Adeniyi, B. (2012). Galaxy Backbone ICT infrastructure rollout gets $100m boost as FG, China enter $1.1billion loan pact. *Technology Times.* Retrieved 24/06/13 from http://www.technologytimesng.com/galaxy-backbone-ict-infrastructure-rollout-gets-100m-boost-as-fg-china-enter-1-1billion-loan-pact/

Adil, K. M. (2000). *Planning for and Monitoring of Project Sustainability: A Guideline on Concepts, Issues and Tools.* Retrieved May 15, 2013, from http://www.mande.co.uk/docs/khan.htm.

Admin. (2012). *Tweet photos: Nigerians protest fuel subsidy removal Nigerians Abroad.* Retrieved 24/08/13 from http://nigeriansabroadlive.com/tweet-photos-nigerians-protest-fuel-subsidy-removal/

Agostino, D. (2013). Using social media to engage citizens: a study of Italian municipalities. *Public Relations Review, 39*, 232–234. doi:10.1016/j.pubrev.2013.02.009

AIIM. (2012). *Social in the Flow-transforming processes and sharing knowledge.* AIIM Industry Watch.

Ajzen, I. (1991). The theory of planned behavior. *Organizational Behavior and Human Decision Processes, 50*(2), 179–211. doi:10.1016/0749-5978(91)90020-T

Akunyili, D. (2010). *ICT and E-government in Nigeria World Congress on Information Technology.* Retrieved 18/06/2013 at http://goafrit.wordpress.com/2010/06/12/ict-and-e-government-in-nigeria-prof-akunyili/

Alberts, D. S., Huber, R. K., & Moffat, J. (2010). *NATO NEC C2 maturity model.* U.S. Department of Defence.

Alfred, S. S., & Elizabeth, K. (2010). Contribution of Mobile Phones to Rural Livelihoods and Poverty Reduction in Morogoro Region Tanzania. *Electronic Journal on Information Systems in Developing Countries, 42*(3), 1–15.

Alguliev, R. M., & Yusifov, F. F. (2009). *Electronic governance as transformation technology of public management.* Retrieved October 17, 2013, from http://ieeexplore.ieee.org/xpl/articleDetails.jsp?arnumber=5372618

Alguliev, R. M., Imamverdiyev, Y. N., & Yusifov, F. F. (2011). *Some conceptual views on information security of the society.* Retrieved October 17, 2013, from http://ieeexplore.ieee.org/xpl/articleDetails.jsp?arnumber=6110909

Al-Hujran, O., Al-dalahmeh, M., & Aloudat, A. (2011). The Role of National Culture on Citizen Adoption of eGovernment Services: An Empirical Study. *Electronic. Journal of E-Government, 9*(2), 93–106.

300

Ali, M., & Bailur, S. (2007). *The challenge of sustainability in ICT4D - Is bricolage the answer?* Paper presented at the 9th international conference on social implications of computers in developing countries. Sao Paulo, Brazil.

Alijerban, M., & Saghafi, F. (2010). M-government Maturity Model with Technological Approach. In *Proceedings of the 4th Conference on New Trends in Information Science and Service Science*. Gyeongju, South Korea: IEEE.

Al-Rababah, B., & Abu-Shanab, E. (2010). E-Government and gender digital divide: The case of Jordan. *International Journal of Electronic Business Management, 8*(1), 1–6.

Alshawi, S., & Alalwany, H. (2009). E-government evaluation: Citizen's perspective in developing countries. *Information Technology for Development, 15*(3), 193–208. doi:10.1002/itdj.20125

Alter, S. (2006). *The Work System Method: Connecting People, Processes and IT for Business Results*. Larkspur, CA: Work System Press.

Amaravadi, C. S. (2005). Digital repositories for e-government. *Electronic Government: An International Journal, 2*(2), 205–218. doi:10.1504/EG.2005.007095

Andersen, K. V., & Henriksen, H. Z. (2006). E-government maturity models: Extension of the Layne and Lee model. *Government Information Quarterly, 23*(2), 236–248. doi:10.1016/j.giq.2005.11.008

Andrus, D. C. (2005). *The Wiki and the Blog: Toward a Complex Adaptive Intelligence*. Central Intelligence Agency.

Anthes, G. (2011). Invasion of the Mobile Apps. *Communications of the ACM, 54*(9), 16–18. doi:10.1145/1995376.1995383

Antovski, L., & Gusev, M. (2005). M-Government Framework. In *Proceedings of Euro mGov 2005: The First European Mobile Government Conference* (pp. 36-44). Brighton, UK: Mobile Government Consortium International LLC.

Antovski, L., & Gusev, M. (2006). M-GOV: The Evolution Method. In *Proceedings of Euro mGov 2006: The Second European Conference on Mobile Government* (pp. 26-35). Brighton, UK: Mobile Government Consortium International LLC.

Apostolou, D., Stojanovic, L., Lobo, T. P., Miró, J. C., & Papadakis, A. (2005). Configuring E-Government Services Using Ontologies. In M. Funabashi, & A. Grzech (Eds.), *Challenges of Expanding Internet: E-Commerce* (Vol. 189, pp. 141–155). Springer, US: E-Business, and E-Government. doi:10.1007/0-387-29773-1_10

Apple. (2013). *5 años del App. Store*. Apple. Retrieved from https://itunes.apple.com/WebObjects/MZStore.woa/wa/viewFeature?cc=mx&id=660896396&enlh=1.8321.832.83.8&mt=8&ls=1

Aregbesola, I. (2012). FG partners Foreign Firm to train 5,000 Nigerians on ICT, says Minister. *Business Day*. Retrieved 26/06/13 from www.businessdayonline.com/NG/index.php/tech/78-computing/38556-fg-partners-foreign-firm-to-train-5000-nigerians-on-ict-says-minister

Arizmendi. (2012). Redes Administrativas. *Redes Sociales*. Retrieved on May 20, 2013, from http://www.gobiernolocal.org/docs/publicaciones/RDGL_18_19_baja.pdf

Ashworth, R., Boyne, G., & Delbridge, R. (2007). Escape from the Iron Cage? Organizational Change and Isomorphic Pressures in the public sector. *Journal of Public Administration: Research and Theory, 19*, 165–187. doi:10.1093/jopart/mum038

Australian State Services Authority. (2007). *Joined up government - A review of national and international experiences*. Author.

Baclawski, K., Kokar, M. K., Kogut, P. A., Hart, L., Smith, J., & Holmes, W. S. I. et al. (2002). Extending the Unified Modeling Language for Ontology Development. *Software & Systems Modeling, 1*(2), 142–156. doi:10.1007/s10270-002-0008-4

Balani, N. (2010). *ESB over Service Oriented Architecture frameworks*. Retrieved November 2013, from http://naveenbalani.com/index.php/2010/05/esb-frameworks/

Baltaci-Goktalay, S., & Ozdilek, Z. (2010). Pre-service teachers' perceptions about Web 2.0 technologies. *Procedia Social and Behavioral Sciences, 2*, 4737–4741. doi:10.1016/j.sbspro.2010.03.760

Bannister, F. (2002). Citizen Centricity: A Model of IS Value in Public Administration. *Electronic Journal of Information Systems Evaluation, 5*(2).

Bannister, F., & Connolly, R. (2011). The Trouble with Transparency: A critical view of openess in e-government. *Policy & Internet, 3*(1), Article 8.

Bannister, F., & Connoly, R. (2010). Researching eGovernment: A Review of ECEG in its Tenth Year. In *Proceedings of European Conference on E-Government* (ECEG) (pp. 53-62). Academic Publishing Limited.

Barret, K., & Green, R. (2001). *Powering Up. How Public Managers Can Take Control of Information Technology*. Washington, DC: CQ Press.

Bassara, A., Wisniewski, M., & Zebrowski, P. (2005). *USE-ME.GOV – A Requirements-driven Approach for M-GOV Services Provisioning*. Business Information Systems.

Baum, C. H., & Di Maio, A. (2000). *Gartner's Four Phases of E-Government Model*. Retrieved on June 6, 2013, from http://www.gartner.com/DisplayDocument?id=317292

BBC News Africa. (2010). *Facebook influences Nigeria football team ban U-turn*. Retrieved 26/08/13 from http://www.bbc.co.uk/news/10525699

Beath, C. M., & Walker, G. (1998). Outsourcing of Application Software: A Knowledge Management Perspective. In *Proceedings of the ThirtyFirst Hawaii International Conference on System Sciences*, (pp. 666-674). IEEE.

Becker, J., Algermissen, L., & Niehaves, B. (2003). *Implementing e-Government Strategies: A procedural model for process oriented e-Government projects*. Paper presented at the meeting of the International Business Information Management Conference (IBIMA). Cairo, Egypt.

Beer, D., Kunis, R., & Runger, G. (2006). A component based software architecture for e-government applications. In *Proceedings of The First International Conference on Availability, Reliability and Security (ARES 2006)* (pp. 1-8). ARES.

Belanger, F., & Hiller, J. (2006). A framework for e-government: privacy implications. *Business Process Management Journal, 12*(1), 48–60. doi:10.1108/14637150610643751

Berners-Lee, T. (2010, November 22). Long Live the Web: A Call for Continued Open Standards and Neutrality. *Scientific American*. Retrieved from http://www.scientificamerican.com/article.cfm?id=long-live-the-web

Berthold, M., Jürgen, K., Normann, H., Schmeidel, K., Schmutz, G., Trops, M., & Winterberg, T. (2013). *Enterprise Service Bus*. Part of the Industrial SOA article series.

Bertot, J. C., Jaeger, P. T., & Hansen, D. (2012). The impact of polices on government social media usage: Issues, challenges, and recommendations. *Government Information Quarterly, 29*(1), 30–40. doi:10.1016/j.giq.2011.04.004

Blessing, M. M., & Vesper, O. (2006). *Bring M-Government to South African Citizens: Policy Framework, Delivery Challenges and Opportunities*. Paper presented at the Second European Conference on Mobile Government. Brighton, UK.

Blessing, M. M., Vesper, O., & Wallace, T. (2007). Enabling M-government in South Africa. In Mobile Government: An Emerging Direction in E-Government. IGI Global.

Bonham, G., Seifert, J., & Thorson, S. (2001). The transformational potential of e-Government: the role of political leadership. In *Proceedings of the 4th Pan European International Relations Conference*. Kent, UK: Academic Press.

Bonsón, E., Torres, L., Royo, S., & Flores, F. (2012). Local e-government 2.0: Social media and corporate transparency in municipalities. *Government Information Quarterly, 29*(2), 123–132. doi:10.1016/j.giq.2011.10.001

Borras, J. (2004). International Technical Standards for E-Government. *Electronic Journal of E-Government, 2*(2).

Borucki, C., Arat, S., & Kushchu, I. (2005). *Mobile Government and Organizational Effectiveness*. Mobile Government Consortium International LLC.

Bostock, M., Ogievetsky, V., & Heer, J. (2011). D^3: Data-Driven Documents. *IEEE Transactions on Visualization and Computer Graphics, 17*(12), 2301–2309. doi:10.1109/TVCG.2011.185 PMID:22034350

Botha, A., & Makitla, I. et al. (2010). *Mobi4D Platform*. IEEE.

Boynton, A. C., & Zmud, R. W. (1984, Summer). An Assessment of Critical Success Factors. *Sloan Management Review*, 17–27.

Braun, S., Kunzmann, C., & Schmidt, A. (2008). People Tagging & Ontology Maturing: Towards Collaborative Competence Management. In *Proceedings of 8th International Conference on the Design of Cooperative Systems COOP '08*. Carry-le-Rouet, France: COOP.

Braund, P., Frausher, K., Schwittay, A., & Petkoski. (2006). A Report on the Global E-Discussion: Information and Communications Technology for Economic Development, Exploring possibilities for Multi-Sector Technology Collaborations. *World Bank Institute Business, Competitiveness and development Team and RiOS Institute*. Retrieved 27/09/11 from http://www.riosinstitute.org/RiOSWBIediscussio.pdf

Bredin, H., Miguel, A., Witten, I. H., & Chollet, G. (2006). *Detecting Replay Attacks in Audiovisual Identity Verification*. Retrieved October 17, 2013, from http://ieeexplore.ieee.org/xpl/articleDetails.jsp?arnumber=1660097

Budde Com. (2013). *Nigeria - Mobile Market - Overview, Statistics and Forecasts*. Retrieved 18/06/2013 from http://www.budde.com.au/Research/Nigeria-Mobile-Market-Overview-Statistics-and-Forecasts.html

Bughin, J. (2008). The rise of enterprise 2.0. *Journal of Direct. Data and Digital Marketing Practice, 9*, 251–259. doi:10.1057/palgrave.dddmp.4350100

Bughin, J., Chui, M., & Miller, A. (2009). How companies are benefiting from Web 2.0 - McKinsey Global Survey Results. *The McKinsey Quarterly*.

Bunt, L., & Harris, M. (2010). *Mass Localism: A way to help small communities solve big social challenges*. Retrieved February 28th, 2013, from http://www.nesta.org.uk/publications/reports/assets/features/mass_localism

Burnett, K., Ng, K. B., & Park, S. (1999). A Comparison of the Two Traditions of Metadata Development. *Journal of the American Society for Information Science American Society for Information Science, 50*(13), 1209–1217. doi:10.1002/(SICI)1097-4571(1999)50:13<1209::AID-ASI6>3.0.CO;2-Y

Caffrey, L. (1998). *Information Sharing between and within Governments: A Study Group Report*. London: Commonwealth Secretariat.

Camarinha-Matos, L. M., & Afsarmanesh, H. (2003). Elements of a base VE infrastructure. *Computers in Industry, 51*, 139–163. doi:10.1016/S0166-3615(03)00033-2

Camenisch, J., & Pfitzmann, B. (2007). Federated Identity Management. In M. Petkovic, & W. Jonker (Eds.), Security, Privacy and Trust in Modern Data Management. Springer.

Carlson, V., & Eyler-Werve, K. (2012). *Civic Apps Competition Handbook* (Vol. 1). O'Reilly Media. Retrieved from http://shop.oreilly.com/product/0636920024484.do

Carmen, H., & Gabriela, G. (2011).. . *Mobile Learning Through Microblogging Procedia Social and Behavioral Sciences, 15*, 4–8. doi:10.1016/j.sbspro.2011.03.039

Carrizales, T., Holzer, M., Kim, S. T., & Kim, C. G. (2006). Digital governance worldwide: A longitudinal assessment of municipal web sites. *International Journal of Electronic Government Research, 2*(4), 1–23. doi:10.4018/jegr.2006100101

Carromeu, C., et al. (2010). *Component-based Architecture for e-Gov Web Systems Development*. Paper presented at the 17th IEEE International Conference and Workshops on Engineering of Computer Based Systems (ECBS). Oxford, UK.

Cartwright, I., & Doernenburg, E. (2006, May). Time to jump on the SOA bandwagon. *IT Now*.

Casadesus, M., & Karapetrovic, S. (2005). Has ISO 9000 lost some of its lustre? A longitudinal impact study. *International Journal of Operations & Production Management, 25*(6), 580–596. doi:10.1108/01443570510599737

Castelnovo, W. (2011). Social Computing Tools for Interorganizational Risk Management. In *Proceedings of the 5th European Conference on Information Management Evaluation*. ECIME.

Castelnovo, W. (2013). A Stakeholder Based Approach to Public Value. In *Proceedings of the 13th European Conference on E-Government*. ACPI.

Castelnovo, W., & Simonetta, M. (2007). The Evaluation of E-Government projects for Small Local Government Organisations. *Electronic Journal of E-Government, 5*(1).

Castelnovo, W. (2012). An Architecture Driven Methodology for Transforming from Fragmented to Connected Government: A Case of a Local Government in Italy. In *Enterprise Architecture for Connected E-Government: Practices and Innovations*. Hershey, PA: IGI Global. doi:10.4018/978-1-4666-1824-4.ch015

CEMR. (2011). *EU Subnational Government – 2010 Key Figures, 2011/2012 Ed.* Council of European Municipalities and Regions. Retrieved June 22, 2013, from http://www.ccre.org/docs/Nuancier2011Web.EN.pdf

Centeno, C., Van Bavel, R., & Burgelman, J. C. (2004). *E-Government in the EU in the Next Decade: The vision and Key Challenges*. Seville, Spain: Inst for Perspective Technological Studies.

Charalabidis, Y., Koussouris, S., & Ramfos, A. (2011). *A Cloud Infrastructure for Collaborative Digital Public Services*. Retrieved October 17, 2013, from http://ieeexplore.ieee.org/xpl/articleDetails.jsp?arnumber=6133162

Chatzinotas, S., Ntaliani, M., Karetsos, S., & Costopoulou, C. (2006). Securing m-Government Services: The Case of Agroporta. In *Proceedings of Euro mGov 2006: The Second European Conference on Mobile Government* (pp. 61-70). Brighton, UK: Mobile Government Consortium International LLC.

Cho, A., Willis, S., & Stewart, M. (2011). *The Resilient Society Innovation, Productivity, and the Art and Practice of Connectedness. Cisco Internet Business Solutions Group*. IBSG.

Chopra, K. R. (2005). *Ecosystems and Human Well-Being: Policy Responses: Findings of the Responses Working Group*. Island Press.

Chui, M., Miller, A., & Roberts, R. P. (2009). Six ways to make Web 2.0 work. *The McKinsey Quarterly*.

Chukwuebuka, U. F. (2013). The Effect of Social Media on Youth Development. *Nigeria Village Square*. Retrieved 24/08/13 from http://nigeriavillagesquare.com/articles/the-effect-of-social-media-on-the-youth-development.html

Chun, S. A., Shulman, S., Sandoval, R., & Hovy, E. (2010). Government 2.0: Making connections between citizens, data and government. *Information Polity*, *15*(1-2), 1–9.

Church, L., Moloney, M., & Bannister, F. (2013). The Sealed Letter: Safeguarding the Public System of Privacy Protection in a Digital World. In *Proceedings of the 46th Hawaii International Conference on System Sciences*. Manoa, HI: IEEE.

Clark, T., & Jones, R. (1999). Organisational Interoperability Maturity Model for C2. In *Proceedings of the Command and Control Research and Tecnology Symposium*. Retrieved June 22, 2013, from http://www.dodccrp.org/events/1999_CCRTS/pdf_files/track_5/049clark.pdf

Claver-Cortes, E., de Juana-Espinosa, S., & Valdes-Conca, J. (2013). The effect of emerging technologies on local e-Government barriers in Spain: a longitudinal perspective. In *Proceedings of the 22nd IBIMA Conference*. Rome, Italy: IBIMA.

Colan, M. (2004, April 21). *Service-oriented architecture expands the vision of web services – part 1*. IBM Corporation.

Coleman, D., & Levine, S. (2008). Collaboration 2.0: technology and best practices for successful collaboration in a Web 2.0 world. Cupertino, CA: Happy About.info Publishing Company.

Colomb, R. M. (2002). *Use of Upper Ontologies for Interoperation of Information Systems: A Tutorial*. Padova, Italy: Academic Press.

Conca, F. J., Llopis, J., & Tarí, J. J. (2004). Development of a Measure to Assess Quality Management in Certified Firms. *European Journal of Operational Research*, *156*, 683–697. doi:10.1016/S0377-2217(03)00145-0

Coordinating, N. E. C. (2001). *M-Government: The convergence of wireless technologies and e-Government*. Retrieved May 21, 2013, from www.ec3.org/Downloads/2001/m-Government_ED.pdf

Cotterill, S., & King, S. (2007). Public sector Partnerships to deliver local e-government: a social network study. In *Proceedings of EGOV 2007* (LNCS), (vol. 4656, pp. 240-251). Berlin: Springer-Verlag.

Cotton, B. (2012). *Century Governance Transforming IT Infrastructures to Meet Critical Imperatives. Frost & Sullivan*.

Crawford, T., & Lester, W. (2004). *Information Management challenges and opportunities for community based organisations serving people living with HIV/AIDS.* Academic Press.

Criado Grando, J. I. (2012). *Redes Sociales y Open Government.* Retrieved on June 2, 2013, from http://www.gobiernolocal.org/docs/publicaciones/RDGL_18_19_baja.pdf

Criado, J. I., & Ramilo, M. C. (2003). E-Government in practice: An analysis of web site orientation to the citizens in Spanish municipalities. *International Journal of Public Sector Management, 126*(3), 191–218. doi:10.1108/09513550310472320

Cronbach, L. J. (1951). Coefficient alpha and the internal structure of tests. *Psychometrika, 16*(3), 297–334. doi:10.1007/BF02310555

CSTransform. (2011). *E-Government Interoperability: A comparative analysis of 30 countries.* London: CS Transform Limited.

Culbertson, S. (2005). E-Government and Organizational Change. In M. Khosrow-Pour (Ed.), *Practicing E-Government: A Global Perspective* (pp. 83–131). Hershey, PA: Idea Group Publishing. doi:10.4018/978-1-59140-637-2.ch005

Cullen, R., & Hernon, P. (2004). *Wired For Well-Being: Citizens' Response to E-Government.* Academic Press.

Curristine, T., Lonti, Z., & Joumard, I. (2007). Improving Public Sector Efficiency: Challenges and Opportunities. *OECD Journal on Budgeting, 7*(1), 4–9. doi:10.1787/budget-v7-art6-en

Dadić, J., Despotović-Zrakić, M., Barać, D., Paunović, L., & Labus, A. (2012). Managing E-Government Information Resources Using Faceted Taxonomy. In M. Gasco (Ed.), *12th European Conference on eGovernment (ECEG 2012)* (pp. 169–175). Barcelona: Academic Publishing International Limited.

Dadić, J., Bogdanović, Z., Radenković, M., Mazinjanjin, Đ., & Jovanić, B. (2012). Developing a Multifaceted Model for Scaffolding Information Management in E-Government Systems. *Metalurgia International, 17*(12), 140–146.

Dadić, J., Labus, A., Simić, K., Radenković, B., & Despotović-Zrakić, M. (2012). A Model for Structuring Information Resources in E-Government. *Innovative Issues and Approaches in Social Sciences, 5*(2), 104–117. doi:10.12959/issn.1855-0541.IIASS-2012-no2-art07

Dadzie, A.-S., & Rowe, M. (2011). Approaches to Visualising Linked Data: A Survey. *Semantic Web, 2*(2), 89–124. doi: doi:10.3233/SW-2011-0037

Da-li, H., Hua-lin, W., & Chang-nan, W. (2008). *Research on Framework of Public Crisis Management System under the Circumstance of E-Governance.* Retrieved October 17, 2013, from http://ieeexplore.ieee.org/xpl/articleDetails.jsp?arnumber=4680352

Damianos, E. L., Cuomo, D., Griffith, J., Hirst, D. M., & Smallwood, J. (2007). Exploring the Adoption, Utility, and Social Influences of Social Bookmarking in a Corporate Environment. In *Proceedings of the 40th Hawaii International Conference on System Sciences.* IEEE.

Daud, M. Y., & Zakaria, E. (2011). Web 2.0 application to cultivate creativity in ICT literacy. *Procedia - Social and Behavioral Sciences, 59*, 459 – 466.

Dawes, S. S., & Pardo, T. (2002). Building collaborative digital government systems. In W. J. McIver, & A. K. Elmagarmid (Eds.), *Advances in Digital Government: Technology, Human Factors, and Policy* (pp. 259–273). Norwell, MA: Kluwer Academic Publishers. doi:10.1007/0-306-47374-7_16

Dawson, R. (2009). *Implementing Enterprise 2.0: A Practical Guide To Creating Business Value Inside Organizations With Web Technologies.* Advanced Human Technologies.

de Bruijn, J. A., & Heuvelhof, E. F. (2000). *Networks and Decision Making.* Utrecht, The Netherlands: Lemma.

De Bruin, T., Freeze, R., & Uday, K. (2005). Understanding the Main Phases of Developing a Maturity Assessment Model. In *Proceedings of Australasian Conference on Information Systems.* Sydney: Academic Press.

De Kool, D., & van Wamelen, J. (2008). Web 2.0: A New Basis for E-Government? In Proceedings of Information and Communication Technologies: From Theory to Applications, (pp. 1-7). ICTTA.

Dean, T., & Boutilier, M. (2011). *Joint service delivery in federal countries: Research report prepared for the forum of federation.* The Global Network on Federalism.

DelloiteResearch. (2000). *At the Dawn of e-Government - eGovernment Resource Centre.* Retrieved 15th September, 2012, from http://www.egov.vic.gov.au/pdfs/e-government.pdf

Developer, A. (2013). *Android, the world's most popular mobile platform.* Google. Retrieved from http://developer.android.com/intl/es/about/index.html

Dewhurst, F., Martínez Lorente, A. R., & Dale, B. G. (1999). TQM in public organisations: an examination of the issues. *Managing Service Quality, 9*(4), 265–273. doi:10.1108/09604529910273210

Di Maio, A. (2004). *Move 'Joined-Up Government' From Theory to Reality.* Gartner Industry Research, ID Number: G00123844.

DiMicco, J. M., Geyer, W., Dugan, C., Brownholtz, B., & Millen, D. R. (2009). People Sensemaking and Relationship Building on an Enterprise Social Networking. In *Proceedings of HICSS 2009.* IEEE.

Disaster-Report. (2012). *Natural Disasters In Nigeria 2012.* Retrieved 24/08/13 from http://www.disaster-report.com/2013/04/natural-disasters-in-nigeria-2012.html

Disaster-Report. (2013). *Natural Disasters In Nigeria 2013.* Retrieved 24/08/13 from http://www.disaster-report.com/2013/04/natural-disasters-in-nigeria-2013.html

Dixon, B. E. (2010). Towards E-Government 2.0: An Assessment of Where E-Government 2.0 is and Where It Is Headed. *Public Administration & Management, 15*(2), 418–454.

Đokić, D., Labus, A., Jevremović, S., Stokić, A., & Milić, A. (2012). Portal for the Management of Digitally Signed Electronic Documents. *Metalurgia International, 17*(9), 120–128.

Doz, Y. L. (1996). The Evolution of Cooperation in Strategic Alliances: Initial Conditions or Learning Processes? *Strategic Management Journal, 17*, 55–83. doi:10.1002/smj.4250171006

Dublin Core Metadata Initiative. (2012). *Dublin Core Metadata Element Set, Version 1.1.* Retrieved June 22, 2013, from http://dublincore.org/documents/2012/06/14/dces/

Ebbers, W. E., Pieterson, W. J., & Noordman, H. N. (2008). Electronic government: rethinking channel management strategies. *Government Information Quarterly, 25*(2), 181–201. doi:10.1016/j.giq.2006.11.003

Ebrahim, Z., & Irani, Z. (2005). E-Government adoption: architecture and barriers. *Business Process Management Journal, 11*(5), 589–611. doi:10.1108/14637150510619902

EC. (2002). *eEurope 2005: An information society for all.* Communication from the Commission - COM(2002) 263 final. Retrieved June 22, 2013, from http://eur-lex.europa.eu/LexUriServ/LexUriServ.do?uri=COM:2002:0263:FIN:EN:PDF

EC. (2010). *EUROPE 2020 - A strategy for smart, sustainable and inclusive growth.* Communication from the Commission - COM(2010) 2020. Retrieved June 22, 2013, from http://ec.europa.eu/eu2020/pdf/COMPLET%20EN%20BARROSO%20%20%20007%20-%20Europe%202020%20-%20EN%20version.pdf

EC. (2012). *Single Market Act II - Together for new growth.* Communication from the Commission - COM(2012) 573. Retrieved June 22, 2013, from http://ec.europa.eu/internal_market/smact/docs/single-market-act2_en.pdf

EC. (2013). *Study on Analysis of the Needs for Cross-Border Services and Assessment of the organisational, Legal, Technical and Semantic Barriers.* European Commission. Retrieved June 22, 2013, from http://ec.europa.eu/digital-agenda/en/news/final-report-study-analysis-needs-cross-border-services-and-assessment-organisational-legal

Edwards, L., & Waelde, C. (2000). *Law and the Internet.* London, UK: Hart Publishing.

Eggers, W. D. (2013, March 13). *Getting Mobile Right: Six Steps to Success in Government.* Retrieved from http://www.governing.com/columns/mgmt-insights/col-mobile-technology-six-steps-success-implementation-government.html

Eggers, W. D., & Jaffe, J. (2013). *Gov on the go.* Deloitte. Retrieved from http://dupress.com/articles/gov-on-the-go/

Ehrlich, K., & Shami, N. S. (2010). Microblogging inside and outside the workplace. In *Proceedings of 4th International AAAI Conference on Weblogs and Social Media*. Washington, DC: AAAI.

EIF. (2010). *European Interoperability Framework (EIF) for European public services*. Retrieved June, 22, 2013 from http://eur-lex.europa.eu/LexUriServ/LexUriServ.do?uri=COM:2010:0744:FIN:EN:PDF

El Haddad, J. (2009). Service-Oriented Architecture and Web Services. In *State of the art: languages for services interface description and for services composition*. Paris: INRIA Paris-Rocquencourt.

El Kiki, T., & Lawrence, E. (2006). *Government as a mobile enterprise: real-time, ubiquitous government*. Paper presented at the Information Technology: New Generations, 2006. New York, NY.

El-Kiki, T., Lawrence, E., & Steele, R. (2005). A management framework for mobile government services. In Proceedings of CollECTeR. Sydney, Australia: CollECTeR.

EPC. (2010). *Digital Single Market Newsletter*. Retrieved June 22, 2013, from http://www.epc.eu/dsm/1/Digital_Single_Market.pdf

Erl, T. (2005). *Core principles for service-oriented architectures*. Retrieved Oct 2013, from www.looselycoupled.com/opinion/2005/erl-core-infr0815.html

Estalella, A., & Ardevol, E. (2011). E-research: Challenges and opportunities for social sciences. *Convergencia Revista de Ciencias Sociales, 18*(55), 87–111.

Estevez, E., & Janowski, T. (2007). *Programmable Messaging for Electronic Government - Building a Foundation*. Academic Press. doi:10.1007/978-3-540-75221-9_9

Etzioni, A. (1999). *The Limits of Privacy*. New York: Basic Books.

European Commission. (2010). A Digital Agenda for Europe (COM(2010) 245 final/2). Author.

European Commission. (2011). *Digital Agenda: Turning government data into gold*. Retrieved 11 3, 2012, from http://europa.eu/rapid/press-release_IP-11-1524_en.htm

European Union. (2007). Retrieved on October 17, 2010, from www.egovbarriers.org/downloads/deliverables/solutions_report/Solutions_for_eGovernment.pdf

Eyob, E. (2004). E-Government: breaking the frontiers of inefficiencies in the public sector. *Electronic Government, 1*(1), 107–114. doi:10.1504/EG.2004.004140

Fasanghari, M., & Samimi, H. (2009). A Novel Framework for M-Government Implementation. In *Proceedings of International Conference on Future Computer and Communications*. Los Alamitos, CA: IEEE.

Federal Trade Commission. (2010). *Protecting Consumer Privay in an Era of Rapid Change: A Proposed Framework for Businesses and Policymakers*. Washington, DC: Federal Trade Commission.

Fedotova, O., Teixeira, L., & Alvelos, H. (2012). E-participation in Portugal: evaluation of government electronic platforms. *Procedia Technology, 5*, 152–161. doi:10.1016/j.protcy.2012.09.017

Ferguson, J., & Ballantyne, P. (2002). *Sustaining ICT-Enabled Development: Practice Makes Perfect?* Academic Press.

Fidel, R., Scholl, H. J., Liu, S., & Unsworth, K. (2007). Mobile government fieldwork: a preliminary study of technological, organizational, and social challenges. In *Proceedings of the 8th annual international conference on Digital government research: bridging disciplines & domains* (Vol. 228, pp. 131-139). Philadelphia, PA: Digital Government Research Center.

Fu, B., Lin, J., Li, L., Faloutsos, C., Hong, J., & Sadeh, N. (2013). Why people hate your app: making sense of user feedback in a mobile app. store. In *Proceedings of the 19th ACM SIGKDD international conference on Knowledge discovery and data mining* (pp. 1276–1284). New York, NY: ACM. doi:10.1145/2487575.2488202

Fundación Orange. (2005). *Informe Anual sobre el Desarrollo de la Sociedad de la Información en España 2005*. Retrieved on June 3, 2013, from http://fundacionorange.es/areas/25_publicaciones/EESPA_A2005_COMPLETO_V3.pdf

Fundación Orange. (2012). *Informe Anual sobre el Desarrollo de la Sociedad de la Información en España 2012*. Retrieved on June 3, 2013, from http://www.proyectos-fundacionorange.es/docs/eE2012.pdf

Fundación Telefónica. (2012). *La sociedad de la información es España 2011*. Retrieved on April 15, 2013, from http://e-libros.fundacion.telefonica.com/sie11/aplicacion_sie/ParteA/pdf/SiE_2011.pdf

Garshol, L. M. (2004). Metadata? Thesauri? Taxonomies? Topic Maps! Making Sense of it all. *Journal of Information Science, 30*(4), 378–391. doi:10.1177/0165551504045856

Garson, G. D. (2003). *Public Information Technology: Policy and Management Issues*. Hershey, PA: Idea Group Publishing.

Garson, G. D. (2003). Toward an information technology research agenda for public administration. In G. D. Garson (Ed.), *Public Information Technology: Policy and Management Issues* (pp. 331–357). Hershey, PA: Idea Group Publishing.

GartnerResearch. (2000). *Gartner's Four Phases of E-Government Model*. Retrieved 15 Spetember, 2012, from http://www.gartner.com/id=317292

Gašević, D., Djuric, D., & Devedžic, V. (2009). The Ontology UML Profile. In *Model Driven Engineering and Ontology Development* (2nd ed., pp. 235–243). Berlin: Springer. doi:10.1007/978-3-642-00282-3_9

Gauld, R., Gray, A., & McComb, S. (2009). How responsive is E-Government? Evidence from Australia and New Zealand. *Government Information Quarterly, 26*(1), 69–74. doi:10.1016/j.giq.2008.02.002

Geiger, C. P., & von Lucke, J. (2012). Open Government and (Linked) (Open) (Government) (Data). *JeDEM - eJournal of eDemocracy and Open Government, 4*(2), 265-278.

Georgiadis, C. K., & Stiakakis, E. (2010). Extending an e-Government Service Measurement Framework to m-Governement Services. In *Proceedings of Mobile Business and 2010 Ninth Global Mobility Roundtable (ICMB-GMR), 2010 Ninth International Conference on*. Athens, Greece: IEEE.

Ghyasi, A. F., & Kushchu, I. (2004). *m-Government: Cases of Developing Countries*. Paper presented at the European Conference on e-Government. London, UK.

Goggin, G. (2011). Ubiquitous apps: politics of openness in global mobile cultures. *Digital Creativity, 22*(3), 148–159. doi:10.1080/14626268.2011.603733

Gompert, D. C., & Nerlich, U. (2002). *Shoulder to Shoulder - The Road to U.S. European Military Co-operability: A German American Analysis*. RAND Corporation. Retrieved June, 22, 2013, from http://www.rand.org/content/dam/rand/pubs/monograph_reports/2005/MR1575.pdf

Goteza, J. et al. (2009). Cross-National Interoperability and Enterprise Architecture. *Informatica, 20*(3), 369–396.

Grönlund, Å., & Horan, T. A. (2004). Introducing e-gov: History, definitions, and issues. *Communications of the Association for Information Systems, 15*, 713–729.

Groves, D. (2005, September 11). Successfully planning for SOA. *BEA Systems Worldwide*.

Gruber, T. R. (1993). A Translation Approach to Portable Ontology Specifications. *Knowledge Acquisition, 5*(2), 199–220. doi:10.1006/knac.1993.1008

Guardian Newspaper. (2013, June 23). *Edward Snowden and the NSA files – Timeline*. Retrieved July 15th, 2013, from http://www.guardian.co.uk/world/2013/jun/23/edward-snowden-nsa-files-timeline

Hamzat, A. O. (2013). *Cybercrime, Social Media, Customs And The Nigeria Trade-Hub*. Retrieved 24/08/2013 from http://elombah.com/index.php/articles-mainmenu/17288-cybercrime-social-media-customs-and-the-nigeria-trade-hub

Hassan, H. S., Shehab, E., & Peppard, J. (2011). Recent advances in e-service in the public sector: state-of-the-art and future trends. *Business Process Management Journal, 17*(3), 526–545. doi:10.1108/14637151111136405

Haughwout, J. (2009). *Meeting the New Requirements for Transparency, Participation and Collaboration with Neighborhood America's Solutions*. Retrieved May 2011, from http://www.ingagenetworks.com/docs/mgmt/Wht_paper_Gov_JHaughwout_May09.pdf

Heeks, R., & Bailur, S. (2007). Analyzing e-government research: Perspectives, philosophies, theories, methods, and practice. *Government Information Quarterly, 24*(2), 243–265. doi:10.1016/j.giq.2006.06.005

Heer, J., Bostock, M., & Ogievetsky, V. (2010). A Tour Through the Visualization Zoo. *Communications of the ACM, 53*(6), 59–67. doi:10.1145/1743546.1743567

Hentrich, C., & Zdun, U. (2012). *Process-Driven SOA: Patterns for Aligning Business and IT.* Taylor & Francis. Retrieved November 2013, from http://books.google.co.uk/books?id=HRHk6nG5Z3QC

Herrick, D. R. (2009). Google this! using Google apps for collaboration and productivity. In *Proceedings of the 37th annual ACM SIGUCCS fall conference* (pp. 55–64). New York, NY: ACM. doi:10.1145/1629501.1629513

Hewson, C. (2008). Internet-mediated research as an emergent method and its potential role in facilitating mixed methods research. In *Handbook of Emergent Methods* (pp. 543–570). New York: Guilford Press.

Higuera, R. P., Gluch, D. P., Dorofee, A. J., Murphy, R. L., Walker, J. A., & Williams, R. C. (1994). *An Introduction to Team Risk Management.* (Version 1.0), Special Report CMU/SEI-94-SR-1, Carnegie Mellon University.

Ho, A. T. (2002). Reinventing Local Government and the E-Government Initiative. *Public Administration Review, 62*(4), 434–444. doi:10.1111/0033-3352.00197

Hoda, A., Roy, A., & Karforma, S. (2012). Application of ECDSA for security of transaction in E-Governance. In *Proceedings of Second National Conference on Computing and Systems (NaCCS)* (pp. 281-286). The University of Burdwan.

Hohpe, G. (2002, May). Stairway to Heaven. *Software Development.*

Holwerda, T. (2011, June 24). *The History of «App» and the Demise of the Programmer.* Retrieved from http://www.osnews.com/story/24882/The_History_of_App_and_the_Demise_of_the_Programmer

Hood, C. (1995). The New Public Management in the 1980s: variations on a theme. *Accounting, Organizations and Society, 20*(2/3), 93–109. doi:10.1016/0361-3682(93)E0001-W

Horman, W. (1440). *William_Horman.* Retrieved October 14, 2013, from http://en.wikipedia.org/wiki/William_Horman

Hosmer, C., Jeffcoat, C., Davis, M., & McGibbon, T. (2011). Use of mobile technology for information collection and dissemination. Data & Analysis Center for Software, 77.

Housing Technology & Race Online. (2011). *Digital by Default 2012.* London, UK: The Intelligent Business Company and Race Online.

Howe, J. (2006). Your Web, Your Way. *Time Magazine, 168*(26), 60–63.

Hui, G., & Hayllar, M. R. (2010). Creating Public Value in E-Government: A Public-Private-Citizen Collaboration Framework in Web 2.0. *Australian Journal of Public Administration, 69*(s1), S120–S131. doi:10.1111/j.1467-8500.2009.00662.x

Hutchison, B., Johnson, K., & Schmidt, M. (2005). *Increasing IT flexibility with IBM WebSphere ESB software.* BM Software Group.

Hyde, J. (2008). How to make the rhetoric of joined-up government really work. *Australia and New Zealand Health Policy.* doi:10.1186/1743-8462-5-22 PMID:18983680

Iansiti, M., & Levien, R. (2004). *The keystone advantage: What the new dynamics of business ecosystems mean for strategy, innovation, and sustainability.* Boston: Harvard Business School Press. Retrieved from http://books.google.ca/books/about/The_Keystone_Advantage.html?id=T_2QFhjzGPAC

IBM. (2009). *Developer Works Digital Library.* Retrieved November 2013, from www.ibm.com/developerworks/library/ws-coor/

IBM. (2013). *IBM Cloud Services and Solutions.* Retrieved from www-03.ibm.com/systems/cloud/

Independent Commission for Good Governance in Public Services. (2004). *The Good Governance Standard for Public Services.* Joseph Rowntree Foundation.

India, T. R. A. O. (2011). *Highlights of Telecom Subscription Data.* Retrieved September 16, 2013, from www.trai.gov.in/WriteReadData/trai/upload/PressReleases/835/Press%20Release%20June11.pdf

InfoDev. (2003). *Improving Health, connecting people: The role of ICTs in the health sector in developing countries.* Author.

InfoDev. (2005). *Harnessing ICTs to fight poverty and promote development: A research strategy and work plan 2005-2007.* InfoDev.

Information and Privacy Commissioner Canada. (2010). *Privacy Risk Management Building privacy protection into a Risk Management Framework to ensure that privacy risks are managed, by default.* Retrieved from www.privacybydesign.ca

International City/County Management Association and Public Technology (ICMA). (2002). *Digital government survey.* Washington, DC: Author.

International Council for IT in Government Administration (ICA). (2006). *Executive Summary. Country Reports.* Author.

Internetworldstatistics.com. (2013). Internet Users in Africa 2012. *Miniwatts Marketing Group.* Retrieved 19/06/2013 from http://www.internetworldstats.com/stats1.htm

Interoperability Solutions for European Public Administrations. (2010). *European Interoperability Framework for European public services, Annex 2* (pp. 1–40). Retrieved from http://ec.europa.eu/isa/documents/isa_annex_ii_eif_en.pdf

Introna, L. D. (1997). Privacy and the Consumer: Why we need privacy in the information society. *Metaphilosophy, 28*(3), 259–275. doi:10.1111/1467-9973.00055

Irani, Z., Love, P. E. D., & Jones, S. (2008). Learning lessons from evaluating eGovernment: reflective case experiences that support transformational government. *The Journal of Strategic Information Systems, 17*(2), 155–164. doi:10.1016/j.jsis.2007.12.005

Iroko, M. (n.d.). FG Abandons $200M Rural Telephony. *Project Zimbio Inc.* Retrieved 22/07/12 from http://www.zimbio.com/Nigeria/articles/GnjWJ99uezp/FG+ABANDONS+200M+RURAL+TELEPHONY+PROJECT

ITU. (2009). *E-government implementation toolkit: A Framework for e-Government Readiness and Action Priorities.* Retrieved from http://www.itu.int/ITU-D/cyb/app/docs/eGovernment%20toolkitFINAL.pdf

Jackson, A., Yates, J., & Orlikowski, W. (2007). Corporate Blogging: Building community through persistent digital talk. In *Proceedings of the 40th Hawaii International Conference on System Sciences.* IEEE.

Jaeger, P. T., & Thompson, K. M. (2003). E-government around the world: lessons, challenges, and future directions. *Government Information Quarterly, 20*(4), 389–394. doi:10.1016/j.giq.2003.08.001

Janowski, T., Estevez, E., & Ojo, A. (2012). *Conceptualizing Electronic Governance Education.* Retrieved October 17, 2013, from http://ieeexplore.ieee.org/xpl/articleDetails.jsp?arnumber=6149289

Janssen, M., Gortmaker, J., & Wagenaar, R. W. (2006). Web service orchestration in public administration: challenges, roles and growth stages. *Information Systems Management, 23*(2), 44–55. doi:10.1201/1078.10580530/45925.23.2.20060301/92673.6

Jidaw. (2013). ICT development Advisory Support and Consulting. *Nigeria Computers.* Retrieved 26/06/2013 from http://nigeriacomputers.com/technews/development-advisory-support-and-consulting-nigeria/#more-2046

Johan, H. (2008). *Mobile phones for good governance – challenges and way forward.* Retrieved September 19, 2013, from http://www.w3.org/2008/10/MW4D_WS/papers/hellstrom_gov.pdf

Johnson, R. C. (2010). Apps Culture reinventing mobile Internet. *Electronic Engineering Times, (1588),* 24-26, 28, 30.

Johnston, P. (2006). *21st Century Networked Local Government.* Cisco Systems. Retrieved June, 22, 2013 from http://www.cisco.com/web/about/ac79/docs/wp/21st_Century_Networked_Local_Government.pdf

Joia, L. A. (2004). Developing Government-to-Government enterprises in Brazil: a heuristic model drawn from multiple case studies. *International Journal of Information Management, 24*(2), 147–166. doi:10.1016/j.ijinfomgt.2003.12.013

Jones, S. (1999). *Doing Internet Research.* Academic Press.

Jordanian e-Government Program. (2006). *Jordan e-Governmen Strategy*. Retrieved November 2013, from www.thieswittig.eu%2Fdocs%2FMPC_Strategies%2FJordan%2FJordan_e-GovernmenStrategy.pdf&ei=_AaJUp_0IpGthQe3kYDoCw&usg=AFQjCNF6vGV2J9PB9F6vzAeozWhoiswKbw&sig2=AOwpOup_MMQhZ7TqQlZKeQ

Kaczorowski, W. (Ed.). (2004). *Connected Government*. London: Premium Publishing.

Kailasam, R. (2010). m-Governance - Leveraging Mobile Technology to extend the reach of e-Governance, 2010. Academic Press. English, M. (2011). Government –shaping or catching the new wave? IBM Global Business Services.

Kamal, M. M., & Alsudairi, M. (2009). Investigating the importance of factors influencing integration technologies adoption in local government authorities. *Transforming Government: People. Process and Policy, 3*(3), 302–331.

Kapogiannis, G., Touzos, M., & Kreps, D. (2006). M-Business & M-Government: Co-operation, The Greek case study. In *Proceedings of Euro mGov 2006: The Second European Conference on Mobile Government* (pp. 154-159). Brighton, UK: Mobile Government Consortium International LLC.

Karantjias, & Papastergiou. (2009). Design principles of secure federated e/m-government framework. *International Journal of Electronic Governance, 2*(4), 402-423.

Kavanaugh, A. L., Fox, E. A., Sheetz, S. D., Yang, S., Li, L. T., & Shoemaker, D. J. et al. (2012). Social media use by government: From the routine to the critical. *Government Information Quarterly, 29*, 480–491. doi:10.1016/j.giq.2012.06.002

Kenneth, K. (2009). Moving towards integrated public governance: improving service delivery through community engagement. *International Review of Administrative Sciences, 75*(2), 239–254. doi:10.1177/0020852309104174

Kersten, G. E., Kersten, M. A., & Rakowski, W. M. (2001). Application software and culture: Beyond the surface of software interface. *Culture (Canadian Ethnology Society)*, 1–17.

Kesavarapu, S., & Choi, M. (2012). M-government - A framework to investigate killer applications for developing countries: An Indian case study. *Electronic Government: An International Journal, 9*(2), 200–219. doi:10.1504/EG.2012.046269

Kim. (2009). *Access Control Service Oriented Architecture Security*. Retrieved November 2013, from http://www.cs.wustl.edu/~jain/cse571-09/ftp/soa/

Kimaro, H. C. & Nhampossa, J.L. (2005). Analysing the problem of unsustainable health information systems in less-developed countries: case studies from Tanzania and Mozambique. *Information technology for development 11*(3), 273-298.

Kim, D. Y. (2010). E-government maturity model using the capability maturity model integration. *Journal of Systems and Information Technology, 12*(3), 230–244. doi:10.1108/13287261011070858

King, J., Lampinen, A., & Smolen, A. (2011). *Privacy: Is there an app. for that?* Academic Press. doi:10.1145/2078827.2078843

Klijn, E., & Koppenjan, J. F. M. (2000). Politicians and Interactive Decision Making: Institutional Spoilsports or Playmakers. *Public Administration, 78*(2), 365–387. doi:10.1111/1467-9299.00210

KMPG International. (2010). *Dynamic Technologies for Smarter Government. Unlocking Knowledge in the Web 2.0 Age*.

Kodali, R. (2005). *An introduction to SOA, What is service-oriented architecture?* Retrieved November 2013, from http://www.javaworld.com/javaworld/jw-06-2005/jw-0613-soa.html

Koh, C. E., Prybutok, V. R., Ryan, S., & Ibragimova, B. (2006). The importance of strategic readiness in an emerging e-Government environment. *Business Process Management Journal, 12*(1), 22–33. doi:10.1108/14637150610643733

Kolari, P., Finin, T., Lyons, K., Yesha, Y., Perelgut, S., & Hawkins, J. (2007). On the structure, properties and utility of internal corporate blogs. In *Proceedings of the International Conference on Weblogs and Social Media ICWSM 2007*. ICWSM.

Kraemer, K., Danziger, J., & King, J. (1978). Local Government and Information Technology in the United States. *OECD Informatics Studies, 12.*

Kramer, R. M. (1999). Trust and Distrust in Organizations: Emerging Perspectives, Enduring Questions. *Annual Review of Psychology, 50*, 569–598. doi:10.1146/annurev.psych.50.1.569 PMID:15012464

Kumar, M., & Sinha, O. P. (2007). *M-government–mobile technology for e-government.* Paper presented at the International conference on e-government. New Delhi, India.

Kumar, A., Sukanta, M., & Sahu, K. (2008). *Challenge of Wireless and Mobile Tecnologies in Government.* Academic Press.

Kushchu, I. & Kuscu. (2003). *From E-Government to M-government: Facing the inevitable.* Dublin, Ireland: MGovLab.

Labs, I. (2012). *Apps for Democracy* (Vol. 2012). Academic Press.

Landsbergen, D. Jr, & Wolken, G. Jr. (2001). Realizing the Promise: Government Information Systems and the Fourth Generation of Information Technology. *Public Administration Review, 61*(2), 206–220. doi:10.1111/0033-3352.00023

Laufer, R., & Wolfe, M. (1977). Privacy as a concept and a social issue: A multidimensional developmental theory. *The Journal of Social Studies, 33*(3), 22–42.

Layne, K., & Lee, J. (2001). Developing Fully Functional E-government: A four-stage model. *Government Information Quarterly, 18*(2), 122–136. doi:10.1016/S0740-624X(01)00066-1

Lee, B., Li, J., & Kim, K. (2009). *Identity-Based Secret Signature Scheme.* Retrieved October 17, 2013, from http://ieeexplore.ieee.org/xpl/articleDetails.jsp?arnumber=5369964

Legg, C. (2007). Ontologies on the Semantic Web. *Annual Review of Information Science & Technology, 41*(1), 407–451. doi:10.1002/aris.2007.1440410116

Legislation.gov.uk. (2011). *Localism Act.* Retrieved April 3rd, 2013, from http://www.legislation.gov.uk/ukpga/2011/20/contents/enacted

Lerman, K. (2007). User Participation in Social Media: Digg Study. In *Proceedings of 2007 IEEE WIC ACM International Conferences on Web Intelligence and Intelligent Agent Technology Workshops,* (pp. 255-258). IEEE.

Lientz, B. P., & Swanson, E. B. (1981). Problems in application software maintenance. *Communications of the ACM, 24*, 763–769. doi:10.1145/358790.358796

Lientz, B. P., Swanson, E. B., & Tompkins, G. E. (1978). Characteristics of application software maintenance. *Communications of the ACM, 21*, 466–471. doi:10.1145/359511.359522

Li, F. (2003). Implementing E-Government strategy in Scotland: current situation and emerging issues. *Journal of Electronic Commerce in Organizations, 1*(2), 44–65. doi:10.4018/jeco.2003040104

Liu, J. K., Wei, V. K., Siu, C., Chan, R. L., & Choi, T. (2001). *Multi-application smart card with elliptic curve cryptosystem certificate.* Retrieved October 17, 2013, from http://ieeexplore.ieee.org/xpl/articleDetails.jsp?arnumber=938143

Lugano, G. (2008). Mobile social networking in theory and practice. *First Monday, 13*(11). doi:10.5210/fm.v13i11.2232

Mabbutt, A. (2010). *Big Society.* Retrieved April 3rd, 2013, from http://www.conservatives.com/Policy/Where_we_stand/Big_Society.aspx

Mahmood, Z. (2007a). Service Oriented Architecture: Potential Benefits and Challenges. In *Proceedings of WSEAS Int Conference.* WSEAS.

Mahmood, Z. (2007b). Service oriented architecture: tools and technologies. In *Proceedings of WSEAS Int Conference.* WSEAS.

Mahmood, Z. (2007c). Service Oriented Architecture: A New Paradigm for Enterprise Application Integration. In *Proceedings of WSEAS Int Conference.* WSEAS.

Manjunath, G., Muthanna, R., Thara, S., Dhanasekaran, H., Tiwari, P., & Das, P. (2012). *Delivering Mobile e-Governance on Low-End-Phones.* Retrieved October 17, 2013, from http://ieeexplore.ieee.org/xpl/articleDetails.jsp?arnumber=6341417

Mansoor, A., & Rohan, D. (2010). *An M-government Solution Proposal for Dubai Government*. Paper presented at the 9th WSEAS International Conference on Telecommunications and Informatics. Sicily, Italy.

Marche, S., & McNiven, D. (2003). *E-Government and E-Governance: The Future Isn't What It Used To Be*. Canadian Journal of Administrative Sciences / Revue Canadienne des Sciences de l'Administration.

Maria Emmanouilidou, D. K. (2010). A framework for accessible m-government implementation. *Electronic Government, an International Journal, 7*(3), 252 - 269.

Maumbe, B. M., & Owei, V. (2006). Bringing M-government to South African Citizens: Policy Framework, Delivery Challenges and Opportunities. In *Proceedings of Euro mGov 2006: The Second European Conference on Mobile Government* (pp. 160-173). Brighton, UK: Mobile Government Consortium International LLC.

McAfee, P. A. (2006). Enterprise 2.0: The Dawn of the Emergent Collaboration. *MIT Sloan Management Review, 47*(3), 21–28.

McAfee, P. A. (2009). *Enterprise 2.0: New Collaborative Tools for Your Organization's Toughest Challenges*. Cambridge, MA: Harvard Business Publishing.

MCIT. (2003). *e-Government Lembaga*. MCIT.

McKenzie, R., Crompton, M., & Wallis, C. (2008). *Use Cases for Identity Management in E-Government*. Retrieved October 17, 2013, from http://ieeexplore.ieee.org/xpl/articleDetails.jsp?arnumber=4489850

McKinsey Global Institute. (2012). *The Social Economy: Unlocking value and productivity through social technologies*. Retrieved on May 26, 2013, from http://www.mckinsey.com/insights/high_tech_telecoms_internet/the_social_economy

Melitski, J. (2003). Capacity and e-government performance: An analysis based on early adopters of Internet technologies in New Jersey. *Public Performance and Management Review, 26*(4), 376–390. doi:10.1177/1530957603026004005

Mengistu, D., Hangjung, Z., & Jae, J. R. (2009). M-government: Opportunities and Challenges to Deliver Mobile Government Services in Developing Countries. In *Proceedings of Computer Sciences and Convergence Information Technology, 2009. ICCIT '09. Fourth International Conference on*. Seoul: IEEE.

Mergel, I., Schweik, C., & Fountain, J. (2009). *The Transformational Effect of Web 2.0 Technologies on Government*. Retrieved June, 22, 2013 from http://papers.ssrn.com/sol3/papers.cfm?abstract_id=1412796

Mergel, I. (2013). Social media adoption and resulting tactics in the US federal government. *Government Information Quarterly, 30*, 123–130. doi:10.1016/j.giq.2012.12.004

Meyers, B. C. (2006). *Risk Management Considerations for Interoperable Acquisition, TECHNICAL NOTE, CMU/SEI-2006-TN-032*. Carnegie Mellon University.

Microsoft. (2011). *Connected Government in a Connected World*. Microsoft Corp. Retrieved June, 22, 2013 from http://www.microsoft.com/download/en/details.aspx?id=8295

Milam, L., & Avery, E. J. (2012). Apps4Africa: A new State Department public diplomacy initiative. *Public Relations Review, 38*(2), 328–335. doi:10.1016/j.pubrev.2011.12.013

Misuraca, G., Alfano, G., & Viscusi, G. (2011). Interoperability Challenges for ICT-enabled Governance: Towards a pan-European Conceptual Framework. *Journal of Theoretical and Applied Electronic Commerce Research, 6*(1), 95–111. doi:10.4067/S0718-18762011000100007

Mittal, P. A., Kumar, M., Mohania, M. K., Nair, M., Batra, N., Roy, P., et al. (2004). *A framework for e-Governance solutions*. Retrieved October 17, 2013 from http://ieeexplore.ieee.org/xpl/articleDetails.jsp?arnumber=5388839

Mohammed, E., Emarah, A. E., & El-Shennawy, K. (2001). *Elliptic curve cryptosystems on smart cards*. Retrieved October 17, 2013, from http://ieeexplore.ieee.org/xpl/articleDetails.jsp?arnumber=962835

Mohandes, M. (2010). *A smart card management and application system*. Retrieved October 17, 2013, from http://ieeexplore.ieee.org/xpl/articleDetails.jsp?arnumber=5687971

Mongo, D. B. (2013). *MongoDB is an open-source document database and the leading NoSQL database.* Retrieved from www.mongodb.org

Morgeson, F. V., VanAmburg, D., & Mithas, S. (2011). Misplaced Trust? Exploring the Structure of the E-Government-Citizen Trust Relationship. *Journal of Public Administration: Research and Theory, 21*(2), 257–283. doi:10.1093/jopart/muq006

Morisonn, J. (2010). Gov 2.0: Towards a User Generated State? *The Modern Law Review, 4.*

Munzner, T. (2009a). Visualization. In *Fundamentals of Computer Graphics* (pp. 675–720). Academic Press.

Munzner, T. (2009b). A Nested Model for Visualization Design and Validation. *IEEE Transactions on Visualization and Computer Graphics, 15*(6), 921–928. doi:10.1109/TVCG.2009.111 PMID:19834155

Murthy, D. N., & Kumar, R. V. P. (2003). *Software architectural design model for e-governance systems.* Retrieved October 17, 2013, from http://ieeexplore.ieee.org/xpl/articleDetails.jsp?arnumber=1273310

Murugesan, S., Rossi, G., Wilbanks, L., & Djavanshir, R. (2011a). The Future of Web Apps. *IT Professional Magazine, 13*(5), 12–14. doi:10.1109/MITP.2011.89

Naccache, D., & M'Raihi, D. (1996). *Cryptographic smart cards.* Retrieved October 17, 2013, from http://ieeexplore.ieee.org/xpl/articleDetails.jsp?arnumber=502402

Nam, T. (2012). Suggesting frameworks of citizen-sourcing via Government 2.0. *Government Information Quarterly, 29*(1), 12–20. doi:10.1016/j.giq.2011.07.005

NAN 1. (2013). FG to launch Fibre-optic Network connecting 27 universities. *News Agency of Nigeria.* Retrieved 24/06/2013 from http://www.nanngronline.com/section/technology/fg-to-launch-fibre-optic-network-connecting-27-universities

NAN 2. (2013). 2 mobile phone coys to begin production in Nigeria in 2013, says Minister. *News Agency of Nigeria.* Retrieved 24/06/2013 from http://www.nanngronline.com/section/technology/2-mobile-phone-coys-to-begin-production-in-nigeria-in-2013-says-minister

Nava, A. S., & Dávila, I. L. (2005). M-Government for Digital Cities: Value Added Public Services. In *Proceedings of the First European Mobile Government Conference Mobile Government Consortium International.* Academic Press.

Newman, J., & Clarke, J. (2009). *Publics, Politics and Power: Remaking the Public in Public Services.* London: Sage Publications.

Nigeria Intel. (2013). Why Nigeria must embrace e-governance. *Nigeria Intel.* Retrieved 18/06/2013 from http://www.nigeriaintel.com/2013/03/28/why-nigeria-must-embrace-e-governance/

Nimalaprakasan, S., Ramanan, S., Malalasena, B. A., Shayanthan, K., Gamage, C., & Fernando, M. S. D. (2009). *Privacy enhanced data management for an electronic identity system.* Retrieved October 17, 2013, from http://ieeexplore.ieee.org/xpl/articleDetails.jsp?arnumber=5224184

Noreng, O. (1980). *The oil industry and government strategy in the North Sea.* Retrieved November 2013, from http://books.google.co.uk/books?id=jZIOAAAAQAAJ&printsec=frontcover#v=onepage&q&f=false

Norris, D. F., Fletcher, P. D., & Holden, S. H. (2001). *Is your local government plugged in? Highlights of the 2000 electronic government survey. Paper prepared for the International City/County Management Association (ICMA) and Public Technology, Inc.* Baltimore, MD: PTI.

Noy, N. F., & McGuinness, D. L. (2001). *Ontology Development 101 : A Guide to Creating Your First Ontology.* Academic Press.

Ntaliani, M., Costopoulou, C., Manouselis, N., & Karetsos, S. (2009). M-government services for rural SMEs. *International Journal of Electronic Security and Digital Forensic, 2*(4), 407–423. doi:10.1504/IJESDF.2009.027672

Nugultham. (2012). Using Web 2.0 for Innovation and Information Technology in Education course. *Procedia - Social and Behavioral Sciences, 46,* 4607 – 4610.

NUS Institute of Systems Science, National University of Singapore. (2010). Retrieved June 22, 2013, from http://unpan1.un.org/intradoc/groups/public/documents/unpan/unpan039390.pdf

O'Hara, K. (2012). Transparency, open data and trust in government: shaping the infosphere. In *Proceedings of the 3rd Annual ACM Web Science Conference* (pp. 223–232). New York, NY: ACM. doi:10.1145/2380718.2380747

O'Reilly, T. (2005). *What is Web 2.0: design patterns and business models for the next generation of software.* Retrieved on Sept 30th, 2013, from http://oreilly.com/web2/archive/what-is-web-20.html

Obiozor, W. E. (2013). ICT and National Development. *Academia.Edu.* Retrieved 26/06/13 from http://academia.edu/2392046/ICT_and_NATIONAL_DEVELOPMENT

OECD, & ITU. (2011). *M-Government.* Paris: Organisation for Economic Co-operation and Development. Retrieved from http://www.oecd-ilibrary.org/content/book/9789264118706-en

OECD. (2009). *e-Government Studies Rethinking e-Government Services User-Centred Approaches: User-Centred Approaches.* OECD Publishing. Retrieved November 2013, from http://books.google.co.uk/books?id=Y2f1N2HVtLsC

OECD/International Telecommunication Union. (2011). *M-Government – Mobile Technologies for Responsive Governments and Connected Societies.* Author.

Oghogho, I., Odikayor, D. C., Adebayo, A. A., & Wara, S. T. (2012). VoIP vs GSM Technology: The Way of the Future for Communication In Ekekwe & Islam (Ed.), Disruptive Technologies, Innovation and Global Redesign: Emerging Implications (pp. 280-298). Hershey, PA: IGI Global.

Oghogho, I., & Ezomo, P. I. (2013). ICT for National Development in Nigeria: Creating an Enabling Environment. *International Journal of Engineering and Applied Sciences, 3*(2).

Ollus, M. (2007). Approaches and solutions supporting collaboration in networks. In *Proceedings of the International Cluster Conference.* Venice, Italy: Academic Press.

Ollus, M. (2005). A Holistic Approach towards Collaborative Networked Organizations. In *Innovation and the Knowledge Economy: Issues, Applications, Case Studies.* IOS Press.

ONTSI. (2011). *Informe Anual de los Contenidos Digitales en España 2011.* Retrieved on May 20, 2013, from www.ontsi.red.es/ontsi/es/estudios-informes/informe-anual-de-los-contenidos-digitales-en-espa%C3%B1a-2011

Onwuemele, A. (2011). Impact of Mobile Phones on Rural Livelihoods Assets in Rural Nigeria: A Case Study of Ovia North East Local Government Area. *JORIND, 9*(2), 223–236.

Open Group. (2009). *OSIMM.* Retrieved Oct 2013, from https://www2.opengroup.org/ogsys/jsp/publications/PublicationDetails.jsp?publicationid=12450

Orimisan, B. (2012). Fuel subsidy protests and power of social media. *Nigerian Best Forum.* Retrieved 24/08/13 from http://www.nigerianbestforum.com/generaltopics/fuel-subsidy-protests-and-power-of-social-media/

Osborne, D., & Gabler, T. A. (1992). *Reinventing Government: How the Entrepreneurial Spirit is Transforming the Public Sector.* Reading, MA: Addison-Wesley.

Osimo, D. (2008). *Web 2.0 in government: Why and how.* Institute for Prospectice Technological Studies (IPTS), JRC, European Commission, EUR, 23358.

Osimo, D. (2008). *Web 2.0 in Government: Why and How?* JRC Scientific and Technical Reports, European Communities.

Pade, C., Mallinson, B., & Sewry, D. (2006). *An Exploration of the Critical Success Factors for the Sustainability of Rural ICT Projects – The Dwesa Case Study.* SAICSIT.

Pallab, S. (2010). *Understanding the Impact of Enterprise Architecture on Connected Government.* Academic Press.

Papazoglou, M., & Willem-Jan, H. (2007). *Service oriented architectures: approaches, technologies and research Issues, 16*(3), 389-415.

Pappel, I., Pappel, I., & Saarmann, M. (2012). Digital records keeping to information governance in Estonian local governments. In *Proceedings of 2012 International Conference* (pp. 199–204). London: Information Society (i-Society).

Parameswaran, M., & Whinston, A. B. (2007). Social Computing: An Overview. *Communications of the Association for Information Systems, 19,* 762–780.

Pardo, T. A., Cresswell, M. A., Dawes, S. S., & Burke, G. B. (2004). Modeling the Social & Technical Processes of Interorganizational Information Integration. In *Proceedings of the 37th Hawaii International Conference on System Sciences*. IEEE.

Pardo, T. A., Gil-Garcia, R., & Burke, G. B. (2006). Building Response Capacity through Cross-boundary Information Sharing: The Critical Role of Trust. In *Exploiting the Knowledge Economy: Issues, Applications, Case Studies*. Amsterdam: IOS Press.

Parrado, S. (2008). Failed policies but institutional innovation through layering and diffusion in Spanish central administration. *International Journal of Public Sector Management, 21*(2), 230–252. doi:10.1108/09513550810855672

Pascu, C. (2008). *An Empirical Analysis of the Creation, Use and Adoption of Social Computing Applications*. JRC Scientific and Technical Reports. European Communities.

Paunović, L., Simić, K., Dadić, J., Jovanić, B., & Barać, D. (2012). The Impact of Applying the Concept of the Semantic Web In E-Government. *Innovative Issues and Approaches in Social Sciences, 5*(2), 161–179. doi:10.12959/issn.1855-0541.IIASS-2012-no2-art11

Pavlou, P. A. (2011). State of the Information Privacy Literature: Where are we now and Where should we go? *Management Information Systems Quarterly, 35*(4), 977–988.

Peterson, R. A. (1994). A Meta-analysis of Cronbach's Coefficient Alpha. *The Journal of Consumer Research, 21*(2), 381–391. doi:10.1086/209405

PICOS. (2007). *Privacy and Identity Management for Community Services*. Retrieved March 23, 2011, from http://www.picos-project.eu/

Pieterson, W., Ebbers, W., & Van Dijk, J. (2005). The opportunities and barriers of user profiling in the public sector. In *Proceedings of the 2005 EGOV*. Copenhagen, Denmark: EGOV.

Pocatilu, P. (2010). Developing Mobile Learning Applications for Android using Web Services. *Informatica Economica, 14*(3), 106–115.

Pollitt, C. (2000). Is the emperor in his underwear? An analysis of the impacts of public management reform. *Public Management, 2*(2), 181–199. doi:10.1080/146166700411229

Potekar, S. D., & Giragaonkar, K. P. (2004). *Application of ICT for better decision making in e-governance*. Retrieved October 17, 2013, from http://ieeexplore.ieee.org/xpl/abstractAuthors.jsp?arnumber=1307597

Pratt, W., Hearst, M. A., & Fagan, L. M. (1999). A Knowledge-Based Approach to Organizing Retrieved Documents. In *Proceedings of the sixteenth national conference on Artificial intelligence and the eleventh Innovative applications of artificial intelligence conference innovative applications of artificial intelligence* (pp. 80–85). AAAI.

Primelife. (2008). *Primelife*. Retrieved March 23, 2011, from http://www.primelife.eu/

Privacy Rights Clearinghouse. (2011, December 16). *Data Breaches: A Year in Review*. Retrieved January 26, 2012, from http://www.privacyrights.org/data-breach-year-review-2011

Putnik, G., & Cunha, M. M. (2007). *Knowledge and Technology Management in Virtual Organizations: Issues, Trends, Opportunities and Solutions*. Retrieved November 2013, from http://books.google.co.uk/books?id=nXn5BgqYdcUC

Raeth, P., Smolnik, S., Urbach, N., & Butler, B. (2009). Corporate Adoption of Web 2.0: Challenges, Success, and Impact. In *Proceedings of the pre-ICIS 2009 SIM Academic Workshop Enterprise and Industry Applications of Web 2.0*. Phoenix, AZ: SIM.

Rahman, M. M., & Rajon, S. A. A. (2011). *An effective framework for implementing electronic governance in developing countries: Bangladesh perspective*. Retrieved October 17, 2013, from http://ieeexplore.ieee.org/xpl/articleDetails.jsp?arnumber=6164814

Raj, P. (2012). *Cloud Enterprise Architecture*. CRC Press. Retrieved from http://www.peterindia.net/peterbook.html

Raja, T. N., Ramana, V. V. V., & Damodharam, A. (2012). *A framework for CSC oriented eGovernance applications*. Retrieved October 17, 2013, from http://ieeexplore.ieee.org/xpl/articleDetails.jsp?arnumber=6414944

Rajon, S. A. A., & Rahman, M. M. (2012). *Issues in implementing electronic governance: Bangladesh perspective.* Retrieved October 17, 2013, from http://ieeexplore.ieee.org/xpl/articleDetails.jsp?arnumber=6509710

Ranganathan, S. R. (1965). *The Colon Classification.* New Brunswick, NJ: Rutgers University Press.

Ranganathan, S. R. (1967). *Prolegomena to Library Classification* (3rd ed.). New York: Asia Publishing House.

Raths, D. (2011). Cities add citizen engagement mobile apps to their portfolios. *KM World, 20*(5), 10–11.

Reddick, C., & Turner, M. (2012). Channel choice and public service delivery in Canada: Comparing e-government to traditional service delivery. *Government Information Quarterly, 29*(1), 1–11. doi:10.1016/j.giq.2011.03.005

Richter, A., & Riemer, K. (2009). Corporate Social Networking Sites - Modes of Use and Appropriation through Co-Evolution. In *Proceedings of 20th Australasian Conference on Information Systems Corporate Social Networking Sites.* Melbourne, Australia: Academic Press.

Riege, A., & Lindsay, N. (2006). Knowledge management in the public sector: stakeholder partnerships in the public policy development. *Journal of Knowledge Management, 10*(3), 24–39. doi:10.1108/13673270610670830

Robertson, S. P., Vatrapu, R. K., & Medina, R. (2010). Off the wall political discourse: Facebook use in the 2008 U.S. presidential election. *Information Polity, 15*(1-2), 11–31.

Rodríguez Bolívar, M. P., Alcaide Muñoz, L., & López Hernández, A. M. (2010). Trends of e-Government research: contextualization and research opportunities. *The International journal of digital accounting research, 10*(16), 6.

Rodriguez, M., Steinbock, D., Watkins, J., Gershenson, C., Bollen, J., Grey, V., & Degraf, B. (2007). Smartocracy: Social Networks for Collective Decision Making. In *Proceeding of the 40th Annual Hawaii International Conference on System Sciences* (pp. 90-100). IEEE.

Rosa, J., Teixeira, C., & Pinto, J. S. (2013). Risk factors in e-justice information systems. *Government Information Quarterly, 30*, 241–256. doi:10.1016/j.giq.2013.02.002

Roy, A., & Karforma, S. (2011a). Risk and Remedies of E-Governance Systems. *Oriental Journal of Computer Science & Technology (OJCST), 4*(2).

Roy, A., & Karforma, S. (2012a). Object Oriented approach of Digital certificate based E-Governance mechanism. In *Proceedings of Computational Intelligence and Communication Engineering, International Joint Conference on CIIT, CENT, CSPE and CIITCom 2012* (pp. 360-366). Chennai, India: Springer.

Roy, A., & Karforma, S. (2013b). UML based modelling of ECDSA for secured and smart E-Governance system. In *Proceedings of National Conference on Advancement of Computing in Engineering Research (ACER13)* (pp. 207 - 222). WB India.

Roy, A., Banik, S., Karforma, S., & Pattanayak, J. (2010). Object Oriented Modelling of IDEA for E-Governance Security. In *Proceedings of International Conference on Computing and Systems 2010 (ICCS 2010)* (pp. 263-269). The University of Burdwan.

Roy, A., Sarkar, S., Mukherjee, J., & Mukherjee, A. (2012). Biometrics as an authentication technique in E-Governance security. In *Proceedings of UGC sponsored National Conference on Research And Higher Education In Computer Science And Information Technology, RHECSIT-2012* (pp. 153-160). WB India.

Roy, A., Banik, S., & Karforma, S. (2011). Object Oriented Modelling of RSA Digital Signature in E-Governance Security.[IJCEIT]. *International Journal of Computer Engineering and Information Technology, 26*(1).

Roy, A., & Karforma, S. (2011b). A Survey on E-Governance Security.[IJCECA]. *International Journal of Computer Engineering and Computer Applications, 8*(1).

Roy, A., & Karforma, S. (2012b). A Survey on digital signatures and its applications. *Journal of Computer and Information Technology, 3*(1&2).

Roy, A., & Karforma, S. (2013a). Object oriented metrics analysis for implementation of authentication in smart card based E-Governance mechanism. *Researchers World – Journal of Arts. Science and Commerce, 4*(2), 103–109.

Roy, A., Karforma, S., & Banik, S. (2013). *Implementation of authentication in E-Governance – An UML Based Approach.* LAP Lambert Academic Publishing.

Roy, J., & Langford, J. (2008). *Integrating Service Delivery across Levels of Government: Case Studies of Canada and Other Countries*. IBM Centre for the Business of Government.

Sacco, G. M. (2000). Dynamic Taxonomies: A Model for Large Information Bases. *IEEE Transactions on Knowledge and Data Engineering*, *12*(3), 468–479. doi:10.1109/69.846296

Sadeh & Norman. (2002). *M-Commerce: Technologies, Services and Business Models*. John Wiley and Sons.

Sandoval-Almazan, R. (2012). Open Government in Mexico: An Assessment Preview 2007-2010. In *Proceedings of Conference for E-democracy and Open GOvernment* (Vol. 1, pp. 255-266). Krems, Austria: Austrian Institute of Technology.

Sandoval-Almazan, R., Gil-Garcia, J. R., Luna-Reyes, L. F., Luna, D. E., & Rojas-Romero, Y. (2012). Open government 2.0: citizen empowerment through open data, web and mobile apps. In *Proceedings of the 6th International Conference on Theory and Practice of Electronic Governance* (pp. 30–33). New York, NY: ACM. doi:10.1145/2463728.2463735

Sandoval-Almazan, R., & Gil-Garcia, J. R. (2012). Are E-Government Portals Becoming Central Components for Public Information Sharing Networks? An Initial Exploration of Local Governments in Mexico. *Government Information Quarterly*, *29*, S72–S81. doi:10.1016/j.giq.2011.09.004

Sandoval-Almazan, R., Gil-Garcia, J. R., Luna-Reyes, L.F., & Luna, D. E., & Diaz-Murillo. (2011). The use of Web 2.0 on Mexican State Websites: A Three Year Assessment. *Electronic. Journal of E-Government*, *9*(2), 107–121.

Sandy, G. A., & McMillan, S. (2005). A Success Factors Model For M-Government. In *Proceedings of Euro MGOV 2005*. Brighton, UK: MGOV.

Sari, B., Schaffers, H., Kristensen, K., Loh, H., & Slagter, R. (2008). Collaborative knowledge workers: Web tools and workplace paradigms enabling enterprise collaboration 2.0. In *ECOSPACE IP-eProfessional Collaborative Workspace*. Dienstag.

Sarkar, S., & Roy, A. (2012). A Study on Biometric based Authentication. In *Proceedings of Second National Conference on Computing and Systems (NaCCS)* (pp. 263-268). WB India.

Sarkar, S., & Roy, A. (2013). Survey on Biometric applications for implementation of authentication in smart Governance. *Researchers World – Journal of Arts. Science and Commerce*, *4*(1), 103–114.

Savari, M., Montazerolzohour, M., & Thiam, Y. E. (2012). *Comparison of ECC and RSA algorithm in multipurpose smart card application*. Retrieved October 17, 2013, from http://ieeexplore.ieee.org/xpl/articleDetails.jsp?arnumber=6246121

Schinfeld, E. (2010, October 18). *Steve Jobs: Open Systems Don't Always Win*. Retrieved March 26th, 2013, from http://techcrunch.com/2010/10/18/steve-jobs-open-dont-win/

Scholl, H. J., & Luna-Reyes, L. F. (2011). Transparency and openness in government: a system dynamics perspective. In *Proceedings of the 5th International Conference on Theory and Practice of Electronic Governance* (pp. 107-114). Tallinn, Estonia: ACM.

Scholl, H. J., Fidel, R., Mai, J. E., & Unsworth, K. (2006). Seattle's Mobile City Project. In *Proceedings of Euro mGov 2006: The Second European Conference on Mobile Government* (pp. 144-153). Brighton, UK: Mobile Government Consortium International LLC.

Scholl, H. J., & Klischewski, R. (2007). E-Government Integration and Interoperability: Framing the Research Agenda. *International Journal of Public Administration*, *30*(8-9), 889–920. doi:10.1080/01900690701402668

Scott, M., Golden, W., & Hughes, M. (2004). Implementation strategies for e-Government: a stakeholder analysis approach. In *Proceedings of ECIS: The European Information Systems Profession in the Global Networking Environment*. Turku, Finland: ECIS.

Sellami, M., & Jmaiel, M. (2007). *A Secured Service-Oriented Architecture for E-government in Tunisia*. ReDCAD research unit. 8-9.

Semy, S. K., Pulvermacher, M. K., & Obrst, L. J. (2004). *Toward the Use of an Upper Ontology for U. S. Government and U. S. Military Domains: An Evaluation*. Retrieved from http://www.dtic.mil/cgi-bin/GetTRDoc?AD=ADA459575

Sharma, M. K., & Vaisla, K. S. (2011). E-governance applications in public healthcare for rural areas of Uttarakhand. *Elixir Comp. Sci., &. Engg., 41*, 5583–5586.

Sharma, S. K., & Gupta, J. (2004). Web Services Architecture for M-government: Issues and Challenges. *International Journal of Electronic Government, 1*(4), 462–474. doi:10.1504/EG.2004.005921

Sheedy, C., & Moloney, M. (2013). *Leveraging the Postal Infrastructure for the Authentication of Individuals Towards an Online Government Service Provision*. Paper presented at CRRI 21st Conference on Postal and Delivery Economics. Dublin, Ireland.

Sheng, H., & Trimi, S. (2006). Towards a Framework to Understand M-Government. In *Proceedings of AMCIS 2006*. Acapulco, Mexico: AIS.

Shneiderman, B. (1996). The Eyes Have It: A Task by Data Type Taxonomy for Information Visualizations. In *Proceedings of IEEE Symposium on Visual Languages* (pp. 336–343). IEEE.

Shoeman, F. (1984). Privacy: Philosophical Dimensions. *American Philosophical Quarterly, 21*(39), 199–213.

Siau, K., & Long, Y. (2005). Synthesizing e-government stage models - a meta-synthesis based on meta-ethnography approach. *Industrial Management & Data Systems, 105*(3), 443–458. doi:10.1108/02635570510592352

Signore, O., Chesi, F., & Pallotti, M. (2005). *E-government: challenges and opportunities*. Paper presented at the Proceedings of the CMG Italy XIX annual conference. Rome, Italy.

Simpson, M. C. S. (1996). Smart Power, a smart card electricity payment system. In *Proceedings of IEE Colloquium on UK Electricity Prepayment Systems*. IEE.

Smith, S., Macintosh, A., & Millard, J. (2009). *Major factors shaping the development of eParticipation*. Retrieved on May 15, 2013, from http://islab.uom.gr/eP/index2.php?option=com_docman

Smith, A., & Peck, B. (2010). The teacher as the 'digital perpetrator': Implementing Web 2.0 technology activity as assessment practice for higher education Innovation or Imposition? *Procedia Social and Behavioral Sciences, 2*, 4800–4804. doi:10.1016/j.sbspro.2010.03.773

Smith, H. J., Dinev, T., & Xu, H. (2011). Information Privacy Research: An Interdisciplinary Review. *Management Information Systems Quarterly, 35*(4), 989–1015.

SOCITM. (2006). *Modern public services: a role for change: The CIO as agent of transformation*. Retrieved on January 19, 2010, from http://www.socitm.gov.uk/socitm/Services/Socitm+Insight/News/Role+of+CIO.htm

Sprecher, M. H. (2000). Racing to e-government: using the Internet for citizen service delivery. *Government Finance Review, 16*(5), 21–22.

Stadlhofer, B., Salhofer, P., & Tretter, G. (2009). Ontology Driven E-Government. *2009 Fourth International Conference on Systems, 7*(4), 251–255. doi:10.1109/ICONS.2009.20

Stegarescu, D. (2006). *Decentralised government in an integrating world: quantitative studies for OECD countries*. Springer Science & Business.

Stewart, K., Clarke, H., Goillau, P., Verrall, N., & Widdowson, W. (2004). Non-technical Interoperability in Multinational Forces. In *Proceedings of 9th International Command and Control Research and Technology Symposium*. Copenhagen, Denmark: Academic Press.

Sundberg, H. K., & Sandberg, K. W. (2006). Towards e-Government: a survey of problems in organizational processes. *Business Process Management Journal, 21*(2), 146–161.

Sur, C., Roy, A., & Ranik, S. (2010). A Study of the State of E-Governance in India. In *Proceedings of National Conference on Computing and Systems (NACCS)*. WB India.

Surowiecki, J. (2004). *The Wisdom of Crowds: Why the Many Are Smarter Than the Few and How Collective Wisdom Shapes Business, Economies, Societies and Nations*. Anchor Books.

Taleb, Z., & Sohrabi, A. (2012). Learning on the move: the use of mobile technology to support learning for university students. *Procedia- Social and Behavioral Sciences, 69*, 1102–1109.

Tan, C., Pan, S., & Lim, E. (2005). Managing Stakeholder Interests in e-Government Implementation: lessons learned from a Singapore e-Government Project. *Journal of Global Information Management, 13*(1), 31–53. doi:10.4018/jgim.2005010102

Tang, L.-Y., Hsiu, P.-C., Huang, J.-L., & Chen, M.-S. (2013). iLauncher: an intelligent launcher for mobile apps based on individual usage patterns. In *Proceedings of the 28th Annual ACM Symposium on Applied Computing* (pp. 505–512). New York, NY: ACM. doi:10.1145/2480362.2480461

Tapia, R. S. (2009). *Assesing business-IT alignment in networked organizations.* (PhD Thesis). University of Twente, Eschende, The Netherlands.

Technosite. (2011). *Monitoring eAccessibility in Europe 2011: Annual Report.* Retrieved on June 6, 2013, from http://www.eaccessibility-monitoring.eu/researchResult.aspx

Teicher, J., Hughes, O., & Dow, N. (2002). E-Government: a new route to public sector quality. *Managing Service Quality, 12*(6), 384–393. doi:10.1108/09604520210451867

Thomas, J. C., & Streib, G. (2005). E-Democracy, E-Commerce, and E-Research: Examining the Electronic Ties Between Citizens and Governments. *Administration & Society, 37*(3), 259–280. doi:10.1177/0095399704273212

Tolk, A. (2003). Beyond Technical Interoperability - Introducing a Reference Model for Measures of Merit for Coalition Interoperability. In *Proceedings of the 8th International Command and Control Research and Technology Symposium (ICCRTS).* Washington, DC: ICCRTS.

Torres, L., Pina, V., & Royo, S. (2005). E-Government and the transformation of public administrations in EU countries – beyond NPM or just a second wave of reforms? *Online Information Review, 29*(5), 531–553. doi:10.1108/14684520510628918

Townsend, A. M. (2002). Mobile and wireless technologies: emerging opportunities for digital government. In *Proceedings of the 2002 annual national conference on Digital government research* (pp. 1–5). Los Angeles, CA: Digital Government Society of North America. Retrieved from http://dl.acm.org/citation.cfm?id=1123098.1123153

Traunmüller, R. (2011). Mobile government. In *Proceedings of the Second international conference on Electronic government and the information systems perspective* (pp. 277–283). Berlin: Springer-Verlag. Retrieved from http://dl.acm.org/citation.cfm?id=2033665.2033694

Triangle Management Services Limited. (2011, March). *Post Offices and Local Government Services – An International Literature Review.* Retrieved February 18th, 2013, from http://www.consumerfocus.org.uk/scotland/files/2011/08/POs-Government-Services-International-Comparisons-Final-Triangle-Report.pdf

Trimi, S., & Sheng, H. (2008). Emerging trends in M-government. *Communications of the ACM, 51*(5), 53–58. doi:10.1145/1342327.1342338

Turban, E., Liang, T. P., & Wu, S. P. J. (2011). A Framework for Adopting Collaboration 2.0 Tools for Virtual Group Decision Making. *Group Decision and Negotiation, 20,* 137–154. doi:10.1007/s10726-010-9215-5

Ubaldi, B. (2011). The impact of the Economic and Financial crisis on e-Government in OECD Member Countries. *European Journal of ePractice, 11,* 5-18.

Uddin, M. N., & Janecek, P. (2007). The Implementation of Faceted Classification in Web Site Searching and Browsing. *Online Information Review, 31*(2), 218–233. doi:10.1108/14684520710747248

UK Government. (2010). *Government ICT Strategy: Smarter, cheaper, greener.* UK Government.

UN DESA. (2012). *United Nations E-government Survey 2012: E-Government for the People.* Department of Economic and Social Affairs (United Nations). Retrieved 18/06/2013 from www.un.org/en/development/desa/publications/connecting-governments-to-citizens.html

UNDESA. (2008). e-Government Survey 2008: From e-Government to Connected Governance. New York: United Nations Department of Economic and Social Affairs.

United Nations E-Government Survey. (2012). *Executive Summary.* Retrieved 24/06/2013 from http://unpan3.un.org/egovkb/

United Nations. (2005). *Global e-Government Readiness Report 2005.* Retrieved on June 4, 2013, from http://unpan1.un.org/intradoc/groups/public/documents/un/unpan021888.pdf

United Nations. (2010). *United Nations e-Government Survey 2010*. Retrieved on November 25, 2011, from http://unpan1.un.org/intradoc/groups/public/documents/un/unpan038851.pdf

United Nations. (2012). *United Nations e-Government Survey 2012*. Retrieved on June 4, 2013, from http://unpan1.un.org/intradoc/groups/public/documents/un/unpan048065.pdf

UnitedNations. (2001). *Benchmarking e-Government: A Global perspective: Assesing the UN member states*. New York: United Nations Department of Economic and Social Afairs.

UnLtd.org.uk. (2003). *About UnLtd*. Retrieved April 3rd, 2013, from http://unltd.org.uk/about_unltd/

US Postal Service Office of the Inspector General. (2013). *e-Government and the Postal Service*. Retrieved February 18th, 2013, from http://www.uspsoig.gov: http://www.uspsoig.gov/foia_files/RARC-WP-13-003.pdf

US-CREST. (2000). *Coalition Military Operations - The Way Ahead Through Co-operability*. U.S. Center for Research and Education on Strategy and Technology. Retrieved June 22, 2013, from http://www.uscrest.org/CMOfinalReport.pdf

Valente, A. (2005). Types and Roles of Legal Ontologies. In V. R. Benjamins, P. Casanovas, J. Breuker, & A. Gangemi (Eds.), *Law and the Semantic Web* (Vol. 3369, pp. 65–76). Springer. doi:10.1007/978-3-540-32253-5_5

Valentina, N. (2004). E-Government for Developing Countries: Opportunity and Challenges. *Information Systems in Developing Countries, 18*(1), 1–24.

Van Dyke, T. P., Midha, V., & Nemati, H. (2007). The Effect of Consumer Privacy Empowerment on Trust and Privacy Concerns in E-Commerce. *Electronic Markets, 17*, 68–81. doi:10.1080/10196780601136997

van Wamelen, J., & de Kool, D. (2008). Web 2.0: a basis for the second society? In *Proceedings of the 2nd international conference on Theory and practice of electronic governance* (pp. 349-354). ACM. doi:10.1145/1509096.1509169

Vassilakis, C., & Lepouras, G. (2006). Ontology for e-Government Public Services. In *Encyclopedia of E-Commerce* (pp. 865–870). E-Government, and Mobile Commerce.

Vassilakis, C., Lepouras, G., Fraser, J., Haston, S., & Georgiadis, P. (2005). Barriers to electronic service development. *E-service Journal, 4*(1), 41–63. doi:10.2979/ESJ.2005.4.1.41

Vigoda-Gadot, E., Shoham, A., Schwabsky, N., & Ruvio, A. (2005). Public Sector Innovation for the Managerial and the Post-Managerial Era: Promises and Realities in a Globalizing Public Administration. *International Public Management Journal, 8*(1), 57–81.

W3C. (2004). *Web Services Addressing (WS-Addressing)*. Retrieved from http://www.w3.org/Submission/ws-addressing/

Wagner, C. (2004). Wiki: A Technology For Conversational Knowledge Management And Group Collaboration. *Communications of the Association for Information Systems, 14*, 265–289.

Waisanen, B. (2002). The future of E-Government: Technology fuelled management toll. *Public Management, 84*(5), 6–9.

Walker, R. M., & Enticott, G. (2004). Using Multiple Informants in Public Administration: Revisiting the Managerial Values and Actions Debate. *Journal of Public Administration: Research and Theory, 14*(3), 417–434. doi:10.1093/jopart/muh022

Walravens, N., & Ballon, P. (2011). The City as a Platform: Exploring the Potential Role(s) of the City in Mobile Service Provision through a Mobile Service Platform Typology. In *Proceedings of 2011 Tenth International Conference on Mobile Business (ICMB)* (pp. 60-67). doi:10.1109/ICMB.2011.22

Walravens, N. (2012). Mobile business and the smart city: Developing a business model framework to include public design parameters for mobile city services. *J. Theor. Appl. Electron. Commer. Res., 7*(3), 121–135. doi:10.4067/S0718-18762012000300011

Walsham, G., & Sahay, S. (2006). Research on Information Systems in developing countries: Current Landscape and Future Prospects. *Information Technology for Development, 12*(1), 7–24. doi:10.1002/itdj.20020

Wang, S., Li, C., Liu, J., & Wang, Z. (2011). Design of identity verification unit and management system. In *Proceedings of IET International Conference on Communication Technology and Application (ICCTA 2011)* (pp. 792–795). IET.

Ward, M. (2006). Information Systems Technologies: a public-private sector comparison. *Journal of Computer Information Systems*, *46*(3), 50–56.

Warner, M. (2010). The future of local government: twenty-first-century challenges. *Public Administration Review*, *70*, S145–S147. doi:10.1111/j.1540-6210.2010.02257.x

Warr, A. W. (2008). Social software: fun and games, or business tools? *Journal of Information Science*, *34*(4), 591–604. doi:10.1177/0165551508092259

Wauters, P., Declercq, K., van der Peijl, S., & Davies, P. (2011). *Study on cloud and service-oriented architectures for e-government*. FP7 report ref. Ares(2012)149022 - 09/02/2012.

Wei, Z., Gao, X., Jia, D., & Yang, Y. (2010). Research of mobile government based on multi-modal platform with unified engine. In *Proceedings of Intelligent Computing and Integrated Systems (ICISS), 2010 International Conference on* (pp. 786-789). ICISS. doi:10.1109/ICISS.2010.5657109

Weibel, S. (2005a). The Dublin Core: A Simple Content Description Model for Electronic Resources. *Bulletin of the American Society for Information Science and Technology*, *24*(1), 9–11. doi:10.1002/bult.70

Weibel, S. (2005b). The State of the Dublin Core Metadata Initiative: April 1999. *Bulletin of the American Society for Information Science and Technology*, *25*(5), 18–22. doi:10.1002/bult.127

Welch, E. W., Hinnant, C. C., & Moon, M. J. (2005). Linking Citizen Satisfaction with E-Government and Trust in Government. *Journal of Public Administration: Research and Theory*, *15*(3), 371–391. doi:10.1093/jopart/mui021

Welch, E., & Wong, W. (2001). Global information technology pressure and government accountability: The mediating effect of domestic context on Website openness. *Journal of Public Administration: Research and Theory*, *11*(4), 509–538. doi:10.1093/oxfordjournals.jpart.a003513

West, D. M. (2004). E-Government and the Transformation of Service Delivery and Citizen Attitudes. *Public Administration Review*, *64*(1), 48–60. doi:10.1111/j.1540-6210.2004.00343.x

Westin, A. (1967). *Privacy and Freedom*. New York: Atheneum.

Whitley, E. A. (2009). *Informational Privacy, Consent and the Control of Personal Data*. London: Elsevier.

Woods, E. (2009, May 3). *Web 2.0 and the public sector - Public Sector - Breaking Business and Technology*. Academic Press.

Wright, A., Bates, D. W., Middleton, B., Hongsermeier, T., Kashyap, V., Thomas, S. M., & Sittig, V. F. (2009). Creating and Sharing Clinical Decision Support Content with Web 2.0: Issues and Examples. *Journal of Biomedical Informatics*, *42*, 334–346. doi:10.1016/j.jbi.2008.09.003 PMID:18935982

Wu, H., Ozok, A. A., Gurses, A. P., & Wei, J. (2009). User aspects of electronic and mobile government: Results from a review of current research. *Electronic Government, an International Journal*, *6*(3), 233-251.

Xu, Q., Erman, J., Gerber, A., Mao, Z., Pang, J., & Venkataraman, S. (2011). Identifying diverse usage behaviors of smartphone apps. In *Proceedings of the 2011 ACM SIGCOMM conference on Internet measurement conference* (pp. 329-344). New York, NY: ACM. doi:10.1145/2068816.2068847

Xu, Z., Ni, Y., He, W., Lin, L., & Yan, Q. (2012). Automatic Extraction of OWL Ontologies From UML Class Diagrams: a Semantics-Preserving Approach. *World Wide Web (Bussum)*, *15*(5-6), 517–545. doi:10.1007/s11280-011-0147-z

Yamakami, T. (2007). Mobile Web 2.0: Lessons from Web 2.0 and Past Mobile Internet Development. In Proceedings of Multimedia and Ubiquitous Engineering, (pp. 886-890). IEEE.

Yao, L. J., Kam, T. H. Y., & Chan, S. H. (2007). Knowledge sharing in Asian public administration sector: the case of Hong Kong. *Journal of Enterprise Information Management*, *20*(1), 51–69. doi:10.1108/17410390710717138

Yi, M., Oh, S. G., & Kim, S. (2013). Comparison of social media use for the U.S., & the Korean governments. *Government Information Quarterly, 30,* 310–317. doi:10.1016/j.giq.2013.01.004

Yiu, C. (2012). *The Big Data Opportunity, Making government faster smarter and more personal. Policy Exchange Report.*

Yoon, J., & Chae, M. (2009). Varying criticality of key critical success factors national e-strategy along the status of economic development of nations. *Government Information Quarterly, 26,* 5–34. doi:10.1016/j.giq.2008.08.006

Young, O. (2007). *Topic Overview: Web 2.0.* Forrester Research.

Zhai, S., & He, T. (2010). *Design and implementation of password-based identity authentication system.* Retrieved October 17, 2013, from http://ieeexplore.ieee.org/xpl/articleDetails.jsp?arnumber=5623039

Zimmermann, O., Krogdahl, P., & Gee, C. (2004, June 2). *Elements of service-oriented analysis and design.* IBM Corporation.

About the Contributors

Zaigham Mahmood is a published author of ten books, six of which are dedicated to Cloud Computing and the other four focus on the subject of Electronic Government including: *Developing E-Government Projects: Frameworks and Methodologies*; *E-Government Implementation and Practice in Developing Countries*; *IT in the Public Sphere: Applications in Administration, Government, Politics, and Planning*; and this current volume. Additionally, he is developing three new books to appear later in 2014. He has also published more than 100 articles and book chapters and organized numerous conference tracks and workshops. Professor Mahmood is the Editor-in-Chief of *Journal of E-Government Studies and Best Practices* as well as the Series Editor-in-Chief of the IGI book series on E-Government and Digital Divide. He is a Senior Technology Consultant at Debesis Education UK and Associate Lecturer (Research) at the University of Derby UK. He further holds positions as a Foreign Professor at NUST and IIUI universities in Islamabad Pakistan and Professor Extraordinaire at the North West University Potchefstroom South Africa. Professor Mahmood is also a certified cloud computing instructor and a regular speaker at international conferences devoted to Cloud Computing and E-Government. His specialized areas of research include distributed computing, project management, and e-government.

* * *

Carl Adams is a Principal Lecturer/Researcher in the School of Computing, University of Portsmouth, UK. He has over a decade of professional experience in the computing industry as a software engineer and consultant before going into academia. His research interests include e/m-commerce, e/m-government, mobile information systems, social media, Web 2.0 technologies, electronic money, and the impact of technology on society. He has over 100 peer-reviewed publications in journals, international and national conferences, as well as a selection of book chapters. His latest book covers a corporate and social view of money and banking and how the electronic money infrastructure is impacting the economy. He has been a keynote and invited speaker at international conferences and invited speaker at national conferences.

Mohammed Al-Husban is a Senior Lecturer/Researcher in the School of Technology, Southampton Solent University, UK. He has many years of professional experience in the computing industry as a programmer and consultant before going into academia. His research interests include human computer interaction for Web and mobile applications, e-government implementation and practices with emphasis on services integration, electronic business, and electronic commerce. He has many peer-reviewed publications in journals, international conferences and national conferences, as well as a selection of book chapters.

Mihajlo Anđelić is a final year B.Sc. student at the Faculty of Organizational Sciences and is currently employed as a Quality Assurance Engineer on a project concerning the development of next generation medical software. As a student, he participated in projects related to education, student organizing, software research, and development. His workplace assignments are related to quality assurance, cloud technologies, iOS related solutions, Web technologies, and Semantic Web applications. His current professional interests include Internet technologies, visualization, e-business, e-learning, e-government, m-health, programming paradigms, and embedded electronics. Currently, he is dedicated to writing on topics in e-learning and applications of visualization in the field of e-government.

Jean-Paul Van Belle, a Professor at the University of Cape Town, has authored or co-authored about 20 books/chapters, 20 journal articles, and more than 80 peer-reviewed published conference papers. His key research area is the social and organisational adoption of emerging information technologies in a developing world context. The key technologies researched include mobile technologies, e-commerce, m-commerce, e/m-government, and open source software and cloud computing.

Walter Castelnovo is Assistant Professor of Information Systems at the University of Insubria, Italy. His research interests concern technological and organizational innovation in Public Administration and inter-organizational information systems. He is one of the founders of the Research Center for "Knowledge and Service Management for Business Applications" of the University of Insubria, and he is member of the Scientific Committee of the "Interdepartmental Center for Organizational Innovation in Public Administration" of the University of Milan. He served as member of the committee for many international conferences on E-Government and ICT evaluation. He was the General Chair of the 5th European Conference on Information Management and Evaluation (ECIME 2011) and the 13rd European Conference on eGovernment (ECEG 2013). He is co-founder with Danilo Piaggesi and Edson Luiz Riccio of the "ICT for Development International School" (ICT4DEVIS), and he was the Director of the first edition of the School in 2012.

Enrique Claver-Cortés is a Full Professor in Business Management at the University of Alicante, Spain. His Ph.D. dissertation was on Strategic Management Analysis. Dr. Claver's current research includes Strategic Management, International Business Competitiveness and Knowledge Management. His main teaching topic is Strategic Management.

Gary Coyle is a Senior Director with 27 years' experience in the post office, postal, and independent retail sector. He has firsthand experience as a postal operator, having owned and run his own post office for 18 years. He is responsible for founding and developing the first independent company to introduce real competition into the UK post office network. He has expertise and a track record of introducing new ideas/concepts to markets and is an excellent storyteller that encourages large brands and stakeholders to partner and collaborate and most importantly to fund new opportunities. He has built a large network of senior contacts within the global postal and retail sectors and large established brands. He is well known and respected in the postal sector and is seen as an innovator and thought leader. Gary is passionate about driving through change within post office networks, and has previously advised a number of global post office networks and large brands.

Marijana Despotović-Zrakić received her BS degree from the Faculty of Organizational Sciences, University of Belgrade in 2001, and a MSc degree in 2003. She received her PhD degree for the thesis titled "Design of Methods for Postgraduate E-Education Based on Internet Technologies" in 2006. Since 2001, she has been teaching several courses at the Laboratory for Electronic Business, the Faculty of Organizational Sciences: E-Business, E-Education, Simulation and Simulation Languages, Internet Technologies, Internet Marketing, Risk Management in Information Systems, M-Business, and Internet of Things. Since 2011, she is an associate professor. Her current professional and scientific interests include software project management, information systems, Internet technologies, and e-education.

Oghogho Ikponmwosa is a Ph.D. student in Telecommunications Engineering, and has over seven years of teaching and research experience. He is a registered member with the Nigerian Society of Engineers (NSE) and Council for the Regulation of Engineering in Nigeria (COREN). He was the Engineering Representative to the National Council of Science Association of Nigeria (SAN). He developed a roadmap to develop and diffuse microelectronics and embedded systems technologies in Nigeria and to use mobile phones in diffusing innovative farm information to Nigerian farmers. He has a research group currently investigating the dependence of Throughput observed at the transport layer on SNR at the Physical layer in IEEE802.11b/g/n WLAN Systems under a wide variety of environments and QoS traffic. He has attended several international conferences, published several journal articles and book chapters, and serves as a reviewer in book projects and journal houses.

Susana de Juana-Espinosa is a full time Assistant Professor of the Department of Business Organization of the University of Alicante since 2001. Her PhD was on Strategic e-Government. Dr. de Juana's research has mainly focused on using information systems to facilitate transformations in both private and public organizations. Her main teaching areas are Human Resource Management and E-Business.

Sunil Karforma has a B.E (Computer Science and Engineering) and M.E (Computer Science and Engineering) from Jadavpur University, W.B, India. He completed his PhD in the field of Cryptography. Presently, he is working as an Associate Professor in the Department of Computer Science at The University of Burdwan, W.B, India. He has already published his research papers in various reputable national and international journals and conference proceedings. His research interest is in the area of E-Governance and E-Commerce.

Miloš Milutinović received his BS degree from the Faculty of Organizational Sciences in 2011 and MSc degree with Master's thesis titled "Developing a Mobile Application for Learning Japanese Language Based on Learning Objects" in 2012. As a teaching associate at the Laboratory for Electronic Business, he is involved in teaching courses covering the areas of E-business, Internet technologies, Internet marketing, Internet of things, and M-business. As a PhD student, he receives scholarship from the Ministry of Science and Technological Development, the Republic of Serbia. His professional interests include Internet technologies, e-business, m-business, e-education, ontologies, cloud computing, and e-government. He is currently exploring the use of ontologies in government and education, and integration of mobile technologies into existing systems.

Maria Moloney is head of Research in Digital Services at Escher Group (IRL) Ltd, located in Dublin, Ireland. Her research interests span various domains such as e-government, informational privacy, design science research, security, and wireless communication. She has over 10 years of industry experience in the field of Information Systems and conducted her PhD in Trinity College Dublin, Ireland. The topic of her PhD involved the engineering of informational privacy principles into information systems. She has published in various academic journals and conferences in the discipline of Management Information Systems. Her latest publications include a paper presented at the International Conference of Information Systems (ICIS) 2012 in Orlando, Florida and another paper presented at the 46th Hawaii International Conference on System Sciences, in January 2013.

Olalekan Samuel Ogunleye is a mobile application developer and researcher at the Council for Scientific and Industrial Research (CSIR) Meraka Institute since 2008. He is a member of the Next Generation ICT Architectures and Mobile Systems. Since joining the CSIR Meraka Institute, his research interest has been on M-Learning, M-Health, M-Commerce, M-government, e-learning, e-government, Web services and enterprise application development, data mining. Olalekan has published in various peer-reviewed conferences such as IEEE, ACM, IST-Africa, SAICSIT, and M-Government Consortium. He has recently published a book chapter titled *Supporting Mobile Applications Developer through Java IDE Using Contextual Inquiry*.

Pethuru Raj is working as a cloud infrastructure architect at the IBM Global Cloud Center of Excellence (CoE), IBM India Bangalore. He obtained CSIR-sponsored PhD degree from Anna University, Chennai and continued his UGC-sponsored postdoctoral research in the department of Computer Science and Automation at the Indian Institute of Science, Bangalore. Thereafter, he was granted a couple of international research fellowships (JSPS and JST) to work as a research scientist for three years in two leading Japanese universities. His technical competencies lie in the areas of service oriented architecture, cloud computing, enterprise architecture, context-aware computing and machine-to-machine (M2M) integration, big data analytics, and business integration methods. He has been contributing book chapters for a number of high-quality technology books that are being edited by internationally acclaimed professors and professionals. The CRC Press, USA released his book titled *Cloud Enterprise Architecture* in 2012.

Muthu Ramachandran is a Principal Lecturer in the Computing, Creative Technologies, and Engineering School within the Faculty of Arts, Environment, and Technology at Leeds Metropolitan University, UK. Previously, he spent nearly eight years in industrial research (Philips Research Labs and Volantis Systems Ltd, Surrey, UK) where he worked on software architecture, reuse, and testing. His career started as a research scientist where he worked on real-time systems development projects. Muthu is an author and co-author of books including: *Software Components: Guidelines and Applications* (Nova Publishers, 2008) and *Software Security Engineering: Design and Applications* (Nova Publishers, 2011). He has also widely authored published journal articles, book chapters, and conference materials on various advanced topics in software engineering and education. He is a member of various professional organizations and computer societies including IEEE and ACM. Muthu is a Fellow of the BCS and the HEA. He is also invited speaker on several international conferences.

Yaneileth Rojas Romero is Assistant Researcher in the Research Center of Business College in the Autonomous State University of Mexico, in Toluca City. Yaneileth Rojas Romero is the author and co-author of several articles on electronic government, best practices, and innovation in electronic government, open data, transparency, accountability, and social media in government. Her research interests include electronic government, open data, transparency, accountability of social media in government, information technologies and organizations, digital divide technology, online political marketing, new public management, best practices and innovation in electronic government, and multi-method research approaches. Yaneileth Rojas Romero has a Bachelor Degree in Computing Science Administration and is a programming certified and a social media certified practitioner.

Abhishek Roy has completed a B.Sc in Information Technology (Hons) and M.Sc in Computer Technology from The University of Burdwan, W.B. India. Presently, he is working as a registered candidate for the degree of PhD in Computer Science in the Department of Computer Science at the University of Burdwan, W.B. India. As a Life Member of Cryptology Research Society of India (CRSI), Indian Statistical Institute Kolkata, he finds his research interest in the application of cryptography for implementation of information security in E-Governance. He is currently an editorial / reviewer board member of seven reputed international journals. Apart from this, he has worked as an Assistant Professor at a well-known management college, affiliated under the West Bengal University of Technology, W.B. India. He has more than five years of teaching and research experiences in the arena of computer science. He has published several research papers and a book in various reputed national and international platforms.

Rodrigo Sandoval-Almazán is Assistant Professor in the Research Center of Business College in the Autonomous State University of Mexico, in Toluca City. Dr. Sandoval-Almazán is the author or co-author of articles published in *Government Information Quarterly, Information Polity, Electronic Journal of Electronic Ggovernment, Journal of Information Technology for Development, American Journal of Business, Espiral: Estudios sobre Estado y Sociedad*, and *Espacios Públicos*. His research interests include electronic government, social media in government, information technologies and organizations, digital divide technology, online political marketing, new public management, and multi-method research approaches. Dr. Sandoval-Almazán is also editor of the *Academic Journal RECAI (Journal of Studies on Accountability, Management, and Informatics)* supported by the University of Mexico. Professor Sandoval-Almazan has a Bachelor's Degree in Political Science and Public Administration, a Masters in Management focused on Marketing, and a Ph.D. in Management with Information Systems.

Konstantin Simić received his B.Sc. degree from the Faculty of Organizational Sciences in 2010, and M.Sc. degree with Master thesis titled "Usage of Mobile Technologies in the Development of Application for Cloud Computing Infrastructure in E-Education" in 2011. As teaching associate at the Laboratory for Electronic Business, he is involved in teaching courses covering the areas of e-business, Internet technologies, Internet marketing, Internet of things, m-business, and concurrent programming. As a PhD student, he receives scholarship from the Ministry of Science and Technological Development, the Republic of Serbia. His current professional interests include Internet technologies, cloud computing, e-business, e-education, digital identities, e-government, and social media. He is currently experimenting with embedded devices, M2M communication, and intelligent environments.

Jorge Valdés-Conca is a full-time professor in the Department of Business Organization of the University of Alicante since 2000. He holds a PhD in Business Organization. His main teaching areas are Human Resource Management and Business Administration. Dr. Valdés's research has been centrally focused on Competency-based Human Resource Models.

Muhammad Yusuf is a PhD student in the School of Computing, University of Portsmouth, UK. He is also a lecturer in Informatics Engineering Department, Informatics Management Department and the Multimedia and Network Engineering Department at the Universty of Trunojoyo, Madura, Indonesia. His research interests are e-government, information system, Web 2.0, e-commerce, cloud computing, social computing, and research philosophy. He is conducting PhD research in e-government, specifically in e-participation, which develops a novel framework of e-participation and related theory of e-participation as well as investigating roles of different technology in e-participation. His PhD research uses Actor Network Theory (ANT). He is also a member of Association of Computing Machinery (ACM) and Special Interest Group on Applied Computing (SIGAPP) – ACM.

Index

A

Accountability 50, 127, 149, 152-153, 187, 198, 207-208, 210, 215, 218, 222, 228, 261, 279

Administration Requirements 57-58, 74

Authentication 6-7, 20, 43, 57-58, 75, 131, 280-281, 286-291, 293, 298-299

B

Bandwidth 84-85, 87, 101

Barriers 23, 54, 87, 89-90, 161, 250-251, 260, 262-263, 270-272

Big Data Analytics 149, 165-166, 177

C

Citizen Participation 125, 127, 145, 206

Citizen Requirements 57, 74

Cloud Computing 11, 17, 109, 162, 164, 169-171, 174, 176, 286-287

Component Based Design 2, 20

Connected service delivery 51-54, 56-61, 74

Controlled Vocabulary 26, 29, 33, 35, 49

Co-Operability 231, 233-234, 239-243, 250-251, 257

Cooperation Platform 234, 240-242, 257

Critical Success Factor 191, 244, 278

Cross-administrative services 235

Cross-border services 231-235, 244, 250

D

Data categorization 22

Data Flow Diagram (DFD) 282, 290-291, 298

Data Integrity 20, 58, 60

Data modelling 280, 291, 295

Data protection 57, 124, 129, 170, 263

D

Development index 82, 85, 265

DFD 282, 290-292, 298

Digital Identity 125, 131, 141, 145

Domain Ontology 31, 38, 49

Drivers 52, 157, 167, 175, 232, 260-261, 268, 271-272

E

E-awareness 87-88

ECEG 104-113, 119, 123

Enterprise Mobility 159

Enterprise Service Bus (ESB) 8, 54, 71, 74

Entity Relationship Diagram (ERD) 282, 290, 293, 298

E-Participation 85, 108-112, 119-120, 124-125, 128, 130, 145, 265, 267, 279

E-Readiness 77, 79-81, 90, 97, 101

E-service 6, 70, 89, 131, 265, 267, 298

E-Service Server 298

European Conference on E-Government 104-105, 119, 123

F

Faceted classification 27

Flexibility 1-3, 5, 12-13, 15, 17, 20, 56, 80, 116, 163, 166-167, 206, 208, 210, 240, 250

Focus Group Discussion (FGD) 104, 111, 123

G

G-Cloud 178

Granularity 20, 35, 43, 199

H

Horizontal Service Integration 75

HTTP (Hypertext Transfer Protocol) 75

CPSIA information can be obtained at www.ICGtesting.com
Printed in the USA
BVOW06*1309170614

356538BV00002B/2/P